283 Current Topics in Microbiology and Immunology

Springer-Verlag Berlin Heidelberg GmbH

Y. Kawaoka (Ed.)

Biology of Negative Strand RNA Viruses: The Power of Reverse Genetics

With 19 Figures and 7 Tables

Springer

Professor Yoshihiro Kawaoka, DVM, PhD
University of Wisconsin-Madison
School of Veterinary Medicine
Department of Pathobiological Sciences
2015 Linden Drive West
Madison, Wisconsin 535706
USA
e-mail: Kawaokay@svm.vetmed.wisc.edu

Cover Illustration by Yuko Kawaoka:
Generation of negative strand RNA virus from cloned cDNA.

ISSN 0070-217X
ISBN 978-3-642-07375-5 ISBN 978-3-662-06099-5 (eBook)
DOI 10.1007/978-3-662-06099-5

Library of Congress Catalog Card Number 72-152360

springeronline.com

Originally published by Springer-Verlag Berlin Heidelberg New York in 2004

Cover design: design & production GmbH, Heidelberg

Preface

Viruses consist of genetic information and components that help to maintain the integrity of this information through replication, packaging, and transmission. Each major class of viruses occupies its own niche, distinguished by the shape, host range, and type of genetic information common to its members. Negative-strand RNA viruses are characterized by the polarity of their genomic RNA to mRNA and by their involvement in diseases of medical and veterinary importance. Moreover, many representatives of this class, exemplified by Ebola and Nipah viruses, are considered emerging pathogens, and other viruses are awaiting discovery.

Remarkable progress in understanding negative-strand RNA viruses has come through reverse genetics, a technology that allows one to generate viruses possessing genes derived from cloned cDNA. First developed for negative-strand RNA viruses in 1989, reverse genetics systems are now established for members of most genera of the families Rhabdoviridae and Paramyxoviridae, influenza A and B viruses (Orthomyxoviridae), as well as for Ebola virus (Filoviridae). This powerful research tool, once available in only a few laboratories worldwide, has become a routine aid in dissecting the assembly and life cycle of negative-strand RNA viruses, as well as the pathogenic role of viral proteins and their interplay with host components. Finally, reverse genetics has opened the way to the development of live attenuated vaccines and of vectors for vaccine and gene delivery.

This volume presents a historical perspective on reverse genetics for negative-strand RNA viruses and reviews advances in this technology for segmented and nonsegmented viruses within this class. There is also a series of reviews on the replication of these viruses, ranging from viral RNA synthesis to the roles of viral and host proteins in viral replication, and finally to the assembly of viral components. The volume concludes with discussion of reverse genetics in the development of novel vaccines and gene delivery vectors. Considered together, these sections provide state-of-the-art knowledge in this maturing field, thus serving as a foundation for future research and a stimulus for interaction among scientists from diverse disciplines.

Y. Kawaoka

List of Contents

List of Contributors

(Their addresses can be found at the beginning of their respective chapters.)

Reverse Genetics of *Mononegavirales*

K. K. Conzelmann

Max von Pettenkofer-Institut and Genzentrum,
Ludwig-Maximilians-Universität München, Feodor-Lynen-Str. 25,
81377 Munich, Germany
E-mail: conzelma@lmb.uni-muenchen.de

Abstract "Reverse genetics" or de novo synthesis of nonsegmented negative-sense RNA viruses (*Mononegavirales*) from cloned cDNA has become a reliable technique to study this group of medically important viruses. Since the first generation of a negative-sense RNA virus entirely from cDNA in 1994, reverse genetics systems have been established for members of most genera of the *Rhabdo-*, *Paramyxo-*, and *Filoviridae* families. These systems are based on intracellular transcription of viral full-length RNAs and simultaneous expression of viral proteins required to form the typical viral ribonucleoprotein complex (RNP). These systems are powerful tools to study all aspects of the virus life cycle as well as the roles of virus proteins in virus-host interplay and pathogenicity. In addition, recombinant viruses can be designed to have specific properties that make them attractive as biotechnological tools and live vaccines.

1 Introduction

Mononegavirales, or "nonsegmented negative-strand RNA viruses" (NNSV), are enveloped viruses that have genomes consisting of a single RNA molecule of negative (anti-mRNA) sense. The *Mononegavirales* order includes the *Rhabdoviridae*, *Paramyxoviridae*, *Filoviridae*, and *Bornaviridae* families and thus viruses with high medical and economical relevance. Although they have different hosts, ranging from plants to mammals, and distinct morphological and biological properties, the elements essential for the typical mode of replication and gene expression have been retained throughout the *Mononegavirales*, illustrating that they have originated from a common ancestor (for review see Pringle 1997).

Although some negative-strand RNA viruses such as vesicular stomatitis virus (VSV) have long served as model organisms for specific aspects of virology and cell biology, such as RNA synthesis or glycoprotein transport and sorting, analysis of negative-strand RNA virus genetics and biology has been hampered by the lack of systems allowing directed genetic manipulation of their genomes. In contrast to positive-strand RNA viruses, or most DNA viruses, nucleic acids isolated from negative-strand RNA viruses or virus-infected cells are not able to initiate an infectious cycle on introduction into an appropriate host cell. This was indeed a criterion originally used to distinguish "positive" -and "negative"-strand RNA viruses (Baltimore et al. 1970). The minimal infectious unit of a negative-strand RNA virus is rather a nucleocapsid or ribonucleoprotein (RNP) complex, in which the RNA is properly associated with the viral RNA polymerase. The development of techniques for generation of recombinant negative-strand RNA viruses from cDNA thus involved reconstitution of active RNPs from individual components, that is, RNA and proteins. To date, recombinant viruses are available from all *Mononegavirales* families except the *Bornaviridae.* Genetic engineering of their genomes has become as easy as with positive-strand RNA viruses. Apart from the analysis of the molecular biology of each of the viruses, the great potential of *Mononegavirales* as tools for studying cell biology and as biomedicals is being exploited now. After a brief outline on the principles of *Mononegavirales* gene expression that are important with regard to reverse genetics (for details see chapter by Neumann et al., this volume), this chapter provides an overview of the approaches taken to establish methods for engineering NNSV and an overview of the currently available recombinant systems.

2 Principles of Replication and Transcription of *Mononegavirales*: Obstacles to Genetic Engineering

Viruses belonging to the *Mononegavirales* consist of two major functional units, an RNP complex, which represents the genetic information and which is active in replication and gene expression, and an envelope involved in getting the RNPs in and out of target cells. Simple members of the *Rhabdoviridae* like rabies virus can manage with only five virus-encoded proteins. In the RNP, the genomic or antigenomic single-strand RNA is tightly encapsidated in a nucleoprotein N (NP for *Paramyxoviridae*; throughout this chapter also designated N and associated with the viral RNA-dependent RNA polymerase. The latter consists of a large catalytic subunit (L), which shows typical RNA polymerase motifs, and a noncatalytic cofactor, the phosphoprotein P. The envelope is composed of a matrix (M) protein lattice lining the inner side of the host cell-derived lipid bilayer and a viral transmembrane glycoprotein (G) making up viral spikes. The RNP protein genes N, P, and L and the envelope protein M together with one or two glycoprotein genes make up also the basic, essential gene repertoire of most other *Mononegavirales.* "Accessory" proteins encoded by specific groups of *Mononegavirales* may encode proteins with regulatory function in virus replication or "luxury" functions to cope with host reactions and to provide for an optimal environment (see chapter by Nagai and Kato, this volume).

After infection of a cell, the RNP serves as a template for two distinct RNA synthesis functions, namely, successive transcription of subgenomic mRNAs as well as replication of full-length RNAs. The RNP genomes appear to possess a single entry site/promoter at the 3′ end of the RNA at which the viral RNA-dependent RNA polymerase enters for both mRNA transcription and genome replication. Thus the polymerase may act in a processive mode for synthesis of full-length RNAs and a nonprocessive mode for transcription, in which internal sequences define the gene or cistron borders. The two types of RNA synthesis are mostly explained by proposing the presence of a "transcriptase" and of a "replicase" form of the polymerase complex. The "transcriptase" is able to recognize gene borders and successively gives rise to a short leader RNA and several capped and polyadenylated mRNAs. Gene borders are defined by gene end sequences (GE) containing an oligo(U) stretch that constitutes the template for a stuttering synthesis of the mRNA's poly(A) tail, a putatively nontranscribed "intergenic region," and conserved gene start signals (GS) complementary to the 5′ end of the next mRNA. In contrast to transcription, the product of replication is not a free RNA but an RNP with an encapsidated full-length RNA. Because constant supply of N/P and P/L complexes is a prerequisite for replication of all negative-strand RNA viruses, it is assumed that RNA polymerization and encapsidation of the growing RNA chain into nucleoprotein are linked (Horikami et al. 1992; Gubbay et al. 2001). Because the genomic 3′ region would thus simultaneously represent the promoter and the encapsidation signal, mechanisms to prevent encapsidation of mRNAs should exist. In *Mononegavirales* this might be achieved through the release of the leader RNA carrying the encapsidation signal. Reinitiation at the junction between the leader and the first protein-coding gene (usually N/NP) appears to be necessary to convert the polymerase to the nonprocessive transcription mode in which elongation is independent of concurrent assembly with N protein (Vidal and Kolakofsky 1989). Transcription is stopped at the GE signal of the most downstream gene (L) by a stop/polyadenylation signal that is not followed by a restart sequence, whereas during replication polymerization proceeds to the 5′ terminus of the template. The 3′ terminal region of the resulting antigenomic RNA (RNP) is highly similar in sequence to that of the genomic promoter and functions as a strong promoter (antigenomic promoter) for synthesis of genome sense RNPs but not for transcription of subgenomic RNAs.

According to the conserved genome organization and expression strategy, the RNA of *Mononegavirales* cannot be infectious. Proteins required to make up an RNP are expressed neither from the genomic RNA, because of its anti-mRNA sense, nor from the complementary positive (antigenomic) strand (at least not effectively and not all) because the modular organization of the RNA does not allow translation of all viral proteins by the cellular machinery. Therefore, to render *Mononegavirales* RNA infectious, at least N, P, and L proteins must be supplied. The latter is easily accomplished by standard recombinant DNA techniques. However, encapsidation of RNA by N protein is not so easily achieved, because concurrent encapsidation by the viral polymerase is functional only on an RNP template and not on a naked RNA. Approaches toward functional RNPs containing cDNA-derived RNA thus involve artificial encapsidation, also called "illegitimate" encapsidation. This is so far a very inefficient process and indeed represented the technical challenge in NNSV reverse genetics.

3
Assembly of cDNA-Derived RNA into RNPs by *Paramyxoviridae* Helper Viruses

Initial work encouraging to the entire *Mononegavirales* research field came from a group working on the segmented influenza A virus. Peter Palese and colleagues reported the first approach allowing successful generation of biologically active RNPs containing cDNA-derived RNA. In vitro RNA transcripts that contained the authentic terminal sequences from an influenza genome segment flanking an internal chloramphenicol acetyl transferase (CAT) reporter gene were assembled with purified influenza virus nucleoprotein (NP) and the influenza virus polymerase proteins PA, PB1, and PB2 in vitro (Luytjes et al. 1989). After transfection of the reconstituted recombinant RNP into cells infected with influenza virus, the construct was replicated and the CAT reporter gene was expressed. The helper influenza virus not only provided the proteins needed for further RNA synthesis but, because of the segmented nature of its genome, also allowed packaging of the synthetic genome segment into progeny virus particles and passage of CAT activity in tissue culture. In the following, such "reassortant" or "transfectant" viruses possessing specific alterations in different genome segments could be generated (Enami et al. 1990). This achievement has rendered influenza virus accessible to detailed experimental investigations and has initiated great progress as illustrated in the chapters on segmented negative-strand

RNA viruses (chapters by Neumann and Kawaoka, Neumann et al., and Subbarao and Katz, this volume).

Unfortunately, and in contrast to influenza virus, in vitro encapsidation of *Mononegavirales* RNA is ineffective, which is most likely the result of a tighter RNP structure (Baudin et al. 1994; Iseni et al. 1998; Mavrakis et al. 2002). Nevertheless, a few synthetic nucleotides corresponding to the genome ends of VSV could be associated in vitro with RNP proteins (Smallwood and Moyer 1993; Moyer et al. 1991). No reports on successful encapsidation of larger constructs containing both termini are available. Mark Krystal and colleagues were then the first in demonstrating that an artificial *Paramyxoviridae*-derived RNA could be rescued inside a cell into a form that allows recognition and amplification by the polymerase of a *Paramyxoviridae* helper virus (Park et al. 1991). The artificial RNA corresponded to a negative-strand Sendai virus minigenome in which the entire coding region was replaced with the coding region of the CAT reporter gene, and in this respect resembled the above influenza minigenome segment. The model minigenome possessed the viral 3′ terminal sequence including the suspected promoter for the L polymerase as well as the signal directing leader RNA transcription and initiation of mRNA transcription. The 5′ end contained the GE signal derived from the most 5′-located cistron (L) and downstream terminal sequences supposed to contain a complete antigenomic promoter for replication. The RNA was generated in vitro by T7 RNA polymerase from a linearized plasmid to obtain precise RNA ends. After transfection of this RNA into cells, CAT activity was observed after infection of cells with Sendai virus. Furthermore, the artificial RNPs were packaged into infectious virions, as demonstrated by successful passage of CAT activity by transfer of cell-free supernatants. Thus it was confirmed that *cis*-acting sequences required for encapsidation, replication, and transcription of this paramyxovirus reside in the terminal sequences of the Sendai virus genome. Rescue by infectious helper virus of similarly organized, in vitro -transcribed transfected RNAs was then successful also in other *Paramyxoviridae* systems such as RSV (Collins et al. 1991), PIV-3 (Dimock and Collins 1993; De and Banerjee 1993), and measles virus (Sidhu et al. 1995). These early experiments revealed several basic requirements for successful rescue of RNAs. For example, the end of the transcript representing the viral 3′ terminus must be correct and cannot function from an internal site. In contrast, the 5′ end appeared to tolerate additional nucleotides (Collins et al. 1991; De and Banerjee 1993). Indeed, such helper virus-driven rescue of minigenomes provided the tools to study all types of viral *cis*-acting sequences, such as terminal promoters, sequences involved in paramyxovirus RNA editing, transcription signals, and intergenic regions (see chapter by Whelan et al., this volume). The latter has been achieved by the use of bicistronic model genomes in which two or more reporter genes are separated by sequences derived from gene borders (Harty and Palese 1995; Kuo et al. 1996). The efficiency by which *Paramyxoviridae* minigenomes are rescued by helper viruses is illustrated by the successful rescue, after transfection of in vitro-transcribed RNA into RSV-infected cells of a "minigenome" of approximately 50% of the genome length of RSV, comprising the entire L gene of RSV and a CAT reporter gene (Collins et al. 1993). In addition to *Paramyxoviridae*, the *Filoviridae* Marburg virus and Ebola virus have also been reported to support rescue of artificial model RNAs (Muhlberger et al. 1998, 1999).

In striking contrast to *Paramyxoviridae* and *Filoviridae*, rescue of minigenomes by natural mammalian *Rhabdoviridae* helper viruses, such as VSV or rabies virus (RV), has never been reported to be successful. One explanation may again lie in differences in the RNP structures, such that "illegitimate" encapsidation of RNA by nucleoproteins from *Rhabdoviridae* is accomplished less easily. Another feature contributing to the failure of helper virus-driven rescue was identified only after engineered infectious helper rhabdoviruses were made available. Rabies virus CAT-expressing minigenomes that could not be rescued by wt rabies virus were reproducibly rescued by a recombinant 3′ copy-back ambisense rabies virus (Finke and Conzelmann 1999). The successful ambisense virus had the strong antigenome promoter replaced with the weak genome promoter for replication (Finke and Conzelmann 1997). "Strong" and "weak" promoters appear to differ in their ability to bind polymerase and to compete for polymerase. Noteworthy in this respect is the finding that naturally occurring rhabdovirus defective interfering (DI) particles mostly belong to the 5′ copy-back type, containing two "strong" promoters. Competition of helper virus and minigenome promoters may thus contribute to the observed inability of mammalian wt rhabdoviruses to rescue reporter genomes. However, recent reports have revealed that, in contrast to mammalian rhabdoviruses, the fish rhabdoviruses infectious hematopoietic necrosis virus (IHNV) and viral hemorrhagic septicemia virus (VHSV) of the *Novirhabdovirus* genus do support rescue of minigenomes very well (Biacchesi et al. 2000b; Betts and Stone 2001). These reports also documented the first rescue of non-mammalian *Mononegavirales* replicating at low temperature.

4 Helper Virus-Free Systems: Generation of RNPs from cDNA-Encoded Components

Expression of proteins from transfected cDNA rather than from helper virus then not only permitted the recovery of *Rhabdoviridae* minigenomes but also improved the efficiency of *Paramyxoviridae* and *Filoviridae* minigenome recovery. Initially, natural DI RNPs were used to determine the *trans*-acting proteins required for DI propagation. In the first successful approaches, an expression system was used that involves infection of cells with a recombinant vaccinia virus (vv) encoding bacteriophage T7 RNA polymerase (vTF7-3) and transfection of the same cells with multiple expression plasmids driven by a T7 promoter (Fuerst et al. 1986). The vv/T7 system has the advantage of tremendously high expression levels combined with mostly cytoplasmic production of RNAs. The latter circumvented possible problems encountered with nuclear transcription, such as unwanted splicing of transcripts. Transfection of vTF7-3-infected cells with plasmids encoding N, P, and L proteins of VSV and Sendai virus, respectively, was necessary and sufficient to support replication of a VSV DI (Pattnaik and Wertz 1990) and a Sendai virus DI (Curran et al. 1991). Moreover, in case of the VSV DI, transfection of two additional plasmids encoding the VSV envelope proteins M and G resulted in assembly and budding of complete VSV particles (Pattnaik and Wertz 1991).

The intracellular T7 RNA polymerase-driven expression systems allowed not only the expression of virus support proteins from individual "support plasmids" but also the intracellular transcription of the minigenome RNAs from transfected plasmids. The generation of correct 3′ termini of transcripts inside cells, however, represented a problem, because linearized plasmids are not very efficient in transfection and vaccinia enzymes are known to religate free DNA ends very efficiently. A substantial advance was thus the development of transcription vectors designed to yield viruslike RNAs with discrete 3′ termini. This was achieved by exploiting the autolytic activity of ribozyme sequences, as first successfully shown for intracellular generation of functional nodavirus RNA (Ball 1992) and then of an artificial RNA mimicking the natural 2-kb VSV DI-T (Pattnaik et al. 1992). The viral cDNA was cloned between the T7 promoter and a ribozyme sequence from the antigenome of hepatitis delta virus (HDV; Perrotta and Been 1990; Sharmeen et al. 1988) This was followed by a T7 polymerase transcription termination sequence. The HDV ribozyme has the advantage that only sequences downstream of the cleavage site are required for autocatalytic activity and is generally indiscriminate with regard to upstream sequences (Shih and Been 2002). Moreover, the 3′ terminal ribose of the upstream genome analog possesses a cyclic 2′,3′-phosphate instead of a hydroxyl group (Sharmeen et al. 1988), which might prevent polyadenylation of the RNA by vaccinia virus enzymes (Gershon et al. 1991) or protect degradation of the RNA 3′ terminus by cellular exonucleases. This setting resulted in efficient encapsidation and replication of the DI-T RNA analog by VSV proteins expressed from simultaneously transfected "support" plasmids (Pattnaik et al. 1992). The crucial role of a precise 3′ end for rhabdovirus RNA rescue was also confirmed by these authors. Constructs with short heterologous 3′ extensions or deletions were found encapsidated by N protein, but they were not replicated. In contrast, extra nonviral G residues at the 5′ end, introduced to enhance T7 RNA polymerase transcription, were shown to be removed rapidly during replication. Complete infectious DI particles were assembled in cells coexpressing all five VSV proteins, thus allowing for analysis of both *cis*-acting sequences and *trans*-acting factors.

The use of the HDV ribozyme in combination with T7 expression turned out to be highly effective also in other virus systems, including mammalian *Rhabdoviridae* and *Paramyxoviridae* (reviewed in Conzelmann 1996), *Filoviridae* (Muhlberger et al. 1998, 1999), fish *Rhabdoviridae* (Biacchesi et al. 2000b), and even segmented negative-strand RNA viruses such as the *Bunyaviridae* or *Arenaviridae* (Dunn et al. 1995; Lopez et al. 1995, 2001; Cornu and de la Torre 2001).

5
Lessons from Minigenome Experiments

The availability of effective mini- or model genome systems rapidly allowed experimental confirmation of long-standing hypotheses and also promoted new knowledge on the biology of *Mononegavirales*. The minigenome systems are particularly well suited to dissect *cis*-acting sequences including promoters, and transcription signals, or RNA editing signals. One particular *cis*-acting "signal" of *Paramyxovirinae* has only been revealed by reverse genetics experiments with Sendai virus model genomes. Laurent Roux and colleagues observed that artificial Sendai virus genomes replicate much more efficiently when they comprise an exact multimer of six nucleotides in length and established the importance of the so-called "rule of six" (Calain et al. 1992; Calain and Roux 1993). The rule of six is presumably a consequence of the association of each N subunit with exactly six nucleotides, as originally observed by Egelman et al. (Egelman et al. 1989). Flush N/RNA ends at the terminal promoters as well as correct N-phasing of internal *cis*-acting signals is crucial for efficient RNA synthesis. Intriguingly, the rule of six applies to all viruses that are able to edit RNAs, that is, the *Paramyxovirinae* subfamily. Because replication of these viruses may give rise to inadvertently edited, erroneous genomes, the rule of six may be instrumental in giving the correct genomes a selective advantage (Hausmann et al. 1996; Kolakofsky et al. 1998; Vulliemoz and Roux 2001). The rule of six must thus be taken into account in the design of *Paramyxovirinae* model genomes, although there are viruses for which the adherence to the rule of six is less stringent, such as the *Rubulavirus* SV5 (Murphy and Parks 1997). In contrast to the *Paramyxovirinae* no such numerical rules seem to apply to the other *Paramyxoviridae* subfamily, the *Pneumovirinae* (Samal and Collins 1996), or to members of the *Filoviridae* and *Rhabdoviridae* families, although the N protein of VSV has been predicted to be associated with nine nucleotides (Thomas et al. 1985).

Another important finding with regard to the design of model genomes is that *Mononegavirales* "holo"promoters are not necessarily always confined to the nontranslated terminal sequences. Although in all cases analyzed so far the sequences contained in the nontranslated leader and the leader/first gene junction and trailer regions are sufficient for all RNA synthesis, including replication and transcription, paramyxoviruses may contain additional sequences enhancing the activity of the terminal "core" promoters. Sequences internal to the N and L coding regions and requiring proper spacing have been shown to enhance replication of Sendai virus (Tapparel et al. 1998) and SV5 (Murphy et al. 1998; Keller et al. 2001).

In addition to *cis*-acting functions, artificial minigenomes have been used to analyze *trans*-acting factors relevant to gene expression as well as virus particle assembly. The former was particularly interesting with regard to the goal of full-length virus recovery. Intriguingly, for all *Rhab-*

doviridae and *Paramyxoviridae* three proteins, namely, the N, P, and L proteins, were sufficient to perform both replication of RNPs and transcription of mRNAs. However, in the *Paramyxoviridae* additional proteins were identified that are able to modulate RNA synthesis. The small Sendai virus C proteins (C, C′), which are expressed from an alternate, overlapping open reading frame (ORF) of the P gene mRNA, were found to have an inhibitory effect on transcription but not on replication (Curran et al. 1992). C protein coexpression inhibited the amplification of an internally deleted DI genome but had little or no effect on a 5′ copy-back DI genome (DI-H4). The inhibitory effects of C coexpression thus appear to be genome promoter specific (Cadd et al. 1996). A more general inhibitory effect on RNA synthesis in minigenome assays was described for the Sendai virus V and W proteins, which are expressed from edited mRNAs (Curran et al. 1991), and the nonstructural (NS) protein 1 (and NS2 protein) of RSV (Atreya et al. 1998). Interestingly, both the Sendai virus C and the RSV NS proteins were later identified as antagonists of interferon function by analyzing the phenotype of recombinant viruses (Bossert and Conzelmann 2002; Garcin et al. 1999; Schlender et al. 2000), suggesting multiple functions of these "accessory" proteins in virus RNA synthesis and in manipulating the host cell.

A factor positively affecting RNA transcription has been identified in RSV, namely, the M2-1, protein which is encoded by the upstream ORF of the M2 gene. Although RSV N, P, and L proteins define the minimal *trans*-acting requirements for RNA replication and transcription (Yu et al. 1995; Grosfeld et al. 1995), M2-1 supports efficient synthesis of full-length mRNAs as well as synthesis of polycistronic readthrough mRNAs (Collins et al. 1996; Fearns and Collins 1999; Hardy and Wertz 1998; Hardy et al. 1999; Sutherland et al. 2001) by an as yet unknown mechanism. The inclusion of M2-1 support plasmids in reverse genetics experiments is thus helpful in improving gene expression from minigenomes and may also be helpful in recovery of full-length RSV (see below).

An astonishing situation was observed with *Filoviridae* minigenomes. For Marburg virus the three nucleocapsid proteins NP, VP35 (corresponding to P), and L were shown to be necessary and sufficient to drive encapsidation, replication, and transcription of minigenomes. For the closely related Ebola virus, these three proteins were also sufficient for minigenome replication but efficient transcription depended on the presence of a fourth nucleocapsid protein, VP30. Most amazingly, Ebola virus VP30 could be replaced functionally by Marburg virus VP30

(Muhlberger et al. 1999). VP30 seems to act exclusively on a stem loop structure at the 5′ end of the first (N) gene, and destruction of this structure renders transcription of minigenomes VP30 independent. Although the start sequences of downstream genes predict stem loops in the mRNAs, transcription of these appears to be VP30 independent (Weik et al. 2002).

6
Recovery of Infectious Viruses from cDNA: A Positive Approach

The approaches to recovery of infectious viruses could build on a large body of experiments that have revealed basic requirements and pitfalls for rescue, replication, and gene expression of minigenomes. Yet it took some time before the first infectious *Mononegavirales*, the rabies rhabdovirus, was recovered from cDNA in our lab (Schnell et al. 1994). Our approach involved intracellular expression of rabies virus N, P, and L protein, along with a full length RNA whose correct 3′ end was generated by the HDV ribozyme. However, in contrast to approaches in several other laboratories (for review see Roberts and Rose 1998, 1999), we used a transcript corresponding to the viral antigenome (positive strand) rather than to genome sense transcripts. The use of the positive-strand transcript is critical to avoiding a severe antisense problem raised by the presence of N, P, and L sequences in full-length RNAs. If negative-strand transcripts are used, the mRNAs encoding the support proteins N, P, and L can hybridize to the naked genomic RNA (affecting more than half of the genome) and prevent the critical assembly of the genome into the RNP. Starting with the positive-strand, antigenomic transcript, this RNA can form an RNP without any interference from the mRNAs. Once in RNP form, the positive strand can be replicated by the cDNA-encoded support proteins to form full-length minus-strand RNPs that are wrapped into RNPs as nascent RNA chains and thus are immune to interference from mRNAs. From functional genomic RNPs, an infectious cycle is then initiated. In addition to preventing RNA encapsidation by N, the formation of dsRNA hybrids may hamper translation of support proteins. Moreover, dsRNAs may turn on interferon expression in transfected cells, resulting in the establishment of an antiviral state.

Within months of the rabies report, successful recoveries of another rhabdovirus, the *Vesiculovirus* VSV (Lawson et al. 1995; Whelan et al. 1995), and of different *Paramyxoviridae* including the *Respirovirus* Sendai virus (Garcin et al. 1995), the *Pneumovirus* RSV (Collins et al. 1995), and the *Morbillivirus* measles virus (Radecke et al. 1995) confirmed that the use of antigenome rather than genome transcripts tips the balance toward success. Until now, reliable recovery of 18 different *Mononegavirales* species from cDNA has been published (see Table 1 and references therein). The spectrum of recombinant species ranges from cold water fish rhabdoviruses (Johnson et al. 2000; Biacchesi et al. 2000a) to Ebola virus (Volchkov et al. 2001). By far more are the recombinant virus strains, chimeras of different strains or species, mutated viruses, and viruses expressing foreign genes.

Several early studies comparing antigenome to genome RNA rescue in parallel experiments confirmed absolute failure of genome RNA (Lawson et al. 1995; Whelan et al. 1995; Radecke et al. 1995). To date there are two examples of rescue from genome RNA, namely, Sendai virus (Kato et al. 1996) and HPIV-3 (Durbin et al. 1997). Especially in the former work, the magnitude of the antisense problem is illustrated. By optimizing various factors, such as the ratio of support plasmids, and removal of extra nonviral G residues from the T7 promoter directing transcription of the full-length RNA, the rescue efficiency with the antigenome RNA could be increased to one recovery event in 10^4–10^5 transfected cells, the highest efficiency yet reported for any *Mononegavirales*. However, recovery starting with the negative-strand construct was 100-fold lower (Kato et al. 1996). Because in this system negative-strand RNA was synthesized much better than positive-strand RNA, the difference is probably even greater than 100-fold. An advantage, yet less pronounced, of starting with cDNA giving rise to antigenome transcripts was also reported for HPIV-3. Virus was recovered in 100% of cultures transfected with cDNA giving rise to antigenome, whereas genome-directing cDNA yielded virus in 53% of cultures (Durbin et al. 1997).

7
Variations on a Theme: State-of-the-Art Recovery of *Rhabdoviridae*, *Paramyxoviridae*, and *Filoviridae* from cDNA

It turned out that only slight, if any, variations of the original robust "rabies" protocol are required for successful recovery of members of the *Mononegavirales*. Even a segmented negative-strand RNA virus, Bunyamwera virus, has been rescued entirely from cDNA by using intracellular coexpression of antigenome RNAs and support proteins from (vac-

Table 1 Recombinant *Mononegavirales*

Virus	Parent strain	Cells used for recovery	T7 RNAP source	Remarks	Reference
Rhabdoviridae					
Lyssavirus	Attenuated rabies virus	BSR (BHK clone)	vTF7-3		Schnell et al. 1994
Rabies virus	SAD B19	BSR-T7/5	BSR-T7/5 cell line		Finke and Conzelmann 1999
	Attenuated rabies virus RC-HL	BHK-21	vTF7-3		Ito et al. 2001
Vesiculovirus	Serotype Indiana,	BHK-21	vTF7-3	No (−) cDNA rescue	Lawson et al. 1995
VSV, vesicular	San Juan strain	BSR T7/5	BSR T7/5 cell line		Harty et al. 2001
stomatitis virus	Serotype Indiana strain San Juan, G from strain Orsay	BHK-21	vTF7-3	No (−) cDNA rescue	Whelan et al. 1995
Novirhabdovirus 5SHRV, Snakehead rhabdovirus	*Novirhabdovirus* of warm water fish	EPC	vTF7-3		Johnson et al. 2000
Novirhabdovirus IHNV, Infectious hematopoietic necrosis virus	*Novirhabdovirus* of salmonids, European 32/87 strain	EPC	vTF7-3	N support plasmid not required	Biacchesi et al. 2000a

Table 1 (continued)

Virus	Parent strain	Cells used for recovery	T7 RNAP source	Remarks	Reference
Paramyxoviridae					
Respirovirus Sendai virus	Strain Z	HeLa or BHK, egg injection	vTF7-3 +AraC	Vaccinia-mediated recombination	Garcin et al. 1995
	Strain Z	LLC-MK2 egg injection	vTF7-3	No 5′ terminal extra G residue, Rescue of (+) and (−) cDNA; Rescue of transfected RNA	Kato et al. 1996
	Fushimi strain	HeLa	vTF7-3, MVA-T7	HDV genomic ribozyme	Leyrer et al. 1998
	Hamamatsu strain	LLC-MK2 egg injection	vTF7-3		Fujii et al. 2002
Respirovirus HPIV-3, Human parainfluenza virus type 3	Strain 47885	HeLa	vTF7-3 +AraC	No NP support plasmid required, NP expressed from the FL antigenome	Hoffman and Banerjee 1997
	JS wild type (7 mut) 15426	HEp-2	MVA-T7	Rescue of (+) and (−) cDNA; no extra G	Durbin et al. 1997
Respirovirus BPIV-3, Bovine parainfluenza virus type 3	Strain Kansas/15626/84 clone 5-2-4	HEp-2, at 32°C, passage to LLC-MK	MVA-T7		Schmidt et al. 2000
	Strain Kansas/15626/84 vaccine	HeLa	MVA-T7 +AraC		Haller et al. 2000
Respirovirus HPIV-1	Clinical isolate Washington/20993/1964			Rescue by HPIV1 or HPIV3 support plasmids or mixture	Newman et al. 2002

Table 1 (continued)

Virus	Parent strain	Cells used for recovery	T7 RNAP source	Remarks	Reference
Morbillivirus Measles virus	Edmonston, nonlymphatic cell culture-adapted vaccine strain	293-3-46, passage to Vero	293-3-46 cell line (T7RNAP, NP, P)	No (−) cDNA rescue	Radecke et al. 1995
		HeLa	MVA-T7		Schneider et al. 1997
		293-3-46, cocultivation with Vero	293-3-46 cell line (T7RNAP, NP, P)	Heat shock or expression of HSP-72	Parks et al. 1999
	Pathogenic IC-B strain isolated in B95a cells	B95a or 293-3-46 with B5a overlay	vTF7-3 or 293-3-46 cell line	No recovery in 293-3-46 cell line alone	Takeda et al. 2000
Morbillivirus RPV, rinderpest virus	RBOK vaccine strain	HEK 293 passage to Vero	MVA-T7		Baron and Barrett 1997
		Vero or 293	FP-T7 fowlpoxvirus		Das et al. 2000a, b
Morbillivirus CDV	Onderstepoort large plaque (OND-LP) vaccine strain	HeLa, with Vero feeder cells, passage to Vero	MVA-T7	No extra Gs	Gassen et al. 2000
	Onderstepoort small plaque (OS) vaccine strain	HEK 293	MVA-T7	Rescue also by measles virus support plasmids	Von Messling et al. 2001
	avianized Onderstepoort vaccine virus	Hep-2 or A549 cocultured with Vero	MVA-T7, MVA-GKT7	vv promoter-driven N and P support plasmids, heat shock	Parks et al. 2002
Rubulavirus SV5, Siminan virus 5	Strain RAL	A549, passage to CV-1	MVA-T7	V protein not required	He et al. 1997

Table 1 (continued)

Virus	Parent strain	Cells used for recovery	T7 RNAP source	Remarks	Reference
Rubulavirus NDV, Newcastle disease virus	La Sota, lentogenic	CEF with allantoic fluid	FPV-T7 fowlpox-virus		Peeters et al. 1999
	La Sota Clone-30, lentogenic	BSR-T7/5, egg injection	BSR-T7/5		Romer-Oberdorfer et al. 1999
	Beaudette C mesogenic, and avirulent La Sota	HEp-2, egg injection or passage on HEp-2 or DF1 with acetyltrypsin	MVA-T7		Krishnamurthy et al. 2000; Huang et al. 2001
	Hitchner B1, lentogenic	HEp-2, A549, CEF coculture, egg injection	MVA-T7		Nakaya et al. 2001
Rubulavirus Mumps virus	Jeryl Lynn strain	A549	MVA-T7		Clarke et al. 2000
Rubula virus HPIV-2		Vero, coculture with fresh cells	MVA-T7	V important for hPIV2 assembly, maturation, and morphogenesis	Kawano et al. 2001

Table 1 (continued)

Virus	Parent strain	Cells used for recovery	T7 RNAP source	Remarks	Reference
Pneumovirus RSV, Respiratory syncytial virus	strain A2 C4G(+)	HEp-2	MVA-T7	M2-1 support plasmid required for rescue	Collins et al. 1995
		HEp-2		M2-1 support plasmid not required, protein expressed from FL cDNA	Collins et al. 1999
	Strain A2 AVIRON C4G(+)	HEp-2	MVA-T7	M2-1 support plasmid not required for rescue	Jin et al. 1998
	Strain A2		MVA-T7		Oomens et al. 2003
Pneumovirus BRSV	A51908 variant ATue51908	BSRT-7/5	BSR-T7/5 cell line		Buchholz et al. 1999
	A51908	HEp2	MVA-T7		Yunus et al. 2001
Filoviridae					
Ebola virus	Subtype Zaire, Mayinga strain	BSR-T7/5, passage to Vero	BSR-T7/5 cell line	Support plasmids: NP, VP35, VP30, L	Volchkov et al. 2001

cinia borne) T7 RNA polymerase-driven plasmids (Bridgen and Elliott 1996). Variations described so far mostly pertain to optimizing transcription of full-length RNA, the source of T7 RNA polymerase, the cells being used for transfection and /or recovery, and the use of support proteins enhancing the recovery rates.

In summary, if one is able to (a) produce a full-length RNA that corresponds to the viral antigenome RNA, in particular with respect to a correct 3′, (b) express a minimal set of viral support proteins that can make this RNA an mRNA-expressing RNP, and (c) provide a (cellular) environment supporting assembly of RNPs from RNA and support proteins, successful virus rescue is most probable.

7.1 Generation of Antigenome-Like RNA: The Favorite RNA Polymerase and Extra G Residues

For synthesis of defined RNAs in eukaryotic systems bacteriophage RNA polymerases such as T7 RNA polymerase have great advantages, such as short specific promoters, defined initiation sites, and termination sequences coupled with high processivity. Because these single-polypeptide-chain proteins do not need cofactors for enzymatic activity, they can be easily expressed in a functional form in eukaryotes. Although SP6 RNA polymerase-expressing vaccinia virus (Usdin et al. 1993) or cell lines that constitutively express the phage T3 RNA polymerase (Deuschle et al. 1989) are available, only T7 RNA polymerase has so far been used for recovery of *Mononegavirales* (see Table 1). After expression of T7 RNA polymerase in eukaryotic cells, the enzyme is predominantly located and active in the cytoplasm (Fuerst et al. 1986; Elroy-Stein and Moss 1990). However, nuclear localization signals can be added to obtain effective nuclear T7 transcription also (Lieber et al. 1989). Eukaryotic polymerases, such as PolI, that are also able to produce capless RNAs at defined initiation sites are being used for expression of RNAs from segmented viruses replicating in the nucleus, such as influenza virus (Fodor et al. 1999; Hoffmann et al. 2000) and Thogotovirus (Wagner et al. 2001), in combination with PolII or T7 RNA polymerase for expression of virus support proteins (see chapter by Schmitt and Lamb). Also, PolII transcripts have been used to recover virus from cDNA (Inoue et al. 2003). As PolII creates capped RNAs with variable 5′ ends, the use of a 5′

ribozyme is suggested. Moreover, the system relies on the absence of cryptic splice sites in the viral RNA.

As few as 17 nucleotides, present in currently used expression plasmids, specify a highly efficient T7 RNA promoter (5′-pTAATACGACTCACTATAGGG-3′; sense orientation, +1 position of the RNA transcript underlined). T7 RNA polymerase most efficiently initiates transcription with GTP on a DNA template containing a CC motif for transcription initiation (Imburgio et al. 2000; Cheetham and Steitz 2000; Oakley and Coleman 1977). In this regard, *Mononegavirales* cDNAs do not represent the optimal templates, because their RNAs start with an A residue. Yet some *Mononegavirales* full-length RNAs are transcribed with satisfying efficiency with the original A in position +1, such as Sendai virus (Kato et al. 1996), HPIV-3 (Durbin et al. 1997), or CDV (Gassen et al. 2000). In others, the addition of one to five G residues has been shown to be necessary, or at least to considerably enhance transcription initiation of antigenome-like RNAs. It appears that the presence of extra nonviral G residues at the 5′ end is not largely affecting virus recovery, although direct comparison is difficult because of the resulting different RNA transcript levels. As shown for VSV, the extra nucleotides are quickly removed during replication (Pattnaik et al. 1992). It has also been reported that T7 RNA polymerase transcription of full-length genome RNA is much less efficient than transcription of antigenome RNA of VSV. This was attributed to the presence of U-rich sequences of the gene borders leading to transcription termination (Whelan et al. 1995). On the other hand, in vitro transcription of full-length Sendai virus genomic RNA was much more efficient than transcription of antigenomic RNA. In this case, the latter was reported to terminate aberrantly (Kato et al. 1996).

7.2 Viral Proteins Required to Make Antigenome Infectious: Support Proteins

As determined in minigenome experiments, the N, P, and L proteins constitute the necessary and sufficient set of viral proteins bringing about replication of RNPs and transcription from RNPs of *Rhabdoviridae* and *Paramyxoviridae* analyzed so far. According to present knowledge, expression of this set of proteins is also sufficient to recover infectious virus from cDNA. However, this does not necessarily mean that all these proteins must be provided from a separate support plasmid. As first shown for HPIV-3 (Hoffman and Banerjee 1997), and more recently for the fish *Novirhabdovirus* IHNV (Biacchesi et al. 2000a), rescue was achieved after transfection of only P- and L-encoding expression plasmids along with the full-length cDNA directing antigenomic RNA, although with reduced efficiency. In these cases, the N protein can be expressed in sufficient amounts from the antigenome RNA (Hoffman and Banerjee 1997; Biacchesi et al. 2000a) because the N start codons represent the first AUG codon in the antigenome sequence. A similar but less easily explainable situation exists with RSV. The above-described RSV M2-1 protein, which is a potent transcription elongator and antiterminator, is encoded very far downstream in the RSV genome, at the ninth gene position. Especially the elongator function strongly advised the inclusion of M2-1 as a support protein for virus recovery. Initially, inclusion of an M2-1 expression plasmid was reported to indeed be required for successful recovery of infectious RSV (Collins et al. 1995). Another laboratory then reported that only N, P, and L support plasmids were necessary and, moreover, that inclusion of an M2-1 expression plasmid did not significantly increase rescue efficiency (Jin et al. 1998). The former lab then also confirmed rescue after transfection of only N, P, and L support plasmids, albeit at low efficiency. As also shown in this work, transfection of the RSV full-length antigenome-encoding cDNA clone (or of parts containing the M2 ORF) led to M2-1 activity sufficient to support RSV transcription. Thus an extra M2-1 support plasmid is not required for virus recovery. A different question that is so far unsolved is whether M2-1 represents an essential gene product. A recombinant RSV lacking the M2-1 gene has not been described as yet, presuming, but not showing, that M2-1 is essential.

Unlike *Rhabdoviridae* and *Paramyxoviridae*, and unlike its fellow *Filoviridae* family members, minigenome RNA synthesis of Ebola virus subtype Zaire, Mayinga strain, does not manage well with the three proteins N, P (VP35), and L but rather requires its own VP30 or VP30 of Marburg virus for effective transcription initiation (Muhlberger et al. 1999). The recent recovery of infectious Ebola virus thus used support plasmids expressing N, P (VP35), L, and VP30. Omission of any of these plasmids resulted in a lack of development of CPE in the transfected BSR-T7/5 cells or in subsequent passages on Vero cells (Volchkov et al. 2001). As determined in minigenome experiments the dependence of transcription on VP30 correlates with the presence of a certain stem loop structure at the 5′ terminal N mRNA of Ebola Zaire/Mayinga. When the structure of this stem loop is disturbed, transcription is VP30 independent, as it is most probably for the downstream genes (Weik et al. 2002). It will therefore be interesting to clarify whether the VP30 dependence is specific for this particular isolate and what functions VP30 may have in other Ebola and Marburg viruses.

Obviously, the authentic homologous support proteins are the prime choice for rescuing a certain virus. However, with very closely related virus species, there is the option of using heterologous proteins. Successful recovery of CDV by using measles virus N, P-, and L-expressing plasmids in a vaccinia-driven system was reported recently (Von Messling et al. 2001). In addition, HPIV-1 could be recovered by using N, P, and L clones not only from HPIV-1 but also from HPIV-3 or a mixture thereof (Newman et al. 2002). However, in view of the great efficiency with which vaccinia virus mediates homologous recombination (Evans et al. 1988; Garcin et al. 1995; Yao and Evans 2001), careful analysis of the newly generated viruses is needed.

The absolute amounts or ratios of transfected support plasmids used for cDNA rescue cover a broad range. It appears obvious that the N protein required for encapsidation of RNA is needed in the most abundant amounts. Because P acts as a chaperone for N encapsidation (Curran et al. 1995) and is required for replication (Horikami et al. 1992) high amounts of P expression plasmids are also being used for transfection. According to the catalytic functions of the L protein, L-encoding plasmids are mostly used in much lower amounts. Because expression of different gene constructs from identical plasmid vectors, and even more from different vectors, is not equally effective, the plasmid ratios given in the reports do not necessarily reflect the ratios of proteins expressed. So it is not surprising if reported ratios for support plasmids N:P:L range from (rounded) 20:10:1 (Leyrer et al. 1998) to 1:1:2 (Johnson et al. 2000). It is therefore worthwhile to optimize the transfection protocol for each expression plasmid and expression system used as well as for each virus to be rescued.

7.3
Expression Systems for Virus Rescue

7.3.1
Vaccinia Virus and T7 RNA Polymerase, a Working Couple

Systems for transient expression of proteins driven by eukaryotic promoters as well as stable cell lines expressing *Mononegavirales* support proteins (Willenbrink and Neubert 1994), T7 RNA polymerase (Deng et al. 1991; Elroy-Stein and Moss 1990; Lieber et al. 1989), or both (Kaelin et al. 1994), have long been available. Yet *Mononegavirales* "rescue" work was greatly facilitated by the availability of a eukaryotic expression system combining advantages of vaccinia virus gene expression and of bacteriophage T7 RNA polymerase transcription (Fuerst et al. 1986). In vaccinia virus-infected cells vaccinia-encoded enzymes lead to amplification of transfected plasmids, capping of approximately 30% of RNA transcripts, and polyadenylation of RNAs in the cytoplasm (Fuerst et al. 1986; Fuerst and Moss 1989; Gershon et al. 1991; Yao and Evans 2001). Thus both effective translation of *Mononegavirales* support protein from transfected T7 promoter-driven plasmids and accurate transcription of genomes or genome analogs is achieved in cells infected with a recombinant vaccinia virus expressing T7 RNA polymerase. Transfection of separate plasmids each encoding one factor further allows easy optimization of protein and RNA ratios. Moreover, the very broad host range of vaccinia virus makes this a rather universal system, probably applicable immediately to most *Mononegavirales* systems.

The prototype vaccinia virus expressing T7 RNA polymerase, vTF7-3, was developed by Fuerst and colleagues and is based on the "wt" vaccinia strain WR (Fuerst et al. 1986). vTF7-3 has been used successfully for recovery of infectious *Rhabdoviridae*, including rabies viruses (Schnell et al. 1994; Ito et al. 2001), VSV (Lawson et al. 1995; Whelan et al. 1995), and two species of the *Novirhabdovirus* genus from fish (Johnson et al. 2000; Biacchesi et al. 2000a), as well as the *Paramyxoviridae* Sendai virus (Garcin et al. 1995; Kato et al. 1996; Leyrer et al. 1998; Fujii et al. 2002), HPIV-3 (Hoffman and Banerjee 1997) and measles virus (Takeda et al. 2000; Fujii et al. 2002).

The robustness and reliability over a broad range of cell culture systems and conditions of vTF7-3 were illustrated by the successful recovery of fish rhabdoviruses SHRV and IHNV. A *Cyprinus carpio* cell line

(EPC) was used for infection with vTF7-3 and transfection of support plasmids. Permanent incubation at 31°C (Johnson et al. 2000) or even at 14°C after an initial 8-h period at 37°C (Biacchesi et al. 2000a) was sufficient for production of T7 RNA polymerase and support proteins and resulted in recovery of virus. Although Sendai virion assembly in vaccinia virus-infected cells is greatly hampered, Kolakofsky and colleagues succeeded first in recovering Sendai virus with the vTF7-3 system (Garcin et al. 1995). In this case, transfected cells were injected into the allantoic cavity of embryonated chicken eggs. Subsequent further passage in eggs resulted in the loss of vaccinia virus and the recovery of recombinant Sendai virus. These authors experienced the high recombinogenic potential of vaccinia virus, in that a genetic marker of the full-length clone was often replaced by the wild-type sequence originating from a support protein-encoding plasmid. In an ingenious approach they took advantage of vaccinia-mediated homologous recombination (Evans et al. 1988; Yao and Evans 2001) to create a novel virus whose genome was derived from two independent plasmids (Garcin et al. 1995).

Use of the replication-competent vTF7-3 usually requires that the rescued virus be separated from vaccinia virus. This is easily accomplished by simple physical means, such as filtration, as originally used for rabies virus (Schnell et al. 1994) or VSV (Lawson et al. 1995) rescue, or by using inhibitors of vaccinia DNA synthesis, such as 1-β-d-arabinofuranosylcytosine (AraC) (Whelan et al. 1995; Garcin et al. 1995; He et al. 1997; Hoffman and Banerjee 1997) or rifampicin (Kato et al. 1996). Also, inoculation into eggs of transfected cells or supernatant led to the loss of vTF7-3 and recovery of Sendai virus (Garcin et al. 1995; Kato et al. 1996). If available, cells permissive for the rescued virus but not for vTF7-3 can be used. For recovery of Bunyamwera virus (although a segmented virus), Bridgen and Elliott (Bridgen and Elliott 1996) took advantage of this virus growing in mosquito cells. A single passage to mosquito cells eliminated vTF7-3. A very comfortable situation exists with the above-mentioned carp cells used for fish rhabdovirus recovery, because these cells support vaccinia virus gene expression but do not support vaccinia propagation (Johnson et al. 2000; Biacchesi et al. 2000a), so that pure rhabdoviruses can be recovered directly from the transfected cells.

In addition to vTF7-3, a host range-restricted and highly attenuated vaccinia virus strain expressing T7 RNA polymerase, based on a modified vaccinia virus Ankara (MVA), MVA-T7, has been made available more recently (Wyatt et al. 1995; Sutter et al. 1995). This virus is unable to multiply in human and most other mammalian cell lines, with the amazing exception of baby hamster kidney (BHK) cells (Drexler et al. 1998). Because viral gene expression is unimpaired in nonpermissive cells, MVA-T7 is an efficient as well as safe vector. There are currently two MVA-T7 variants available; T7 RNA polymerase expression is driven in one by the vv early/late promoter P7.5 (Sutter et al. 1995) and in the other by the vv late promoter P11 (Wyatt et al. 1995). MVA-T7 has been used immediately for recovery of the first *Pneumovirus*, RSV (Collins et al. 1995), the first *Rubulavirus*, SV5 (He et al. 1997), HPIV-3 (Durbin et al. 1997), rinderpest virus (Baron and Barrett 1997), and measles virus (Schneider et al. 1997). Other virus species followed, including "first" rescues of mumps virus (Clarke et al. 2000), CDV (Gassen et al. 2000), HPIV-2 (Kawano et al. 2001), and BPIV-3 (Schmidt et al. 2000).

Other poxviruses may also be a safe alternative to replication-competent vaccinia virus. Britton et al. (Britton et al. 1996) have generated a recombinant fowlpox virus, fpEFLT7pol, with the T7 RNA polymerase gene under the control of the vv early/late promoter P7.5. This virus supports stable expression of T7 RNA polymerase in both mammalian and avian cells. Only in the latter are infectious fowlpox viruses generated. FpEFLT7pol has been successfully used by Peeters et al. for the first recovery of an avian *Paramyxoviridae* member, NDV (Peeters et al. 1999). FpEFLT7pol was also successfully used for rescue, in Vero cells, of a chimeric rinderpest virus with the glycoprotein genes replaced with those of peste-des-petits ruminants virus (PPRV), for which attempts with MVA-T7 had failed (Das et al. 2000a). This was attributed to a lower cytopathogenicity of FpEFLT7pol.

Other viral vectors expressing T7 RNA polymerase and supporting expression from transfected T7 promoter-driven plasmids have been described, such as adenoviruses (Tomanin et al. 1997; Aoki et al. 1998) or Semliki Forest virus (SFV) (Kohl et al. 1999). The latter was shown to drive transient expression from transfected plasmids in both insect and mammalian cells. Cells expressing T7 RNA polymerase from a noncytopathic, selectable Sindbis virus replicon (Frolov et al. 1996) have been used successfully to express support proteins for RSV minigenomes (Olivo et al. 1998). Importantly, because these viral systems lack the cytoplasmic enzymes for capping (and polyadenylation) of T7 transcripts, the introduction of an internal ribosome entry site (IRES) preceding the coding region is required for high-level protein translation.

7.3.2
Cell Lines Expressing T7 RNA Polymerase and/or Support Proteins

The idea of using the advantages of bacteriophage polymerases for eukaryotic gene expression in the absence of infectious (or defective) viral vectors, such as vaccinia virus, was followed early and has led to the establishment of stable cell lines expressing T3 RNA polymerase (Deuschle et al. 1989) or T7 RNA polymerase, which was directed to the nucleus by adding an NLS (Lieber et al. 1989). For comparison of the vv/T7 system with a pure cell system, a stable mouse cell line generated in the Moss lab, which had also previously provided the vTF7-3 system, was particularly useful (Elroy-Stein and Moss 1990). As with vTF7-3, the original T7 RNA polymerase expressed in this cell line was found in the cytoplasmic compartment and was able to express a reporter gene from transfection plasmids containing T7 promoter and termination signals. However, consistent with the absence of RNA-capping activity in the cytoplasm, effective protein expression was achieved only after the IRES from encephalomyocarditis virus (EMCV) was introduced upstream of the coding region of the reporter gene, allowing cap-independent translation (Jang et al. 1989). Nevertheless, this cap-independent transient expression was increased more than 500-fold when the transfected cells were also infected with vaccinia virus (Elroy-Stein and Moss 1990). Thus T7 polymerase-expressing cell lines were known not only to provide the opportunity to transcribe correct *Mononegavirales* antigenome-like RNAs from transfected plasmids but also to express substantial amounts of *Mononegavirales* support proteins from IRES-containing plasmids. In addition, stable cell lines expressing *Mononegavirales* support proteins from eukaryotic promoters have been successfully used for Sendai virus DI propagation (Willenbrink and Neubert 1994).

A cell line expressing T7 RNA polymerase was initially used to recover the Edmonston vaccine strain of measles virus from cDNA (Radecke et al. 1995). In a human embryo kidney cell line, 293-3-46, both T7 RNA polymerase and measles virus support proteins N and P were expressed from a cytomegalovirus (CMV) promoter. Transfection of plasmids specifying antigenomic RNA and MeV L mRNA allowed rescue of Edmonston virus, as directly monitored by syncytia formation in approximately 30% of the transfected cell culture wells (Radecke et al. 1995). S. Udem and colleagues later showed that rescue efficiency in 293-3-46 cells was increased by two- to threefold by heat shock treatment of the transfected cultures. Another 20-fold increase was observed after additional cocultivation of Vero cells (Parks et al. 1999). The importance of having the appropriate cells for virus recovery available was illustrated in a report from Masato Tashiro's group, who tried to recover the pathogenic measles virus strain IC-B. Whereas IC-B was readily recovered in B95a cells infected with vTF7-3 and transfected with support plasmids, recovery in 293-3-46 cells failed. Only after coculture with B95a cells could virus be recovered (Takeda et al. 2000). To complete the story, measles virus Edmonston virus has also been recovered in the MVA-T7 system (Schneider et al. 1997), making measles virus the record species with respect to the different expression systems used for recovery from cDNA.

Another cell line expressing T7 RNA polymerase and currently in use for virus recovery is based on BSR cells, a BHK clone, which have previously been used for vTF7-3-driven recovery of rabies virus (Schnell et al. 1994). The BSR-T7/5 cell line was selected in our lab from approximately 300 geneticin-resistant clones of BSR transfected with pSC6-T7-NEO (kindly provided by M. Billeter) encoding the T7 RNA polymerase gene under control of the CMV promoter and a neomycin resistance gene for its ability to efficiently express rabies virus N protein from a transfected plasmid (Buchholz et al. 1999). BSR T7/5 cells were first used in our lab for recovery of BRSV (Buchholz et al. 1999) and then also for different rabies viruses (Finke and Conzelmann 1999) including "designer" ambisense gene expression rabies viruses (Finke and Conzelmann 1997). One particular advantage of BSR T7-5 in virus rescue is their inability to express type I interferon, such that even completely interferon-sensitive viruses, such as BRSV lacking the NS1 and/or NS2 gene (Schlender et al. 2000), or partially interferon-resistant viruses, such as chimeric BRSV with HRSV NS genes (Bossert and Conzelmann 2002) could easily be rescued and propagated. In other laboratories, the BSR T7-5 cell line has been used for successful recovery of VSV (Harty et al. 2001), NDV (Romer-Oberdorfer et al. 1999), and also Ebola virus (Volchkov et al. 2001) from transfected cDNA.

8
Rescue of Mutant and Designer NNSV

The use of cDNA-derived RNA transcripts corresponding as closely as possible to the infectious virus's RNA is the most straightforward approach to recovery and the choice for optimizing rescue systems. Once a reliable system is established, the range of possibilities for "genetical manipulation" of *Mononegavirales* is in fact enormous. During the past years, recombinant viruses have been generated that are only subtle variants of their parents, for example, by introducing point mutations, but also heavily designed viruses that have never existed in nature before. Because the *Mononegavirales* recovery systems do not depend on helper virus, recovery of even severely debilitated virus is possible. The basis for the enormous potential in flexibility and stability of recombinant *Mononegavirales* stems in particular from their modular genome organization, their simple and efficient gene expression strategy, and the high stability of their genetic information. The latter is at least in part due to the characteristic structures of *Mononegavirales* RNPs, which make recombination and elimination of nonessential sequences rare events. Here, only a very short digest of approaches is given. More details and examples can be found in the other chapters of this volume dealing with certain aspects of *Mononegavirales* biology, such as the chapter by Whelan et al. (replication and transcription), and the chapters by Schmitt and Lamb, Nagai and Kato, García-Sastre, von Messling and Cattaneo, and Subbarao and Katz (glycoproteins, accessory proteins, interferon antagonists, vaccine vectors, and attenuated vaccines, respectively) in this volume.

8.1 Attenuating Viruses by Mutations, Gene Deletions, Genome Reorganization and Insertion of Foreign Genes

Most mutations introduced into the full-length cDNA clones will change some aspect of the virus, and probably the overwhelming number of them will be attenuating. The identification of mutations attenuating virus in vivo without reducing their capacity of replication in cell culture is the prime goal in live vaccine development. Of recombinant Mononegavirales, RSV is the virus most extensively modified to generate live vaccines. Mutations previously identified in classically attenuated viruses such as temperature-sensitive (ts) or cold-passaged (cp) isolates can now be transferred, combined, and analyzed in defined recombinant backgrounds (for a recent overview on RSV mutants, see Collins and Murphy 2002). Such mutated viruses may also help to identify virus protein functions and their interaction with host cell factors.

Gene deletions or mutations that prevent the expression of a certain gene product have identified accessory genes/proteins in *Mononegavirales* not needed for virus replication in cell culture but important in vivo. In particular, the identification of *Mononegavirales* IFN antagonists is a current topic in reverse genetics (for review, see Young et al. 2000; Garcia-Sastre 2001; Gotoh et al. 2001) and certainly is of great relevance to attenuated vaccine development. The C proteins of *Respirovirus* (Sendai virus and HPIV-3) and V proteins of *Rubulavirus* (SV5, SV41, mumps virus, and HPIV2) have been shown to prevent the IFN response by inhibiting IFN signaling (Parisien et al. 2001; Gotoh et al. 1999, 2001; Basler et al. 2000; Didcock et al. 1999; Garcin et al. 1999). In addition, the V proteins of SV5 and other *Paramyxovirinae* inhibit interferon induction by preventing NF-κB and IRF-3 activation. The highly conserved C-terminal cysteine-rich domain of the V proteins is required for this function (Poole et al. 2002). Also, Ebola virus VP35 (P) affects IFN induction (Basler et al. 2000). Finally, RSV NS1 and NS2 proteins have been shown to cooperatively mediate a complete interferon resistance without affecting IFN signaling (Schlender et al. 2000; Bossert and Conzelmann 2002). With respect to recovery of such interferon-sensitive viruses from cDNA the use of cells that lack intact interferon systems is suggested, such as Vero or BHK.

The modular nature of *Mononegavirales* genomes, made up of a succession of mostly monocistrons, and the well-defined transcription signals at the gene borders make it easy to engineer additional genes into a virus. A multiplicity of foreign sequences have been expressed, including reporter genes making it easy to follow virus traces in their hosts, foreign antigens derived from other viruses such as HIV or HCV, and genes for cellular proteins such as cytokines and growth factors. Because of the typical transcription gradients, adjustment of the rate of foreign gene expression is easily achieved by selecting a more up- or downstream location. Even the 3′ promoter-proximal position has been used (Hasan et al. 1997), although this attenuates transcription of the abundantly needed N protein. Expression levels can be further fine-tuned by modified transcription signals that differ in their ability to direct restart or readthrough (Finke et al. 2000; He and Lamb 1999). Because the relative position of a gene to the 3′ terminal promoter determines its level of transcription, changing the gene order also leads to viruses with altered phenotypes. Stepwise translocation of the N gene of VSV downstream resulted in stepwise attenuation of VSV (Wertz et al. 1998). This strategy can also be exploited to increase the expression of major protective antigens, such as moving the RSV G and F genes to more 3′ proximal positions (Collins and Murphy 2002). An approach allowing the expression of multiple genes without disturbing the expression of virus genes involved the generation of "ambisense" gene expression viruses. On exchange of the rabies virus and Sendai virus antigenome promoter for replication with that of the transcriptionally active genome promoter, expression of genes from the antigenome RNP was possible (Finke and Conzelmann 1997; Le Mercier et al. 2002).

8.2
Chimeric Viruses, Envelope Switching, and Retargeting

NNSV carrying novel proteins in their envelopes may have a role to play as particulate vaccines and as targetable gene delivery vectors. Early experiments with recombinant rhabdoviruses lacking the single glycoprotein G gene confirmed that rhabdovirus particle formation does not require G, although in the presence of G budding efficiency is increased approximately 30-fold (Mebatsion et al. 1996). The major driving force for rhabdovirus budding is indeed the M protein surrounding a condensed helical RNP (Mebatsion et al. 1999), as also determined by analysis of a rabies virus gene deletion mutant. Such deletion of essential genes requires that the lacking product be supplied *in trans*, either by stable cell lines or transiently. An easy way for recovery is by simply expressing the missing gene along with the support proteins in T7 RNA polymerase-expressing cells by transfection of additional plasmids, as described for the above rhabdoviruses, and propagation of the "pseudotyped" viruses in cell lines providing the protein, as described for Sendai virus F deletion mutants (Li et al. 2000). Interestingly, heterologous glycoproteins not only from Mononegavirales but also other viruses, such as HIV Env, or from cells are incorporated in functional form into the envelope of rhabdoviruses. In the absence of G this leads to completely retargeted viruses and/or viruses presenting foreign antigens. (Mebatsion and Conzelmann 1996; Mebatsion et al. 1997; Schnell et al. 1996, 1997; Johnson et al. 1997; Kretzschmar et al. 1997; Kahn et al. 1999). Also, recombinant measles virus and RSV have been described having the VSV G gene as a substitute for their own envelope protein genes (Spielhofer et al. 1998; Oomens et al. 2003). Approaches to retarget measles also include fusion of receptor ligands to the H protein sequence, resulting in a broadened host cell spectrum (Schneider et al. 2000; Hammond et al. 2001). Chimeric viruses with glycoprotein from related species have been described for other paramyxoviruses, such as human or bovine PIV-3, with reciprocal exchanges of the F and HN genes (Schmidt et al. 2000), PIV-1 with F and HN from PIV-1, rinderpest virus with peste-des-petits-ruminants F and H glycoproteins, or bovine RSV with G and F from human RSV or bovine PIV-3 glycoproteins (Buchholz et al. 2000; Stope et al. 2001). These approaches are mostly used to restrict the host range of live vaccines.

NNSVs constitute a rich source of viruses and virus functions from which generation of biomedical tools by reverse genetics is now possible. The exceptional propensity of recombinant NNSV to accommodate functions of other viruses is already being exploited to create really novel artificial viruses are of great value. It is predicted that the overwhelming majority of such novel viruses will be unable to compete with naturally selected viruses in any respect. Nevertheless, it is imperative that appropriate care be taken, as with chimeric DNA and positive-strand RNA viruses.

Acknowledgements. Research in the author's lab is supported by grants from the Deutsche Forschungsgemeinschaft (SFB 455-A3, and SPP1089 CO260/1-1) and the European Commission (QLK2-CT-1999-00443 and QLK2-CT-2002-81399). Critical reading of the manuscript by Stefan Finke and Birgit Bossert is greatly appreciated.

References

Aoki Y, Aizaki H, Shimoike T, Tani H, Ishii K, Saito I, Matsuura Y, Miyamura T (1998) A human liver cell line exhibits efficient translation of HCV RNAs produced by a recombinant adenovirus expressing T7 RNA polymerase. Virology 250:140–150

Atreya PL, Peeples ME, Collins PL (1998) The NS1 protein of human respiratory syncytial virus is a potent inhibitor of minigenome transcription and RNA replication. J Virol 72:1452–1461

Ball LA (1992) Cellular expression of a functional nodavirus RNA replicon from vaccinia virus vectors. J Virol 66:2335–2345

Baltimore D, Huang AS, Stampfer M (1970) Ribonucleic acid synthesis of vesicular stomatitis virus, II. An RNA polymerase in the virion. Proc Natl Acad Sci USA 66:572–576

Baron MD, Barrett T (1997) Rescue of rinderpest virus from cloned cDNA. J Virol 71:1265–1271

Basler CF, Wang X, Muhlberger E, Volchkov V, Paragas J, Klenk HD, Garcia-Sastre A, Palese P (2000) The Ebola virus VP35 protein functions as a type I IFN antagonist. Proc Natl Acad Sci USA 97:12289–12294

Baudin F, Bach C, Cusack S, Ruigrok RW (1994) Structure of influenza virus RNP. I. Influenza virus nucleoprotein melts secondary structure in panhandle RNA and exposes the bases to the solvent. EMBO J 13:3158–3165

Betts AM, Stone DM (2001) Rescue of viral haemorrhagic septicaemia virus minigenomes by helper virus. Virus Res 77:19–23

Biacchesi S, Thoulouze MI, Bearzotti M, Yu YX, Bremont M (2000a) Recovery of NV knockout infectious hematopoietic necrosis virus expressing foreign genes. J Virol 74:11247–11253

Biacchesi S, Yu YX, Bearzotti M, Tafalla C, Fernandez-Alonso M, Bremont M (2000b) Rescue of synthetic salmonid rhabdovirus minigenomes. J Gen Virol 81:1941–1945

Bossert B, Conzelmann KK (2002) Respiratory syncytial virus (RSV) nonstructural (NS) proteins as host range determinants: a chimeric bovine RSV with NS genes from human RSV is attenuated in interferon-competent bovine cells. J Virol 76:4287–4293

Bridgen A, Elliott RM (1996) Rescue of a segmented negative-strand RNA virus entirely from cloned complementary DNAs. Proc Natl Acad Sci USA 93:15400–15404

Britton P, Green P, Kottier S, Mawditt KL, Penzes Z, Cavanagh D, Skinner MA (1996) Expression of bacteriophage T7 RNA polymerase in avian and mammalian cells by a recombinant fowlpox virus. J Gen Virol 77 (Pt 5):963–967

Buchholz UJ, Finke S, Conzelmann KK (1999) Generation of bovine respiratory syncytial virus (BRSV) from cDNA: BRSV NS2 is not essential for virus replication in tissue culture, and the human RSV leader region acts as a functional BRSV genome promoter. J Virol 73:251–259

Buchholz UJ, Granzow H, Schuldt K, Whitehead SS, Murphy BR, Collins PL (2000) Chimeric bovine respiratory syncytial virus with glycoprotein gene substitutions from human respiratory syncytial virus (HRSV): effects on host range and evaluation as a live-attenuated HRSV vaccine. J Virol 74:1187–1199

Cadd T, Garcin D, Tapparel C, Itoh M, Homma M, Roux L, Curran J, Kolakofsky D (1996) The Sendai paramyxovirus accessory C proteins inhibit viral genome amplification in a promoter-specific fashion. J Virol 70:5067–5074

Calain P, Curran J, Kolakofsky D, Roux L (1992) Molecular cloning of natural paramyxovirus copy-back defective interfering RNAs and their expression from DNA. Virology 191:62–71

Calain P, Roux L (1993) The rule of six, a basic feature for efficient replication of Sendai virus defective interfering RNA. J Virol 67:4822–4830

Cheetham GM, Steitz TA (2000) Insights into transcription: structure and function of single-subunit DNA-dependent RNA polymerases. Curr Opin Struct Biol 10:117–123

Clarke DK, Sidhu MS, Johnson JE, Udem SA (2000) Rescue of mumps virus from cDNA. J Virol 74:4831–4838

Collins PL, Camargo E, Hill MG (1999) Support plasmids and support proteins required for recovery of recombinant respiratory syncytial virus. Virology 259:251–255

Collins PL, Hill MG, Camargo E, Grosfeld H, Chanock RM, Murphy BR (1995) Production of infectious human respiratory syncytial virus from cloned cDNA confirms an essential role for the transcription elongation factor from the 5′ proximal open reading frame of the M2 mRNA in gene expression and provides a capability for vaccine development. Proc Natl Acad Sci USA 92:11563–11567

Collins PL, Hill MG, Cristina J, Grosfeld H (1996) Transcription elongation factor of respiratory syncytial virus, a nonsegmented negative-strand RNA virus. Proc Natl Acad Sci USA 93:81–85

Collins PL, Mink MA, Hill MG, III, Camargo E, Grosfeld H, Stec DS (1993) Rescue of a 7502-nucleotide (49.3% of full-length) synthetic analog of respiratory syncytial virus genomic RNA. Virology 195:252–256

Collins PL, Mink MA, Stec DS (1991) Rescue of synthetic analogs of respiratory syncytial virus genomic RNA and effect of truncations and mutations on the expression of a foreign reporter gene. Proc Natl Acad Sci USA 88:9663–9667

Collins PL, Murphy BR (2002) Respiratory syncytial virus: reverse genetics and vaccine strategies. Virology 296:204–211

Conzelmann KK (1996) Genetic manipulation of non-segmented negative-strand RNA viruses. J Gen Virol 77 (Pt 3):381–389

Cornu TI, de la Torre JC (2001) RING finger Z protein of lymphocytic choriomeningitis virus (LCMV) inhibits transcription and RNA replication of an LCMV S-segment minigenome. J Virol 75:9415–9426

Curran J, Boeck R, Kolakofsky D (1991) The Sendai virus P gene expresses both an essential protein and an inhibitor of RNA synthesis by shuffling modules via mRNA editing. EMBO J 10:3079–3085

Curran J, Marq JB, Kolakofsky D (1992) The Sendai virus nonstructural C proteins specifically inhibit viral mRNA synthesis. Virology 189:647–656

Curran J, Marq JB, Kolakofsky D (1995) An N-terminal domain of the Sendai paramyxovirus P protein acts as a chaperone for the NP protein during the nascent chain assembly step of genome replication. J Virol 69:849–855

Das SC, Baron MD, Barrett T (2000a) Recovery and characterization of a chimeric rinderpest virus with the glycoproteins of peste-des-petits-ruminants virus: homologous F and H proteins are required for virus viability. J Virol 74:9039–9047

Das SC, Baron MD, Skinner MA, Barrett T (2000b) Improved technique for transient expression and negative strand virus rescue using fowlpox T7 recombinant virus in mammalian cells. J Virol Methods 89:119–127

De BP, Banerjee AK (1993) Rescue of synthetic analogs of genome RNA of human parainfluenza virus type 3. Virology 196:344–348

Deng H, Wang C, Acsadi G, Wolff JA (1991) High-efficiency protein synthesis from T7 RNA polymerase transcripts in 3T3 fibroblasts. Gene 109:193–201

Deuschle U, Pepperkok R, Wang FB, Giordano TJ, McAllister WT, Ansorge W, Bujard H (1989) Regulated expression of foreign genes in mammalian cells under the control of coliphage T3 RNA polymerase and lac repressor. Proc Natl Acad Sci USA 86:5400–5404

Didcock L, Young DF, Goodbourn S, Randall RE (1999) The V protein of simian virus 5 inhibits interferon signalling by targeting STAT1 for proteasome-mediated degradation. J Virol 73:9928–9933

Dimock K, Collins PL (1993) Rescue of synthetic analogs of genomic RNA and replicative-intermediate RNA of human parainfluenza virus type 3. J Virol 67:2772–2778

Drexler I, Heller K, Wahren B, Erfle V, Sutter G (1998) Highly attenuated modified vaccinia virus Ankara replicates in baby hamster kidney cells, a potential host for virus propagation, but not in various human transformed and primary cells. J Gen Virol 79 (Pt 2):347–352

Dunn EF, Pritlove DC, Jin H, Elliott RM (1995) Transcription of a recombinant bunyavirus RNA template by transiently expressed bunyavirus proteins. Virology 211:133–143

Durbin AP, Hall SL, Siew JW, Whitehead SS, Collins PL, Murphy BR (1997) Recovery of infectious human parainfluenza virus type 3 from cDNA. Virology 235:323–332

Egelman EH, Wu SS, Amrein M, Portner A, Murti G (1989) The Sendai virus nucleocapsid exists in at least four different helical states. J Virol 63:2233–2243

Elroy-Stein O, Moss B (1990) Cytoplasmic expression system based on constitutive synthesis of bacteriophage T7 RNA polymerase in mammalian cells. Proc Natl Acad Sci USA 87:6743–6747

Enami M, Luytjes W, Krystal M, Palese P (1990) Introduction of site-specific mutations into the genome of influenza virus. Proc Natl Acad Sci USA 87:3802–3805

Evans DH, Stuart D, McFadden G (1988) High levels of genetic recombination among cotransfected plasmid DNAs in poxvirus-infected mammalian cells. J Virol 62:367–375

Fearns R, Collins PL (1999) Role of the M2-1 transcription antitermination protein of respiratory syncytial virus in sequential transcription. J Virol 73:5852–5864

Finke S, Conzelmann KK (1997) Ambisense gene expression from recombinant rabies virus: random packaging of positive- and negative-strand ribonucleoprotein complexes into rabies virions. J Virol 71:7281–7288

Finke S, Conzelmann KK (1999) Virus promoters determine interference by defective RNAs: selective amplification of mini-RNA vectors and rescue from cDNA by a 3′ copy-back ambisense rabies virus. J Virol 73:3818–3825

Finke S, Cox JH, Conzelmann KK (2000) Differential transcription attenuation of rabies virus genes by intergenic regions: generation of recombinant viruses overexpressing the polymerase gene. J Virol 74:7261–7269

Fodor E, Devenish L, Engelhardt OG, Palese P, Brownlee GG, Garcia-Sastre A (1999) Rescue of influenza A virus from recombinant DNA. J Virol 73:9679–9682

Frolov I, Hoffman TA, Pragai BM, Dryga SA, Huang HV, Schlesinger S, Rice CM (1996) Alphavirus-based expression vectors: strategies and applications. Proc Natl Acad Sci USA 93:11371–11377

Fuerst TR, Moss B (1989) Structure and stability of mRNA synthesized by vaccinia virus-encoded bacteriophage T7 RNA polymerase in mammalian cells. Importance of the 5′ untranslated leader. J Mol Biol 206:333–348

Fuerst TR, Niles EG, Studier FW, Moss B (1986) Eukaryotic transient-expression system based on recombinant vaccinia virus that synthesizes bacteriophage T7 RNA polymerase. Proc Natl Acad Sci USA 83:8122–8126

Fujii Y, Sakaguchi T, Kiyotani K, Huang C, Fukuhara N, Egi Y, Yoshida T (2002) Involvement of the leader sequence in Sendai virus pathogenesis revealed by recovery of a pathogenic field isolate from cDNA. J Virol 76:8540–8547
Garcia-Sastre A (2001) Inhibition of interferon-mediated antiviral responses by influenza A viruses and other negative-strand RNA viruses. Virology 279:375–384
Garcin D, Latorre P, Kolakofsky D (1999) Sendai virus C proteins counteract the interferon-mediated induction of an antiviral state. J Virol 73:6559–6565
Garcin D, Pelet T, Calain P, Roux L, Curran J, Kolakofsky D (1995) A highly recombinogenic system for the recovery of infectious Sendai paramyxovirus from cDNA: generation of a novel copy-back nondefective interfering virus. EMBO J 14:6087–6094
Gassen U, Collins FM, Duprex WP, Rima BK (2000) Establishment of a rescue system for canine distemper virus. J Virol 74:10737–10744
Gershon PD, Ahn BY, Garfield M, Moss B (1991) Poly(A) polymerase and a dissociable polyadenylation stimulatory factor encoded by vaccinia virus. Cell 66:1269–1278
Gotoh B, Komatsu T, Takeuchi K, Yokoo J (2001) Paramyxovirus accessory proteins as interferon antagonists. Microbiol Immunol 45:787–800
Gotoh B, Takeuchi K, Komatsu T, Yokoo J, Kimura Y, Kurotani A, Kato A, Nagai Y (1999) Knockout of the Sendai virus C gene eliminates the viral ability to prevent the interferon-alpha/beta-mediated responses. FEBS Lett 459:205–210
Grosfeld H, Hill MG, Collins PL (1995) RNA replication by respiratory syncytial virus (RSV) is directed by the N, P, and L proteins; transcription also occurs under these conditions but requires RSV superinfection for efficient synthesis of full-length mRNA. J Virol 69:5677–5686
Gubbay O, Curran J, Kolakofsky D (2001) Sendai virus genome synthesis and assembly are coupled: a possible mechanism to promote viral RNA polymerase processivity. J Gen Virol 82:2895–2903
Haller AA, Miller T, Mitiku M, Coelingh K (2000) Expression of the surface glycoproteins of human parainfluenza virus type 3 by bovine parainfluenza virus type 3, a novel attenuated virus vaccine vector. J Virol 74:11626–11635
Hammond AL, Plemper RK, Zhang J, Schneider U, Russell SJ, Cattaneo R (2001) Single-chain antibody displayed on a recombinant measles virus confers entry through the tumor-associated carcinoembryonic antigen. J Virol 75:2087–2096
Hardy RW, Harmon SB, Wertz GW (1999) Diverse gene junctions of respiratory syncytial virus modulate the efficiency of transcription termination and respond differently to M2- mediated antitermination 12. J Virol 73:170–176
Hardy RW, Wertz GW (1998) The product of the respiratory syncytial virus M2 gene ORF1 enhances readthrough of intergenic junctions during viral transcription. J Virol 72:520–526
Harty RN, Brown ME, Hayes FP, Wright NT, Schnell MJ (2001) Vaccinia virus-free recovery of vesicular stomatitis virus. J Mol Microbiol Biotechnol 3:513–517
Harty RN, Palese P (1995) Mutations within noncoding terminal sequences of model RNAs of Sendai virus: influence on reporter gene expression. J Virol 69:5128–5131
Hasan MK, Kato A, Shioda T, Sakai Y, Yu D, Nagai Y (1997) Creation of an infectious recombinant Sendai virus expressing the firefly luciferase gene from the 3′ proximal first locus. J Gen Virol 78 (Pt 11):2813–2820

Hausmann S, Jacques JP, Kolakofsky D (1996) Paramyxovirus RNA editing and the requirement for hexamer genome length. RNA 2:1033–1045

He B, Lamb RA (1999) Effect of inserting paramyxovirus simian virus 5 gene junctions at the HN/L gene junction: analysis of accumulation of mRNAs transcribed from rescued viable viruses. J Virol 73:6228–6234

He B, Paterson RG, Ward CD, Lamb RA (1997) Recovery of infectious SV5 from cloned DNA and expression of a foreign gene. Virology 237:249–260

Hoffman MA, Banerjee AK (1997) An infectious clone of human parainfluenza virus type 3. J Virol 71:4272–4277

Hoffmann E, Neumann G, Kawaoka Y, Hobom G, Webster RG (2000) A DNA transfection system for generation of influenza A virus from eight plasmids. Proc Natl Acad Sci USA 97:6108–6113

Horikami SM, Curran J, Kolakofsky D, Moyer SA (1992) Complexes of Sendai virus NP-P and P-L proteins are required for defective interfering particle genome replication in vitro. J Virol 66:4901–4908

Huang Z, Krishnamurthy S, Panda A, Samal SK (2001) High-level expression of a foreign gene from the most 3′-proximal locus of a recombinant Newcastle disease virus. J Gen Virol 82:1729–1736

Imburgio D, Rong M, Ma K, McAllister WT (2000) Studies of promoter recognition and start site selection by T7 RNA polymerase using a comprehensive collection of promoter variants. Biochemistry 39:10419–10430

Inoue K, Shoji Y, Kurane I, Iijima T, Sakai T, Morimoto K (2003). An improved method for recovering rabies virus from cloned cDNA. J Virol Methods 107:229-236.

Iseni F, Barge A, Baudin F, Blondel D, Ruigrok RW (1998) Characterization of rabies virus nucleocapsids and recombinant nucleocapsid-like structures. J Gen Virol 79 (Pt 12):2909–2919

Ito N, Takayama M, Yamada K, Sugiyama M, Minamoto N (2001) Rescue of rabies virus from cloned cDNA and identification of the pathogenicity-related gene: glycoprotein gene is associated with virulence for adult mice. J Virol 75:9121–9128

Jang SK, Davies MV, Kaufman RJ, Wimmer E (1989) Initiation of protein synthesis by internal entry of ribosomes into the 5′ nontranslated region of encephalomyocarditis virus RNA in vivo. J Virol 63:1651–1660

Jin H, Clarke D, Zhou HZ, Cheng X, Coelingh K, Bryant M, Li S (1998) Recombinant human respiratory syncytial virus (RSV) from cDNA and construction of subgroup A and B chimeric RSV. Virology 251:206–214

Johnson JE, Schnell MJ, Buonocore L, Rose JK (1997) Specific targeting to CD4+ cells of recombinant vesicular stomatitis viruses encoding human immunodeficiency virus envelope proteins. J Virol 71:5060–5068

Johnson MC, Simon BE, Kim CH, Leong JA (2000) Production of recombinant snakehead rhabdovirus: the NV protein is not required for viral replication. J Virol 74:2343–2350

Kaelin K, Spielhofer P, Schneider H, Radecke F, Kunz C, Sidhu MS, Dowling PC, Udem SA, Billeter MA. Requirements for artificial measles virus mini- and midireplicons. Abstracts of the IXth International Conference on Negative-Strand Viruses, 2–7 October, Estoril, Portugal. 1994. 2-10-0094

Kahn JS, Schnell MJ, Buonocore L, Rose JK (1999) Recombinant vesicular stomatitis virus expressing respiratory syncytial virus (RSV) glycoproteins: RSV fusion protein can mediate infection and cell fusion. Virology 254:81–91

Kato A, Sakai Y, Shioda T, Kondo T, Nakanishi M, Nagai Y (1996) Initiation of Sendai virus multiplication from transfected cDNA or RNA with negative or positive sense. Genes Cells 1:569–579

Kawano M, Kaito M, Kozuka Y, Komada H, Noda N, Nanba K, Tsurudome M, Ito M, Nishio M, Ito Y (2001) Recovery of infectious human parainfluenza type 2 virus from cDNA clones and properties of the defective virus without V-specific cysteine-rich domain. Virology 284:99–112

Keller MA, Murphy SK, Parks GD (2001) RNA replication from the simian virus 5 antigenomic promoter requires three sequence-dependent elements separated by sequence-independent spacer regions. J Virol 75:3993–3998

Kohl A, Billecocq A, Prehaud C, Yadani FZ, Bouloy M (1999) Transient gene expression in mammalian and mosquito cells using a recombinant Semliki Forest virus expressing T7 RNA polymerase. Appl Microbiol Biotechnol 53:51–56

Kolakofsky D, Pelet T, Garcin D, Hausmann S, Curran J, Roux L (1998) Paramyxovirus RNA synthesis and the requirement for hexamer genome length: the rule of six revisited. J Virol 72:891–899

Kretzschmar E, Buonocore L, Schnell MJ, Rose JK (1997) High-efficiency incorporation of functional influenza virus glycoproteins into recombinant vesicular stomatitis viruses. J Virol 71:5982–5989

Krishnamurthy S, Huang Z, Samal SK (2000) Recovery of a virulent strain of newcastle disease virus from cloned cDNA: expression of a foreign gene results in growth retardation and attenuation. Virology 278:168–182

Kuo L, Grosfeld H, Cristina J, Hill MG, Collins PL (1996) Effects of mutations in the gene-start and gene-end sequence motifs on transcription of monocistronic and dicistronic minigenomes of respiratory syncytial virus. J Virol 70:6892–6901

Lawson ND, Stillman EA, Whitt MA, Rose JK (1995) Recombinant vesicular stomatitis viruses from DNA. Proc Natl Acad Sci USA 92:4477–4481

Le Mercier P, Garcin D, Hausmann S, Kolakofsky D (2002) Ambisense sendai viruses are inherently unstable but are useful to study viral RNA synthesis. J Virol 76:5492–5502

Leyrer S, Neubert WJ, Sedlmeier R (1998) Rapid and efficient recovery of Sendai virus from cDNA: factors influencing recombinant virus rescue. J Virol Methods 75:47–58

Li HO, Zhu YF, Asakawa M, Kuma H, Hirata T, Ueda Y, Lee YS, Fukumura M, Iida A, Kato A, Nagai Y, Hasegawa M (2000) A cytoplasmic RNA vector derived from nontransmissible Sendai virus with efficient gene transfer and expression. J Virol 74:6564–6569

Lieber A, Kiessling U, Strauss M (1989) High level gene expression in mammalian cells by a nuclear T7-phase RNA polymerase. Nucleic Acids Res 17:8485–8493

Lopez N, Jacamo R, Franze-Fernandez MT (2001) Transcription and RNA replication of tacaribe virus genome and antigenome analogs require N and L proteins: Z protein is an inhibitor of these processes. J Virol 75:12241–12251

Lopez N, Muller R, Prehaud C, Bouloy M (1995) The L protein of Rift Valley fever virus can rescue viral ribonucleoproteins and transcribe synthetic genome-like RNA molecules. J Virol 69:3972–3979

Luytjes W, Krystal M, Enami M, Pavin JD, Palese P (1989) Amplification, expression, and packaging of foreign gene by influenza virus. Cell 59:1107–1113

Mavrakis M, Kolesnikova L, Schoehn G, Becker S, Ruigrok RW (2002) Morphology of Marburg Virus NP-RNA. Virology 296:300–307

Mebatsion T, Conzelmann KK (1996) Specific infection of CD4+ target cells by recombinant rabies virus pseudotypes carrying the HIV-1 envelope spike protein. Proc Natl Acad Sci USA 93:11366–11370

Mebatsion T, Finke S, Weiland F, Conzelmann KK (1997) A CXCR4/CD4 pseudotype rhabdovirus that selectively infects HIV-1 envelope protein-expressing cells. Cell 90:841–847

Mebatsion T, Konig M, Conzelmann KK (1996) Budding of rabies virus particles in the absence of the spike glycoprotein. Cell 84:941–951

Mebatsion T, Weiland F, Conzelmann KK (1999) Matrix protein of rabies virus is responsible for the assembly and budding of bullet-shaped particles and interacts with the transmembrane spike glycoprotein G. J Virol 73:242–250

Moyer SA, Smallwood-Kentro S, Haddad A, Prevec L (1991) Assembly and transcription of synthetic vesicular stomatitis virus nucleocapsids. J Virol 65:2170–2178

Muhlberger E, Lotfering B, Klenk HD, Becker S (1998) Three of the four nucleocapsid proteins of Marburg virus, NP, VP35, and L, are sufficient to mediate replication and transcription of Marburg virus-specific monocistronic minigenomes. J Virol 72:8756–8764

Muhlberger E, Weik M, Volchkov VE, Klenk HD, Becker S (1999) Comparison of the transcription and replication strategies of Marburg virus and Ebola virus by using artificial replication systems. J Virol 73:2333–2342

Murphy SK, Ito Y, Parks GD (1998) A functional antigenomic promoter for the paramyxovirus simian virus 5 requires proper spacing between an essential internal segment and the 3′ terminus. J Virol 72:10–19

Murphy SK, Parks GD (1997) Genome nucleotide lengths that are divisible by six are not essential but enhance replication of defective interfering RNAs of the paramyxovirus simian virus 5. Virology 232:145–157

Nakaya T, Cros J, Park MS, Nakaya Y, Zheng H, Sagrera A, Villar E, Garcia-Sastre A, Palese P (2001) Recombinant Newcastle disease virus as a vaccine vector. J Virol 75:11868–11873

Newman JT, Surman SR, Riggs JM, Hansen CT, Collins PL, Murphy BR, Skiadopoulos MH (2002) Sequence analysis of the Washington/1964 strain of human parainfluenza virus type 1 (HPIV1) and recovery and characterization of wild-type recombinant HPIV1 produced by reverse genetics. Virus Genes 24:77–92

Oakley JL, Coleman JE (1977) Structure of a promoter for T7 RNA polymerase. Proc Natl Acad Sci USA 74:4266–4270

Olivo PD, Collins PL, Peeples ME, Schlesinger S (1998) Detection and quantitation of human respiratory syncytial virus (RSV) using minigenome cDNA and a Sindbis virus replicon: a prototype assay for negative-strand RNA viruses. Virology 251:198–205

Oomens AGP, Megaw AG, Wertz GW (2003) Infectivity of a human respiratory syncytial virus lacking the SH, G, and F proteins efficiently mediated by the vesicular stomatitis virus G protein. J Virol 77:3785–98.

Parisien JP, Lau JF, Rodriguez JJ, Sullivan BM, Moscona A, Parks GD, Lamb RA, Horvath CM (2001) The V protein of human parainfluenza virus 2 antagonizes type I interferon responses by destabilizing signal transducer and activator of transcription 2. Virology 283:230–239

Park KH, Huang T, Correia FF, Krystal M (1991) Rescue of a foreign gene by Sendai virus. Proc Natl Acad Sci USA 88:5537–5541

Parks CL, Lerch RA, Walpita P, Sidhu MS, Udem SA (1999) Enhanced measles virus cDNA rescue and gene expression after heat shock. J Virol 73:3560–3566

Parks CL, Wang HP, Kovacs GR, Vasilakis N, Kowalski J, Nowak RM, Lerch RA, Walpita P, Sidhu MS, Udem SA (2002) Expression of a foreign gene by recombinant canine distemper virus recovered from cloned DNAs. Virus Res 83:131–147

Pattnaik AK, Ball LA, LeGrone AW, Wertz GW (1992) Infectious defective interfering particles of VSV from transcripts of a cDNA clone. Cell 69:1011–1020

Pattnaik AK, Wertz GW (1990) Replication and amplification of defective interfering particle RNAs of vesicular stomatitis virus in cells expressing viral proteins from vectors containing cloned cDNAs 42. J Virol 64:2948–2957

Pattnaik AK, Wertz GW (1991) Cells that express all five proteins of vesicular stomatitis virus from cloned cDNAs support replication, assembly, and budding of defective interfering particles. Proc Natl Acad Sci USA 88:1379–1383

Peeters BP, de Leeuw OS, Koch G, Gielkens AL (1999) Rescue of Newcastle disease virus from cloned cDNA: evidence that cleavability of the fusion protein is a major determinant for virulence. J Virol 73:5001–5009

Perrotta AT, Been MD (1990) The self-cleaving domain from the genomic RNA of hepatitis delta virus: sequence requirements and the effects of denaturant. Nucleic Acids Res 18:6821–6827

Poole E, He B, Lamb RA, Randall RE, Goodbourn S (2002) The V proteins of Simian virus 5 and other paramyxoviruses inhibit induction of interferon-β. Virology 303:33–46

Pringle CR (1997) The order Mononegavirales—current status. Arch Virol 142:2321–2326

Radecke F, Spielhofer P, Schneider H, Kaelin K, Huber M, Dotsch C, Christiansen G, Billeter MA (1995) Rescue of measles viruses from cloned DNA. EMBO J 14:5773–5784

Roberts A, Rose JK (1998) Recovery of negative-strand RNA viruses from plasmid DNAs: a positive approach revitalizes a negative field. Virology 247:1–6

Roberts A, Rose JK (1999) Redesign and genetic dissection of the rhabdoviruses. Adv Virus Res 53:301–319

Romer-Oberdorfer A, Mundt E, Mebatsion T, Buchholz UJ, Mettenleiter TC (1999) Generation of recombinant lentogenic Newcastle disease virus from cDNA. J Gen Virol 80 (Pt 11):2987–2995

Samal SK, Collins PL (1996) RNA replication by a respiratory syncytial virus RNA analog does not obey the rule of six and retains a nonviral trinucleotide extension at the leader end. J Virol 70:5075–5082

Schlender J, Bossert B, Buchholz U, Conzelmann KK (2000) Bovine respiratory syncytial virus nonstructural proteins NS1 and NS2 cooperatively antagonize alpha/beta interferon-induced antiviral response. J Virol 74:8234–8242

Schmidt AC, McAuliffe JM, Huang A, Surman SR, Bailly JE, Elkins WR, Collins PL, Murphy BR, Skiadopoulos MH (2000) Bovine parainfluenza virus type 3 (BPIV3) fusion and hemagglutinin-neuraminidase glycoproteins make an important contribution to the restricted replication of BPIV3 in primates. J Virol 74:8922–8929
Schneider H, Spielhofer P, Kaelin K, Dotsch C, Radecke F, Sutter G, Billeter MA (1997) Rescue of measles virus using a replication-deficient vaccinia-T7 vector. J Virol Methods 64:57–64
Schneider U, Bullough F, Vongpunsawad S, Russell SJ, Cattaneo R (2000) Recombinant measles viruses efficiently entering cells through targeted receptors. J Virol 74:9928–9936
Schnell MJ, Buonocore L, Kretzschmar E, Johnson E, Rose JK (1996) Foreign glycoproteins expressed from recombinant vesicular stomatitis viruses are incorporated efficiently into virus particles. Proc Natl Acad Sci USA 93:11359–11365
Schnell MJ, Johnson JE, Buonocore L, Rose JK (1997) Construction of a novel virus that targets HIV-1-infected cells and controls HIV-1 infection. Cell 90:849–857
Schnell MJ, Mebatsion T, Conzelmann KK (1994) Infectious rabies viruses from cloned cDNA. EMBO J 13:4195–4203
Sharmeen L, Kuo MY, Dinter-Gottlieb G, Taylor J (1988) Antigenomic RNA of human hepatitis delta virus can undergo self-cleavage. J Virol 62:2674–2679
Shih IH, Been MD (2002) Catalytic strategies of the hepatitis delta virus ribozymes. Annu Rev Biochem 71:887–917
Sidhu MS, Chan J, Kaelin K, Spielhofer P, Radecke F, Schneider H, Masurekar M, Dowling PC, Billeter MA, Udem SA (1995) Rescue of synthetic measles virus minireplicons: measles genomic termini direct efficient expression and propagation of a reporter gene. Virology 208:800–807
Smallwood S, Moyer SA (1993) Promoter analysis of the vesicular stomatitis virus RNA polymerase. Virology 192:254–263
Spielhofer P, Bachi T, Fehr T, Christiansen G, Cattaneo R, Kaelin K, Billeter MA, Naim HY (1998) Chimeric measles viruses with a foreign envelope. J Virol 72:2150–2159
Stope MB, Karger A, Schmidt U, Buchholz UJ (2001) Chimeric bovine respiratory syncytial virus with attachment and fusion glycoproteins replaced by bovine parainfluenza virus type 3 hemagglutinin-neuraminidase and fusion proteins. J Virol 75:9367–9377
Sutherland KA, Collins PL, Peeples ME (2001) Synergistic effects of gene-end signal mutations and the M2-1 protein on transcription termination by respiratory syncytial virus. Virology 288:295–307
Sutter G, Ohlmann M, Erfle V (1995) Non-replicating vaccinia vector efficiently expresses bacteriophage T7 RNA polymerase. FEBS Lett 371:9–12
Takeda M, Takeuchi K, Miyajima N, Kobune F, Ami Y, Nagata N, Suzaki Y, Nagai Y, Tashiro M (2000) Recovery of pathogenic measles virus from cloned cDNA. J Virol 74:6643–6647
Tapparel C, Maurice D, Roux L (1998) The activity of Sendai virus genomic and antigenomic promoters requires a second element past the leader template regions: a motif (GNNNNN)3 is essential for replication. J Virol 72:3117–3128

Thomas D, Newcomb WW, Brown JC, Wall JS, Hainfeld JF, Trus BL, Steven AC (1985) Mass and molecular composition of vesicular stomatitis virus: a scanning transmission electron microscopy analysis. J Virol 54:598–607

Tomanin R, Bett AJ, Picci L, Scarpa M, Graham FL (1997) Development and characterization of a binary gene expression system based on bacteriophage T7 components in adenovirus vectors. Gene 193:129–140

Usdin TB, Brownstein MJ, Moss B, Isaacs SN (1993) SP6 RNA polymerase containing vaccinia virus for rapid expression of cloned genes in tissue culture. Biotechniques 14:222-224

Vidal S, Kolakofsky D (1989) Modified model for the switch from Sendai virus transcription to replication. J Virol 63:1951–1958

Volchkov VE, Volchkova VA, Muhlberger E, Kolesnikova LV, Weik M, Dolnik O, Klenk HD (2001) Recovery of infectious Ebola virus from complementary DNA: RNA editing of the GP gene and viral cytotoxicity. Science 291:1965–1969

von Messling V, Zimmer G, Herrler G, Haas L, Cattaneo R (2001) The hemagglutinin of canine distemper virus determines tropism and cytopathogenicity. J Virol 75:6418–6427

Vulliemoz D, Roux L (2001) "Rule of six": how does the Sendai virus RNA polymerase keep count? J Virol 75:4506–4518

Wagner E, Engelhardt OG, Gruber S, Haller O, Kochs G (2001) Rescue of recombinant Thogoto virus from cloned cDNA. J Virol 75:9282–9286

Weik M, Modrof J, Klenk HD, Becker S, Muhlberger E (2002) Ebola virus VP30-mediated transcription is regulated by RNA secondary structure formation. J Virol 76:8532–8539

Wertz GW, Perepelitsa VP, Ball LA (1998) Gene rearrangement attenuates expression and lethality of a nonsegmented negative strand RNA virus. Proc Natl Acad Sci USA 95:3501–3506

Whelan SP, Ball LA, Barr JN, Wertz GT (1995) Efficient recovery of infectious vesicular stomatitis virus entirely from cDNA clones. Proc Natl Acad Sci USA 92:8388–8392

Willenbrink W, Neubert WJ (1994) Long-term replication of Sendai virus defective interfering particle nucleocapsids in stable helper cell lines. J Virol 68:8413–8417

Wyatt LS, Moss B, Rozenblatt S (1995) Replication-deficient vaccinia virus encoding bacteriophage T7 RNA polymerase for transient gene expression in mammalian cells. Virology 210:202–205

Yao XD, Evans DH (2001) Effects of DNA structure and homology length on vaccinia virus recombination. J Virol 75:6923–6932

Young DF, Didcock L, Goodbourn S, Randall RE (2000) Paramyxoviridae use distinct virus-specific mechanisms to circumvent the interferon response. Virology 269:383–390

Yu Q, Hardy RW, Wertz GW (1995) Functional cDNA clones of the human respiratory syncytial (RS) virus N, P, and L proteins support replication of RS virus genomic RNA analogs and define minimal *trans*-acting requirements for RNA replication. J Virol 69:2412–2419

Yunus AS, Khattar SK, Collins PL, Samal SK (2001) Rescue of bovine respiratory syncytial virus from cloned cDNA: entire genome sequence of BRSV strain A51908. Virus Genes 23:157–164

Reverse Genetics Systems for the Generation of Segmented Negative-Sense RNA Viruses Entirely from Cloned cDNA

G. Neumann[1] · Y. Kawaoka[1,2,3]

[1] Department of Pathobiological Sciences, School of Veterinary Medicine, University of Wisconsin-Madison, 2015 Linden Drive, Madison, WI 53706, USA
E-mail: kawaokay@svm.vetmed.wisc.edu

[2] Department of Microbiology and Immunology, Institute of Medical Science, University of Tokyo, 108-9639 Tokyo, Japan

[3] Core Research for Evolutional Science and Technology, Japan Science and Technology Corporation, Kawaguchi, 332-0012 Saitama, Japan

Abstract Reverse genetics is defined as the generation of virus entirely from cloned cDNA. For negative-sense RNA viruses, whose genomes are complementary to mRNA in their orientation, the viral RNA(s) and the viral proteins required for replication and translation must be provided to initiate the viral replication cycle. Segmented negative-sense RNA viruses were refractory to genetic manipulation until 1989. In this chapter, we review developments in the reverse genetics of segmented negative-sense RNA viruses, beginning with the in vitro reconstitution of viral polymerase complexes in the late 1980s and culminating in the generation of Bunyamwera and influenza virus entirely from plasmid DNA almost a decade later.

1 Introduction

In the late 1970s, the advent of recombinant DNA technology opened the door for the conversion of viral RNA genomes into cDNA clones that can be modified at will by site-directed mutagenesis. This ability soon translated into the de novo synthesis of positive-sense RNA viruses, in which the naked RNA is infectious. Hence, transfection of cells with in vitro-transcribed viral RNA or with a plasmid containing a full-length cDNA copy of the viral genome that is transcribed intracellularly initiates viral replication. The first successful attempt to generate an RNA virus from cloned cDNA was reported in 1978 for the bacteriophage Qβ (Taniguchi et al. 1978). Generation of the first animal virus, poliovirus, from cloned cDNA was reported in 1981 (Racaniello and Baltimore 1981). Infectious clones have since been obtained for a large number of positive-sense RNA viruses (reviewed in Boyer and Haenni 1994).

The genomes of negative-sense RNA viruses are complementary to messenger RNA in their orientation and therefore are not infectious by themselves. To initiate viral replication, negative-sense RNA viruses rely on virion-packaged RNA-dependent RNA polymerase, which transcribes negative-sense viral RNA (vRNA) into positive-sense mRNA. Researchers therefore faced the challenge of providing the vRNA together with components of the viral polymerase complex and the nucleoprotein (which encapsidates the vRNA). In 1989, Luytjes et al. (1989) described the first system for the modification of a negative-sense RNA virus, exemplified by the introduction of a ninth segment (encoding a reporter gene) into the influenza viral genome. This advance was quickly followed by the generation of an influenza virus containing a mutation in one of the viral gene segments (Enami et al. 1990). The generation of a negative-sense RNA virus entirely from cloned cDNA was not achieved until 1994, when Schnell et al. (1994) produced rabies virus, a nonsegmented negative-sense RNA virus belonging to the family *Rhabdoviridae*. A plasmid containing a full-length cDNA copy of the viral antigenome, controlled by the bacteriophage T7 RNA polymerase promoter, was transfected into cells that expressed T7 RNA polymerase, provided from recombinant vaccinia virus. The components of the viral polymerase complex and the NP protein were provided from protein expression plasmids, all controlled by the T7 RNA polymerase promoter. The report by Schnell et al. (1994) was soon followed by the generation of nonsegmented negative-sense RNA viruses of the *Rhabdoviridae, Paramyxoviridae,* and *Filoviridae* families (Collins et al. 1995; Garcin et al. 1995; Lawson et al. 1995; Radecke et al. 1995; Whelan et al. 1995; Kato et al. 1996; Baron and Barrett 1997; Durbin et al. 1997; He et al. 1997; Hoffman and Banerjee 1997; Schneider et al. 1997; Buchholz et al. 1999; Peeters et al. 1999; Romer-Oberdorfer et al. 1999; Clarke et al. 2000; Gassen et al. 2000; Haller et al. 2000; Krishnamurthy et al. 2000; Takeda et al. 2000; Kawano et al. 2001; Volchkov et al. 2001; Neumann 2002).

Producing segmented negative-sense RNA viruses from cloned cDNA is even more challenging because more than one genomic RNA must be provided. In 1996, Bridgen and Elliott (1996) generated a member of the *Bunyaviridae* family, characterized by a genome consisting of three segments of negative-sense viral RNA, entirely from cloned cDNA. The de novo synthesis of influenza virus, which comprises eight segments of negative-sense viral RNA, was finally accomplished in 1999 (Fodor et al. 1999; Neumann et al. 1999). In this chapter, we summarize developments in the reverse genetics of segmented negative-sense RNA viruses, beginning with the in vitro reconstitution of viral polymerase complexes in the late 1980s and culminating in the generation of influenza virus entirely from plasmid DNA a decade later.

2
In Vitro Reconstitution of Influenza Virus Ribonucleoprotein Complexes

The minimal influenza virus replication unit, the viral ribonucleoprotein (vRNP) complex, is composed of three polymerase subunits (PB2, PB1, and PA), nucleoprotein (NP), and viral RNA. Enzymatically active vRNP complexes were first isolated from purified virus (Plotch et al. 1981) and later from infected cells (Beaton and Krug 1986). During the next several years, methods were established to purify the individual components of vRNP complexes, followed by their in vitro reconstitution. Honda et al. (1987) first purified vRNP complexes by glycerol gradient centrifugation and then applied a discontinuous cesium chloride-glycerol gradient to separate the NP from the RNA polymerase-RNA fraction. The latter can be further separated into polymerase proteins and vRNA by cesium trifluoroacetate gradient centrifugation (Honda et al. 1988, 1990).

Parvin et al. (1989) first reconstituted synthetic viral RNA into functional vRNP complexes. Plasmid DNA encoding short model influenza viral RNA encompassing the genomic termini was transcribed in vitro from the T7 RNA polymerase promoter and the resulting RNA transcript was then mixed with purified polymerase and NP proteins, leading to functional vRNP complexes. More importantly, this reconstitution system allowed the synthesis of full-length viral transcripts from purified viral RNA, or from in vitro-synthesized viral transcripts. In a different approach, purified RNP cores were treated with micrococcal nuclease to deplete the viral RNA (Seong and Brownlee 1992; Seong et al. 1992). The micrococcal nuclease-treated complexes were then mixed with in vitro-synthesized short model RNAs to reconstitute vRNP complexes.

3
Cell Culture Systems to Study Influenza Virus Replication

Huang et al. (1990) assembled RNP complexes from purified polymerase and NP proteins and an in vitro-synthesized virus-like RNA that contained the reading frame for the reporter gene chloramphenicol-acetyltransferase (CAT) flanked by the nontranslated regions of the NS segment. The in vitro-assembled RNP complexes were transfected into cells infected with recombinant vaccinia virus expressing the NP and polymerase proteins. CAT expression demonstrated the in vivo formation of functional RNP complexes that replicated and transcribed the virus-like RNA.

For the intracellular reconstitution of vRNP complexes, the polymerase and NP proteins can be provided by recombinant vaccinia virus (Huang et al. 1990), recombinant simian virus 40 (de la Luna et al. 1993), or cell lines stably expressing these proteins (Kimura et al. 1992) or from protein expression plasmids controlled by the T7 RNA polymerase promoter (Mena et al. 1994). For most of these systems, the vRNA was provided by transfection of in vitro-assembled vRNP complexes; however, transfected naked viral RNA was also found to initiate viral replication and transcription (de la Luna et al. 1993; Mena et al. 1994). These systems opened the door to study of all aspects of viral replication intracellularly. The downside of providing the viral polymerase and NP proteins from protein expression systems is that the ratios among the viral proteins may not be identical to those in infected cells, which may affect the fine-tuning of viral replication.

4
Ribonucleoprotein Transfection Method

In 1989, Luytjes et al. (1989) devised the first system to modify the genome of a negative-sense RNA virus, influenza virus (Fig. 1A). These authors generated a plasmid containing the T7 RNA polymerase promoter, a cDNA encoding the 5′ nontranslated region of the influenza A virus NS segment, the coding region for CAT in antisense orientation, a cDNA encoding the 3′ nontranslated region of the NS segment, and a recognition site for a restriction endonuclease to generate the 3′ ends of viral transcripts. In vitro transcription by T7 RNA polymerase thus yielded a virus-like transcript (encoding CAT) with authentic influenza virus 5′ and 3′ ends. The addition of purified NP and polymerase proteins allowed the formation of vRNP complexes that were subsequently transfected into eukaryotic cells. Before or after vRNP transfection, cells were infected with influenza helper virus to provide the remaining viral RNA segments. CAT expression in RNP-transfected and helper virus-infected cells verified the amplification of virus-like RNA by the viral polymerase complex. Virus-like RNAs encoding CAT were packaged into progeny virus particles, as demonstrated by expression of the reporter gene in cells infected with supernatant derived from RNP-transfected and helper virus-infected cells. However, CAT expression was greatly reduced after three passages in cell culture, indicating that the additional segment encoding CAT was not stably maintained.

Using the RNP transfection method, Enami et al. (1990) generated an influenza virus containing a viral gene segment derived from cloned cDNA. In Madin-Darby bovine kidney (MDBK) cells, the NA gene of A/WSN/33 virus renders viral replication independent of trypsin. By

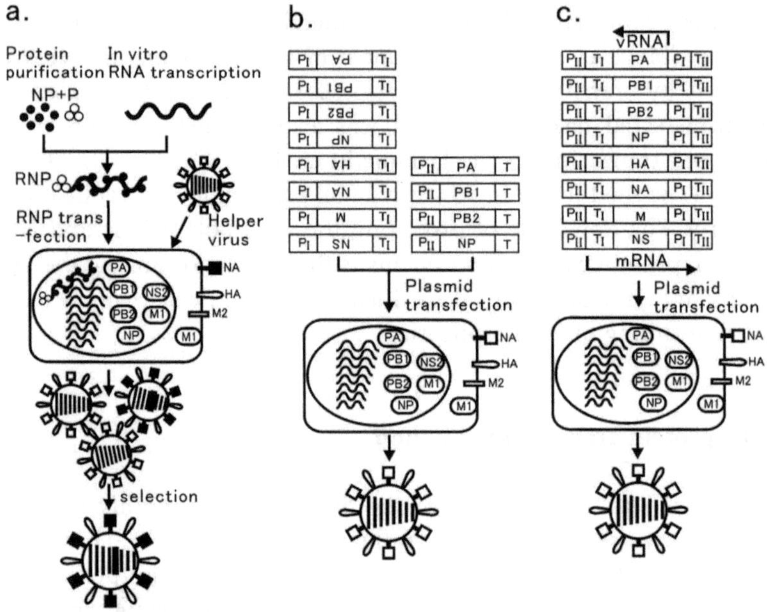

Fig. 1A–C Reverse genetics systems for the modification of influenza virus. **A** RNP transfection method. In vitro-synthesized viral RNA is mixed with purified polymerase and NP proteins to reconstitute vRNP complexes. Artificially assembled vRNP complexes are transfected into eukaryotic cells that are infected with influenza helper virus to provide the remaining seven vRNPs. **B** RNA polymerase I system for the de novo synthesis of influenza virus. Viral cDNAs are cloned in negative-sense orientation between RNA polymerase I promoter and terminator sequences, and after transfection of the resulting constructs into eukaryotic cells, vRNAs are synthesized by the cellular enzyme RNA polymerase I. The polymerase and NP proteins are provided from protein expression plasmids. **C** RNA polymerase I/II system. Viral cDNAs are cloned in negative-sense orientation between RNA polymerase I promoter and terminator sequences, and the resulting cassette is flanked by a RNA polymerase II promoter and polyadenylation signal in positive-sense orientation. In eukaryotic cells, both vRNA and mRNA are synthesized from the same template, eliminating the need for protein expression constructs

contrast, viruses containing NA genes derived from other human strains and HAs uncleavable by ubiquitous proteases (e.g., furin) rely on the addition of trypsin for their amplification. When MDBK cells were transfected with artificially assembled vRNP complexes encoding the A/WSN/33 NA protein, and infected with helper virus encoding NA derived from a different virus, recombinant viruses containing the A/WSN/33 NA gene were selected in the absence of trypsin.

A variation of the RNP transfection method used micrococcal nuclease-treated RNP complexes that were incubated with in vitro-synthesized vRNA encoding CAT. Transfection of in vitro-reconstituted vRNP complexes, followed by influenza helper virus infection, resulted in CAT expression in infected and transfected cells (Kim et al. 1997).

The RNP transfection method allowed the generation of transfectant influenza viruses containing a gene derived from cloned cDNA. Selection of the transfectant virus against the background of (wild type) helper virus required that strong selection systems be established. Thus selection systems based on temperature sensitivity (Enami et al. 1991; Yasuda et al. 1994; Li et al. 1995), host-range restriction (Enami et al. 1990; Subbarao et al. 1993), antibody selection (Enami and Palese 1991; Horimoto and Kawaoka 1994; Barclay and Palese 1995; Rowley et al. 1999), or drug resistance (Castrucci and Kawaoka 1995) were developed to genetically modify the influenza A virus PB2 (Subbarao et al. 1993), HA (Enami and Palese 1991; Horimoto and Kawaoka 1994), NP (Li et al. 1995), NA (Enami et al. 1990), M (Castrucci and Kawaoka 1995), and NS (Enami et al. 1991) segments, as well as the influenza B virus HA (Barclay and Palese 1995) and NA (Rowley et al. 1999) segments. Although these systems were technically demanding, resulting in a low rate of recovery of the transfectant viruses, they conferred the ability to alter the influenza virus genome, thus contributing significantly to our understanding of this virus.

5
A System for the Mutagenesis of Any Influenza Virus Gene Segment

Enami et al. (2000) established a system for the mutagenesis of any influenza virus gene segment. Purified vRNP complexes were incubated with cDNA of the target segment to form an RNA-DNA hybrid, which was digested with RNaseH. vRNP complexes depleted of the target vRNA were then mixed with in vitro-synthesized, genetically modified viral target RNA. Transfection of the full set of influenza A vRNPs into eukaryotic cells yielded transfectant viruses. This approach does not require selection systems, but it is quite technically demanding and has therefore not been adopted by other laboratories.

6 RNA Polymerase I System

The systems described so far rely on the in vitro reconstitution of influenza virus RNP complexes. However, in vitro RNA synthesis, in vitro RNP reconstitution, and RNP transfection are all technically demanding, inefficient, or both; hence, strategies were pursued to generate influenza virus RNA intracellularly. In contrast to most other negative-sense RNA viruses, influenza virus replicates in the nucleus of infected cells. Consequently, the intracellular synthesis of viral RNA requires not only the generation of full-length transcripts with authentic 5′ and 3′ ends, but also their localization to the cell nucleus. This goal was achieved in 1993, when the RNA polymerase I system for influenza vRNA synthesis was established (Zobel et al. 1993; Neumann et al. 1994). RNA polymerase I transcribes ribosomal RNA (rRNA) in the nucleus of eukaryotic cells. Because RNA polymerase I transcripts do not contain 5′ cap or 3′ polyA structures, it seemed feasible to use this enzyme for the nuclear synthesis of (noncapped and nonpolyadenylated) influenza viral RNAs. This concept was tested by generating a plasmid that contained the coding region for HA or CAT (in antisense orientation) flanked by the 5′ and 3′ noncoding regions of HA; this cassette was then fused to RNA polymerase I promoter and terminator sequences (Zobel et al. 1993; Neumann et al. 1994). In vitro and in vivo studies demonstrated the RNA polymerase I-driven synthesis of full-length viral transcripts from such a plasmid. The viral RNA transcripts were not only replicated and transcribed by the viral polymerase complex but also became packaged into virions on infection with influenza helper virus.

In 1996, Pleschka et al. (1996) devised an entirely plasmid-driven minireplicon system. The original RNA polymerase I system was first modified by replacing the RNA polymerase I terminator with the hepatitis delta ribozyme. Cotransfection of an RNA polymerase I-ribozyme construct encoding CAT in antisense orientation, together with protein expression plasmids for the NP and polymerase proteins, resulted in CAT expression. This approach also allowed the generation of transfectant influenza virus containing the NA gene derived from cloned cDNA (Pleschka et al. 1996).

7
Generation of Influenza A Virus Entirely from Plasmids—RNA Polymerase I System

Influenza A virus was finally generated entirely from plasmids in 1999, with the RNA polymerase I system (Fig. 1B; Fodor et al. 1999; Neumann et al. 1999). Neumann et al. (1999) cloned cDNAs encoding all eight segments of A/WSN/33 virus in negative orientation between RNA polymerase I promoter and terminator sequences. On transfection of the eight resulting plasmids into 293T (human embryonic kidney) cells, nuclear RNA polymerase I transcribed negative-sense influenza viral RNAs. The proteins required for transcription and replication of the vRNAs (i.e., the nucleoprotein NP and the polymerase proteins PB2, PB1, and PA) were provided from cotransfected protein expression constructs, controlled by the chicken β-actin promoter. Hence, this approach required cotransfection of 12 plasmids. In an alternative strategy, cells were cotransfected with protein expression plasmids for all nine viral structural proteins (i.e., PB2, PB1, PA, HA, NP, NA, M1, M2, and NS2), together with eight RNA polymerase I plasmids for vRNA synthesis, resulting in the cotransfection of 17 plasmids. Nonetheless, more than 10^7 infectious viruses per milliliter of supernatant were produced (Neumann et al. 1999). Since the initial report, the efficiency of virus recovery has been further improved, now allowing one to routinely generate more than 10^8 infectious viruses per milliliter of supernatant derived from transfected cells.

Fodor et al. (1999) reported a similar approach for the de novo synthesis of influenza virus. In their system, all eight viral RNAs were synthesized from plasmids containing the RNA polymerase I promoter and the hepatitis delta ribozyme. Transfection of Vero cells with all eight RNA polymerase I-ribozyme constructs, and with protein expression constructs for the NP and polymerase protein (controlled by the adenovirus type 2 major late promoter), yielded infectious influenza virus.

With these advances, the once-unattainable goal of generating influenza virus entirely from cloned cDNA had become a reality. The tools were now available to alter the genome of influenza A viruses without any technical limitations. The RNA polymerase I system is undemanding and straightforward, because it requires only DNA cloning, DNA purification, and DNA transfection techniques, which are widely established in molecular biology and virology laboratories. Because this system is entirely plasmid based, helper virus infection and cumbersome selection of transfectant viruses are no longer required. Despite the fact that at least 12 plasmids are required for influenza A virus generation, this approach is the most efficient reverse genetics system for negative-sense RNA viruses described to date, allowing the generation of a spectrum of viruses, even those with severe growth defects.

8
The RNA Polymerase I/II System for the Generation of Influenza A Virus

Hoffmann et al. (2000a,b) devised a modified RNA polymerase I system that reduced the number of plasmids required for influenza virus generation from 12 to 8 (Fig. 1C). In this approach, cDNAs encoding viral RNAs are inserted in positive-sense orientation between an RNA polymerase II promoter and a polyadenylation signal. This cassette is then inserted in negative-sense orientation between the RNA polymerase I promoter and terminator sequences. Hence, capped and polyadenylated mRNAs for the expression of proteins (synthesized by RNA polymerase II) and the negative-sense vRNAs (synthesized by RNA polymerase I) are generated from the same template. Consequently, influenza virus can be generated from eight RNA polymerase I/II plasmids that direct the synthesis of both vRNA and mRNA. The RNA polymerase I/II system may prove advantageous for virus generation in cell lines that cannot be efficiently transfected with 12 plasmids. On the other hand, the flexibility of virus generation is reduced, because both protein expression and vRNA synthesis are achieved from the same template. For example, the generation of virus-like particles lacking one viral gene segment (see below) is not feasible with this approach. Nonetheless, the RNA polymerase I/II system provides an alternative approach to the de novo synthesis of influenza viruses.

9
Robustness of the RNA Polymerase I System

Early reports on RNA polymerase I transcription suggested that both the promoter and terminator elements extend into the transcribed region (reviewed in Paule 1998). Replacement of the ribosomal DNA template with heterologous templates might therefore impair the accuracy and/or efficiency of transcription initiation or termination, or perhaps both. However, the RNA polymerase I transcription system has been used for the synthesis of influenza virus (Fodor et al. 1999; Neumann et al. 1999), Thogotovirus (Wagner et al. 2000, 2001), and Uukuniemi virus-like transcripts (Flick and Pettersson 2001), demonstrating that the RNA polymerase I promoter up to the transcription initiation site and the conserved terminator element are sufficient for the synthesis of authentic viral transcripts.

For RNA polymerase I transcription, species specificity is conferred by a transcription initiation factor that binds to the RNA polymerase I promoter (reviewed in Paule 1998). This specificity might limit the use of the system to cell lines derived from the organism from which the promoter and terminator sequences were obtained. However, the human RNA polymerase I promoter is functional not only in human cells, such as the 293T cell line (Neumann et al. 1999; Hoffmann and Webster, 2000), but also in African green monkey kidney cells such as COS-1 (Hoffmann and Webster 2000) or Vero (Fodor et al. 1999). The mouse RNA polymerase I promoter has so far been used in mouse fibroblasts cell lines (e.g., B82 or NIH3T3 cells) (Neumann et al. 1994), as well as in baby hamster kidney (BHK21) cells (Flick and Pettersson 2001). Moreover, the RNA polymerase I promoter and terminator elements are functional when derived from closely related organisms. Influenza A virus generation was achieved with a human RNA polymerase I promoter, whereas the terminator element was of murine origin (Neumann et al. 1999). With the availability of both human and murine RNA polymerase I transcription cassettes, a number of commonly used cell lines can be exploited for RNA polymerase I-based experiments.

The A/WSN/33 virus, the first influenza virus to be generated entirely from plasmids (Fodor et al. 1999; Neumann et al. 1999), has an extensive history of growth in mice and in cell culture. Subsequent generation of viruses with a limited growth history in cell culture, such as A/Udorn/72 (Takeda et al. 2002) or A/Memphis/8/88 (Hatta et al. 2002), as well as viruses derived from humans (e.g., A/Hong Kong/483/97 and A/Hong Kong/486/97) (Hatta et al. 2001) or animal species (e.g., A/Teal/HK/W312/97) (Hoffmann et al. 2000b) or A/Mallard/New York/6750/78 (Hatta et al. 2002), demonstrated the robustness of the RNA polymerase I systems for influenza virus generation.

10
Generation of Virus-like Particles

With the new-found ability to generate influenza viruses, researchers began to explore the use of virus-like particles (VLPs), that is, particles lacking one or more genome segments, for vaccine and gene delivery purposes. Because these VLPs lack the coding information for one or more viral structural proteins, they do not permit the generation of infectious progeny viruses. VLPs can be generated by expressing all viral structural proteins and a virus-like RNA encoding a reporter gene (Mena et al. 1996; Gomez-Puertas et al. 1999; Neumann et al. 2000b). Neumann et al. (2000b) generated VLPs that contained seven vRNAs, excluding the NS vRNA. In the absence of the NS segment, which encodes the NS1 and NS2 proteins, vRNPs were retained in the nucleus. Further studies with an NS vRNA encoding NS1, but not NS2, identified a critical role for NS2 in vRNP nuclear export (Neumann et al, 2000a). These findings confirmed in vitro studies by O'Neill et al. (1998) that implicated NS2 as the viral nuclear export factor. Watanabe et al. (2002) demonstrated the potential use of VLPs lacking the NS2 coding region as a vaccine.

11
Reverse Genetics of Thogotovirus

Thogotovirus is a tick-transmitted member of the family *Orthomyxoviridae*, whose genome comprises six segments of negative-sense vRNA. As in early studies with influenza A virus, Thogotovirus researchers first focused on the in vitro and in vivo reconstitution of vRNP complexes. Detergent-lysed virions were subjected to cesium chloride-glycerol gradients to obtain RNP fractions devoid of vRNA (Gomez-Puertas et al. 2000). Incubation of these fractions with Thogotovirus-like RNA reconstituted vRNP complexes that were transfected into Thogotovirus-infected cells, where replication of the Thogotovirus-like RNA proved the reconstitution of functional vRNP complexes. Weber et al. (1998) provided the viral NP and polymerase proteins, as well as a virus-like RNA encoding CAT, from plasmids. CAT expression demonstrated the in vivo formation of vRNP complexes and confirmed that the NP and polymerase proteins are sufficient for replication of a Thogotovirus model RNA. In the next step, the virus-like RNA was packaged into VLPs generated by coexpression of all viral structural proteins (Wagner et al.

2000). These successes stimulated the production of Thogotovirus from cloned cDNA in 2001 (Wagner et al. 2001), with the RNA polymerase I system for the synthesis of all six genomic viral RNAs and the vaccinia virus-T7 RNA polymerase system for expression of the structural proteins.

12
Reverse Genetics of *Bunya*- and *Arenaviridae*

Bunyamwera virus, a member of the *Bunyaviridae* family, was the first segmented negative-sense RNA virus generated from cloned cDNA (Bridgen and Elliott 1996). Three cDNAs encoding the positive-sense antigenomic RNAs were cloned between T7 RNA polymerase promoter sequences and the hepatitis delta ribozyme sequence. HeLa cells were infected with recombinant vaccinia virus expressing T7 RNA polymerase and transfected with T7 RNA polymerase-driven plasmids for RNA synthesis and synthesis of viral proteins. Forty-eight hours later the vaccinia virus-infected and plasmid-transfected cells were frozen and thawed, and cell extracts were used to infect insect cells that allowed the selective amplification of recombinant Bunyamwera virus (Bridgen and Elliott 1996). This system has been used to generate transfectant Bunyamwera viruses (e.g., those lacking the NSs gene) (Bridgen et al. 2001).

In one study, Flick and Pettersson (2001) addressed the feasibility of using the RNA polymerase I system to synthesize Uukuniemi *(Bunyaviridae)* virus transcripts. Because RNA polymerase I localizes to the nucleus, transcripts synthesized by this enzyme may not be exported from the nucleus to the cytoplasm, the site of bunyavirus replication. Moreover, the genomes of viruses replicating in the cytoplasm may contain artificial splice signals, resulting in their nuclear retention and splicing. Despite these concerns, RNA polymerase I transcription yielded a functional Uukuniemi virus-like transcript encoding a reporter gene, indicating that RNA polymerase I-derived transcripts are exported from the nucleus to the cytoplasm (Flick and Pettersson 2001).

Viruses in the *Arenaviridae* family, which contain two segments of negative-sense vRNA, remain to be generated from plasmid DNA. Lee et al. (2000) established a minireplicon system in which a virus-like RNA encoding CAT was replicated and transcribed by the L and NP proteins, provided by helper virus infection or T7 RNA polymerase-driven protein expression constructs. Hence, the minimal replication unit of arenaviruses is composed of the NP and L proteins, together with the viral RNA.

13
Concluding Remarks

The ability to alter the genomes of segmented negative-sense RNA viruses in any way desired will undoubtedly have a dramatic impact on the way we perceive these viruses. For example, the availability of new reverse genetics systems now makes it possible to elucidate previously ill-defined steps in viral replication and to identify virulence factors that escaped detection with conventional methods of genome manipulation. Ultimately, it should be possible to produce reassortant viruses for use as inactivated vaccines, as well as safe and effective live attenuated vaccine viruses.

References

Barclay WS, Palese P (1995) Influenza B viruses with site-specific mutations introduced into the HA gene. J Virol 69: 1275-1279

Baron MD, Barrett T (1997) Rescue of rinderpest virus from cloned cDNA. J Virol 71: 1265-1271

Beaton AR, Krug RM (1986) Transcription antitermination during influenza viral template RNA synthesis requires the nucleocapsid protein and the absence of a 5′ capped end. Proc Natl Acad Sci USA 83: 6282-6286

Boyer JC, Haenni AL (1994) Infectious transcripts and cDNA clones of RNA viruses. Virology 198: 415-426

Bridgen A, Elliott RM (1996) Rescue of a segmented negative-strand RNA virus entirely from cloned complementary DNAs. Proc Natl Acad Sci USA 93: 15400-15404

Bridgen A, Weber F, Fazakerley JK, Elliott RM (2001) Bunyamwera bunyavirus nonstructural protein NSs is a nonessential gene product that contributes to viral pathogenesis. Proc Natl Acad Sci USA 98: 664-669

Buchholz UJ, Finke S, Conzelmann KK (1999) Generation of bovine respiratory syncytial virus (BRSV) from cDNA: BRSV NS2 is not essential for virus replication in tissue culture, and the human RSV leader region acts as a functional BRSV genome promoter. J Virol 73: 251-259

Castrucci MR, Kawaoka Y (1995) Reverse genetics system for generation of an influenza A virus mutant containing a deletion of the carboxyl-terminal residue of M2 protein. J Virol 69: 2725-2758

Clarke DK, Sidhu MS, Johnson JE, Udem SA (2000) Rescue of mumps virus from cDNA. J Virol 74: 4831-4838

Collins PL, Hill MG, Camargo E, Grosfeld H, Chanock RM, Murphy BR (1995) Production of infectious human respiratory syncytial virus from cloned cDNA confirms an essential role for the transcription elongation factor from the 5′ proximal open reading frame of the M2 mRNA in gene expression and provides a capability for vaccine development. Proc Natl Acad Sci USA 92: 11563-11567

de la Luna S, Martin J, Portela A, Ortin J (1993) Influenza virus naked RNA can be expressed upon transfection into cells co-expressing the three subunits of the polymerase and the nucleoprotein from simian virus 40 recombinant viruses. J Gen Virol 74: 535-539

Durbin AP, Hall SL, Siew JW, Whitehead SS, Collins PL, Murphy BR (1997) Recovery of infectious human parainfluenza virus type 3 from cDNA. Virology 235: 323-332

Enami M, Luytjes W, Krystal M, Palese P (1990) Introduction of site-specific mutations into the genome of influenza virus. Proc Natl Acad Sci USA 87: 3802-3805

Enami M, Palese P (1991) High-efficiency formation of influenza virus transfectants. J Virol 65: 2711-2713

Enami M, Sharma G, Benham C, Palese P (1991) An influenza virus containing nine different RNA segments. Virology 185: 291-298

Enami M, Enami K (2000) Characterization of influenza virus NS1 protein by using a novel helper-virus-free reverse genetic system. J Virol 74:5556–5561

Flick R, Pettersson RF (2001) Reverse genetics system for Uukuniemi virus (Bunyaviridae): RNA polymerase I-catalyzed expression of chimeric viral RNAs. J Virol 75: 1643–1655

Fodor E, Devenish L, Engelhardt OG, Palese P, Brownlee GG, Garcia-Sastre A (1999) Rescue of influenza A virus from recombinant DNA. J Virol 73: 9679-9682

Garcin D, Pelet T, Calain P, Roux L, Curran J, Kolakofsky D (1995) A highly recombinogenic system for the recovery of infectious Sendai paramyxovirus from cDNA: generation of a novel copy-back nondefective interfering virus. EMBO J 14: 6087-6094

Gassen U, Collins FM, Duprex WP, Rima BK (2000) Establishment of a rescue system for canine distemper virus. J Virol 74: 10737-10744

Gomez-Puertas P, Mena I, Castillo M, Vivo A, Perez-Pastrana E, Portela A (1999) Efficient formation of influenza virus-like particles: dependence on the expression levels of viral proteins. J Gen Virol 80:1635–1645

Gomez-Puertas P, Leahy MB, Nuttall PA, Portela A (2000) Rescue of synthetic RNAs into thogoto and influenza A virus particles using core proteins purified from Thogoto virus. Virus Res 67: 41-48

Haller AA, Miller T, Mitiku M, Coelingh K (2000) Expression of the surface glycoproteins of human parainfluenza virus type 3 by bovine parainfluenza virus type 3, a novel attenuated virus vaccine vector. J Virol 74: 11626-11635

Hatta M, Gao P, Halfmann P, Kawaoka Y (2001) Molecular basis for high virulence of Hong Kong H5N1 influenza A viruses. Science 293: 1840–1842

Hatta M, Halfmann P, Wells K, Kawaoka Y (2002) Human influenza A viral genes responsible for the restriction of its replication in duck intestine. Virology 295: 250–255

He B, Paterson RG, Ward CD, Lamb RA (1997) Recovery of infectious SV5 from cloned DNA and expression of a foreign gene. Virology 237: 249-260

Hoffman MA, Banerjee AK (1997) An infectious clone of human parainfluenza virus type 3. J Virol 71: 4272-4277

Hoffmann E, Neumann G, Hobom G, Webster RG, Kawaoka, Y (2000a) "Ambisense" approach for the generation of influenza A virus: vRNA and mRNA synthesis from one template. Virology 267: 310-317

Hoffmann E, Neumann G, Kawaoka Y, Hobom G, Webster RG (2000b) A DNA transfection system for generation of influenza A virus from eight plasmids. Proc Natl Acad Sci USA 97: 6108-6113

Hoffmann E, Webster RG (2000) Unidirectional RNA polymerase I-polymerase II transcription system for the generation of influenza A virus from eight plasmids. J Gen Virol 81: 2843-2847

Honda A, Ueda K, Nagata K, Ishihama A (1987) Identification of the RNA polymerase-binding site on genome RNA of influenza virus. J Biochem (Tokyo) 102: 1241-1249

Honda A, Ueda K, Nagata K, Ishihama A (1988) RNA polymerase of influenza virus: role of NP in RNA chain elongation. J Biochem (Tokyo) 104: 1021-1026

Honda A, Mukaigawa J, Yokoiyama A, Kato A, Ueda S, Nagata K, Krystal M, Nayak DP, Ishihama A (1990) Purification and molecular structure of RNA polymerase from influenza virus A/PR8. J Biochem (Tokyo) 107:624–628

Horimoto T, Kawaoka Y (1994) Reverse genetics provides direct evidence for a correlation of hemagglutinin cleavability and virulence of an avian influenza A virus. J Virol 68: 3120-3128

Huang TS, Palese P, Krystal M (1990) Determination of influenza virus proteins required for genome replication. J Virol 64: 5669-5673

Kato A, Sakai Y, Shioda T, Kondo T, Nakanishi M, Nagai Y (1996) Initiation of Sendai virus multiplication from transfected cDNA or RNA with negative or positive sense. Genes Cells 1: 569–79

Kawano M, Kaito M, Kozuka Y, Komada H, Noda N, Nanba K, Tsurudome M, Ito M, Nishio M, Ito Y (2001) Recovery of infectious human parainfluenza type 2 virus from cDNA clones and properties of the defective virus without V-specific cysteine-rich domain. Virology 284: 99–112

Kim HJ, Fodor E, Brownlee GG, Seong BL (1997) Mutational analysis of the RNA-fork model of the influenza A virus vRNA promoter in vivo. J Gen Virol 78: 353-357

Kimura N, Nishida M, Nagata K, Ishihama A, Oda K, Nakada S (1992) Transcription of a recombinant influenza virus RNA in cells that can express the influenza virus RNA polymerase and nucleoprotein genes. J Gen Virol 73: 1321-1328

Krishnamurthy S, Huang Z, Samal SK (2000) Recovery of a virulent strain of newcastle disease virus from cloned cDNA: expression of a foreign gene results in growth retardation and attenuation. Virology 278: 168-182

Lawson ND, Stillman EA, Whitt MA, Rose JK (1995) Recombinant vesicular stomatitis viruses from DNA. Proc Natl Acad Sci USA 92: 4477-4481

Lee KJ, Novella IS, Teng MN, Oldstone MB, de La Torre JC (2000) NP and L proteins of lymphocytic choriomeningitis virus (LCMV) are sufficient for efficient transcription and replication of LCMV genomic RNA analogs. J Virol 74: 3470-3477

Li S, Xu M, Coelingh K (1995) Electroporation of influenza virus ribonucleoprotein complexes for rescue of the nucleoprotein and matrix genes. Virus Res 37: 153-161

Luytjes W, Krystal M, Enami M, Parvin JD, Palese P (1989) Amplification, expression, and packaging of foreign gene by influenza virus. Cell 59: 1107-1113

Mena I, de la Luna S, Albo C, Martin J, Nieto A, Ortin J, Portela A (1994) Synthesis of biologically active influenza virus core proteins using a vaccinia virus-T7 RNA polymerase expression system. J Gen Virol 75: 2109-2114

Mena I, Vivo A, Perez E, Portela A (1996) Rescue of a synthetic chloramphenicol acetyltransferase RNA into influenza virus-like particles obtained from recombinant plasmids. J Virol 70: 5016-5024

Neumann G, Zobel A, Hobom G (1994) RNA polymerase I-mediated expression of influenza viral RNA molecules. Virology 202:477–479

Neumann G, Watanabe T, Ito H, Watanabe S, Goto H, Gao P, Hughes M, Perez DR, Donis R, Hoffmann E, Hobom G, Kawaoka Y (1999) Generation of influenza A viruses entirely from cloned cDNAs. Proc Natl Acad Sci USA 96:9345–50

Neumann G, Hughes MT, Kawaoka Y (2000a) Influenza A virus NS2 protein mediates vRNP nuclear export through NES-independent interaction with hCRM1. EMBO J 19:6751–6758

Neumann G, Watanabe T, Kawaoka Y (2000b) Plasmid-driven formation of influenza virus-like particles. J Virol 74:547–51

Neumann G, Feldmann H, Watanabe S, Lukashevich I, Kawaoka Y (2002) Reverse genetics demonstrates that proteolytic processing of the Ebola virus glycoprotein is not essential for replication in cell culture. J Virol 76: 406–410

O'Neill RE, Talon J, Palese P (1998) The influenza virus NEP (NS2 protein) mediates the nuclear export of viral ribonucleoproteins. EMBO J 17: 288-296

Parvin JD, Palese P, Honda A, Ishihama A, Krystal M (1989) Promoter analysis of influenza virus RNA polymerase. J Virol 63: 5142-5152

Paule RME (1998) Transcription of Ribosomal RNA Genes by Eukaryotic RNA Polymerase I. Springer Verlag, Berlin Heidelberg New York

Peeters BP, de Leeuw OS, Koch G, Gielkens AL (1999) Rescue of Newcastle disease virus from cloned cDNA: evidence that cleavability of the fusion protein is a major determinant for virulence. J Virol 73: 5001-5009

Pleschka S, Jaskunas R, Engelhardt OG, Zurcher T, Palese P, Garcia-Sastre A (1996) A plasmid-based reverse genetics system for influenza A virus. J Virol 70: 4188-4192

Plotch SJ, Bouloy M, Ulmanen I, Krug RM (1981) A unique cap(m7GpppXm)-dependent influenza virion endonuclease cleaves capped RNAs to generate the primers that initiate viral RNA transcription. Cell 23: 847-858

Racaniello VR, Baltimore D (1981) Cloned poliovirus complementary DNA is infectious in mammalian cells. Science 214: 916-919

Radecke F, Spielhofer P, Schneider H, Kaelin K, Huber M, Dotsch C, Christiansen G, Billeter MA (1995) Rescue of measles viruses from cloned DNA. EMBO J 14: 5773-5784

Romer-Oberdorfer A, Mundt E, Mebatsion T, Buchholz UJ, Mettenleiter TC (1999) Generation of recombinant lentogenic Newcastle disease virus from cDNA. J Gen Virol 80: 2987-2995

Rowley KV, Harvey R, Barclay WS (1999) Isolation and characterization of a transfectant influenza B virus altered in RNA segment 6. J Gen Virol 80: 2353-2359

Schneider H, Spielhofer P, Kaelin K, Dotsch C, Radecke F, Sutter G, Billeter MA (1997) Rescue of measles virus using a replication-deficient vaccinia-T7 vector. J Virol Methods 64: 57–64

Schnell MJ, Mebatsion T, Conzelmann KK (1994) Infectious rabies viruses from cloned cDNA. EMBO J 13: 4195-4203

Seong BL, Brownlee GG (1992) A new method for reconstituting influenza polymerase and RNA in vitro: a study of the promoter elements for cRNA and vRNA synthesis in vitro and viral rescue in vivo. Virology 186: 247-260

Seong BL, Kobayashi M, Nagata K, Brownlee GG, Ishihama A (1992) Comparison of two reconstituted systems for in vitro transcription and replication of influenza virus. J Biochem 111: 496-499

Subbarao EK, Kawaoka Y, Murphy BR (1993) Rescue of an influenza A virus wild-type PB2 gene and a mutant derivative bearing a site-specific temperature-sensitive and attenuating mutation. J Virol 67: 7223-7228

Takeda M, Takeuchi K, Miyajima N, Kobune F, Ami Y, Nagata N, Suzaki Y, Nagai Y, Tashiro M (2000) Recovery of pathogenic measles virus from cloned cDNA. J Virol 74: 6643-6647

Takeda M, Pekosz A, Shuck K, Pinto LH, Lamb RA (2002) Influenza a virus M_2 ion channel activity is essential for efficient replication in tissue culture. J Virol 76:1391–1399

Taniguchi T, Palmieri M, Weissmann C (1978) QB DNA-containing hybrid plasmids giving rise to QB phage formation in the bacterial host. Nature 274: 223-228

Volchkov VE, Volchkova VA, Muhlberger E, Kolesnikova LV, Weik M, Dolnik O, Klenk HD (2001) Recovery of infectious Ebola virus from complementary DNA: RNA editing of the GP gene and viral cytotoxicity. Science 291: 1965-1969

Wagner E, Engelhardt OG, Weber F, Haller O, Kochs G (2000) Formation of virus-like particles from cloned cDNAs of thogoto virus. J Gen Virol 81:2849–2853

Wagner E, Engelhardt OG, Gruber S, Haller O, Kochs G (2001) Rescue of recombinant Thogoto virus from cloned cDNA. J Virol 75: 9282-9286

Watanabe T, Watanabe S, Neumann G, Kida H, Kawaoka Y (2002) Immunogenicity and protective efficacy of replication-incompetent influenza virus-like particles. J Virol 76: 767-773

Weber F, Jambrina E, Gonzalez S, Dessens JT, Leahy M, Kochs G, Portela A, Nuttall PA, Haller O, Ortin J, Zurcher T (1998) In vivo reconstitution of active Thogoto virus polymerase: assays for the compatibility with other orthomyxovirus core proteins and template RNAs. Virus Res 58: 13–20

Whelan SP, Ball LA, Barr JN, Wertz GT (1995) Efficient recovery of infectious vesicular stomatitis virus entirely from cDNA clones. Proc Natl Acad Sci USA 92: 8388-8392

Yasuda J, Bucher DJ, Ishihama A (1994) Growth control of influenza A virus by M1 protein: analysis of transfectant viruses carrying the chimeric M gene. J Virol 68: 8141-8146

Zobel A, Neumann G, Hobom G (1993) RNA polymerase I catalysed transcription of insert viral cDNA. Nucleic Acids Res 21: 3607-3614

Transcription and Replication of Nonsegmented Negative-Strand RNA Viruses

S. P. J. Whelan[1] · J. N. Barr[2] · G. W. Wertz[2]

[1] Department of Microbiology and Molecular Genetics, Harvard Medical School, 200 Longwood Ave, Boston, MA 02115, USA

[2] Department of Microbiology, University of Alabama School of Medicine, 845 19th Street South, Birmingham, AL 35294, USA

E-mail: gailw@uab.edu

Abstract The nonsegmented negative-strand (NNS) RNA viruses of the order *Mononegavirales* include a wide variety of human, animal, and plant pathogens. The NNS RNA genomes of these viruses are templates for two distinct RNA synthetic processes: transcription to generate mRNAs and replication of the genome via production of a positive-sense antigenome that acts as template to generate progeny negative-strand genomes. The four virus families within the *Mononegavirales* all express the information encoded in their genomes by transcription of discrete subgenomic mRNAs. The key feature of transcriptional control in the NNS RNA viruses is entry of the virus-encoded RNA-dependent RNA polymerase at a single 3′ proximal site followed by obligatory sequential transcription of the linear array of genes. Levels of gene expression are primarily regulated by position of each gene relative to the single promoter and also by *cis*-acting sequences located at the beginning and end of each gene and at the intergenic junctions. Obligatory sequential transcription dictates that termination of each upstream gene is required for initiation of downstream genes. Therefore, termination is a means to regulate expression of individual genes within the framework of a single transcriptional promoter. By engineering either whole virus genomes or subgenomic replicon derivatives, elements important for signaling transcript initiation, 5′ end modification, 3′ end polyadenylation, and transcription termination have been identified. Although the diverse families of NNS RNA virus use different sequences to control these processes, transcriptional termination is a common theme in controlling gene expression and overall transcriptional regulation is key in controlling the outcome of viral infection. The latest models for control of replication and transcription are discussed.

1 Introduction

The nonsegmented negative-strand (NNS) RNA viruses of the order *Mononegavirales* have linear, single-stranded negative-sense RNA as their genetic material. The order includes four families of viruses: the *Rhabdoviridae*, the *Paramyxoviridae*, the *Filoviridae*, and the *Bornaviridae*. These viruses comprise a wide variety of human, animal, and plant pathogens such as rabies virus (RV), measles virus (MV), canine distemper virus (CDV), Rinderpest virus (RPV) and human and bovine respiratory syncytial viruses (HRSV and BRSV) as well as the lethal Ebola (EBO) and Marburg (MBG) viruses and the recently described Nipah (NV) and Hendra virus (HeV).

The four virus families hold in common the fact that the genetic information in their negative-sense RNA genomes is expressed via transcription of a series of discrete monocistronic mRNAs. Transcription originates from a single polymerase entry site near the 3′ end of the genome and is obligatorily sequential, but attenuation (a decrease in product amount, thought to be due to polymerase dissociation) occurs specifically at each gene junction, resulting in a progressive reduction in the transcription of genes that are located further from the promoter (Abraham and Banerjee 1976; Ball and White 1976; Iverson and Rose 1981). This simple method of regulation may account for the observation that the order of the core genes within the *Mononegavirales* is highly conserved with genes whose products are required in large amounts located promoter proximally, whereas those needed in catalytic amounts are more distal. Transcription is carried out by an RNA-dependent RNA polymerase (RdRP) whose major components, the L (large) protein catalytic subunit and phosphoprotein (P) cofactor, plus an additional factor in the *Pneumoviruses* and *Filoviruses*, are encoded by the viral genome and packaged as components of the virion (Baltimore et al. 1970; Emerson and Wagner 1972; Emerson and Yu 1975). The active template for transcription as well as genome replication is the RNA genome encapsidated with the nucleocapsid protein. As a result, viral transcription is the first RNA biosynthetic step after entry and uncoating to release the viral ribonucleocapsid template. Replication of genomes cannot commence until a supply of nucleocapsid (N) protein has been synthesized to encapsidate the nascent antigenomes and their progeny genomes.

The ability to recover first replicable and infectious NNS defective interfering particle RNAs (Pattnaik et al. 1992) and subgenomic replicons (Collins et al. 1991; Park et al. 1991; De and Banerjee 1993; Dimock and Collins 1993; Conzelmann and Schnell 1994; Wertz et al. 1994) and subsequently infectious viruses (Schnell et al. 1994; Collins et al. 1995; Garcin et al. 1995; Lawson et al. 1995; Radecke et al. 1995; Whelan et al. 1995; Conzelmann 1998) from cDNA clones has allowed investigation of the genomic sequences involved in signaling and controlling the processes of mRNA transcription and genome replication. This review focuses on advances made in understanding the *cis*-acting genomic elements involved in control of these RNA synthetic processes since it became possible to engineer changes into the genomes of the NNS RNA viruses and assay their effects on function. The majority of the functional analyses have been carried out with the prototypic *Rhabdoviruses* vesicular stomatitis virus (VSV) or rabies virus (RV), or with various members of the *Paramyxovirinae* including the *rubulavirus* Simian virus 5 (SV5), the *respiroviruses* Sendai virus (SeV) and human parainfluenza virus (HPIV3), the *morbilliviruses* (MV and RPV) or with the *Pneumovirinae*, HRSV or BRSV. The *Bornaviridae* have not yet been recovered from cDNA, so they will not be emphasized. For a discussion of the *trans*-acting protein factors involved in these processes the reader is referred to previous reviews (Emerson, 1976; Banerjee and Chattopadhyay 1990).

1.1
Viral Genomes and Gene Products

The complete genome sequences for representative members of most of the genera of all four families of the *Mononegavirales* have been determined. The genomes range in size from the Borna disease virus (BDV) genome at 8,910 nt to the Ebola virus genome at approximately twice that size (18,957 nt), and they contain from 5 to 10 genes (Fig. 1). All of the virus families contain four core genes: a nucleocapsid protein gene

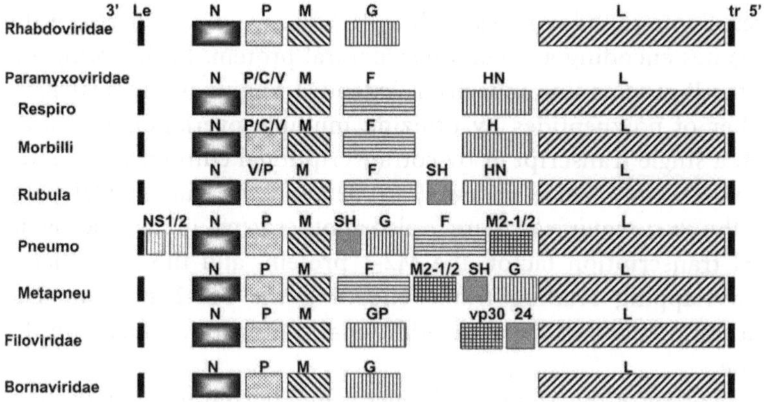

Fig. 1 Diagrammatic representation of the 3′ to 5′ arrangement of transcription units of the genomes of selected representative members of the families of the *Mononegavirales*. Abbreviations: *Le*, leader region; *N*, nucleocapsid or nucleoprotein gene; *P*, phosphoprotein gene; *V/C*, nonstructural protein genes; *M*, matrix gene; *SH*, small hydrophobic protein gene; *F*, fusion protein gene; *G* (or *H* or *HN*), glycosylated (or hemagglutinin or hemagglutinin/neuraminidase) attachment protein gene; *M2*, transcription factor gene; *NS*, nonstructural protein gene; *L*, large protein (polymerase) gene; *tr*, trailer region

(N or NP), a phosphoprotein gene (P), a matrix protein gene (M), and an RNA-dependent RNA polymerase gene (L). There is then either a single additional gene encoding one transmembrane attachment and entry glycoprotein such as the G gene of VSV and rabies, or up to three genes encoding transmembrane glycoproteins with the attachment and entry functions segregated onto two different proteins, G/HN/H and F, respectively, depending on the genus. Additionally a small hydrophobic (SH) gene encoding a protein of unknown function is found in the subfamily *Pneumovirinae* and the genus *Rubulavirus* of the subfamily *Paramyxovirinae*.

Many members of the order *Mononegavirales* encode nonstructural proteins, although the presence of a small number of copies of these proteins within the virion can not be eliminated. These nonstructural proteins are encoded either individually as separate genes or by multiple open reading frames (ORFs) within a single gene. The pneumoviruses such as HRSV and BRSV have separate genes encoding two nonstructural proteins involved in evading the host response (Schlender et al. 2000). Interestingly, these genes are not found in the avian pneumoviruses. Some rhabdoviruses such as IHNV have a gene between the G and L genes encoding a small nonstructural protein. In the *Paramyxovirinae* the P gene coding capacity is extended to give rise to a surprising number of polypeptides by utilizing multiple overlapping reading frames on a single transcript or by cotranscriptional editing. These strategies have recently been reviewed in detail (Lamb and Kolakofsky 2001). In the *Pneumovirinae* there is also another gene, M2, encoding an additional transcription factor, the M2-1 protein, and this gene has a second overlapping reading frame encoding the M2-2 protein. The Filoviruses also encode a separate additional transcription factor in the VP 30 gene.

The gene order of the *Mononegavirales* is highly conserved, with three of the four core genes, the N, P, and M genes, always at or near the 3′ terminus and with the L gene always the most promoter distal. This likely reflects the method for control of gene expression wherein position of a gene relative to the single 3′ promoter is the major determinant of expression levels. Consequently, genes needed in substantial amounts are located close to the promoter, whereas those needed in catalytic amounts are promoter distal. The surface glycoprotein genes are located between the N, P, M and the L genes and their absolute order varies among the various families and genera. The importance of gene order and regulated gene expression in the viral life cycle has been addressed in several functional studies, and these are described in Sect. 5.

1.1.1
Identification of Conserved Sequences at the Genomic Termini

The genomes of all the members of the *Mononegavirales* have a 3′ extracistronic region, the leader region, of approximately 40–50 nt and a 5′ extracistronic region, the trailer, which ranges from 20 nt up to over 600 nt (Table 1). These regions are notable for their high A/U content and the potential for complementarity between the first 10–12 or so most terminal nucleotides of the Le and Tr (Table 1). Major advances have been made in understanding the roles of these regions since it became possible to engineer specific changes into genomes and assay function. As described in detail in the following sections, these regions have been shown to be multifunctional and not only are essential for control of transcription and replication but also contain signals for encapsidation and assembly of newly replicated RNAs into virus particles.

Table 1 Sequences at the genomic termini of members of the order *Mononegavirales* highlighting the presence of complementary nucleotides within the first 20 nt

Family Rhabdoviridae

Genus *Vesiculovirus*

```
VSV     3'   UGCUUCUGUUUGUUUGGUAA    (50)
             |||||||| |  ||||||
        5'   ACGAAGACCACAAAACCAGA    (59)
SVCV    3'   UGCAUCAGUUUUACAGGCAA    (59)
             ||| || | | |  |||
        5'   ACGAAGACUACAAAUCCAG     (19)
```

Genus *Lyssavirus*

```
RV      3'   UGCGAAUUGUUGGUCUAGUU    (58)
             ||||||||||| || |  |
        5'   ACGCUUAACAAAUAAACAAC    (69)
```

Genus *Nucleorhabdovirus*

```
SYNV    3'   UCUCUGUCUUUGAGUCUUUU    (144)
             ||||||| |||||||||| |
        5'   AGAGACAAAAGCUCAGAACA    (159)
```

Genus *Novirhabdovirus*

```
VHSV    3'   CAUAGUAUUUUCUACUACUC    (167)
             |||| | |||  ||    |
        5'   GUAUAGAAAAUAAUACAUAC    (115)
```

Genus *Ephemerovirus*

```
BEFV    3'   UGCUCUUUUUUGUUUUUUUG    (51)
             |||| |||||||||| |||
        5'   ACGAAGAAAAACAAAUAAAA    (69)
```

Family Filoviridae

```
EBO     3'   GCCUGUGUGUUUUUCUUUUU    (55)
              |||||||||||||||||||
        5'    GGACACACAAAAAGGAAAA    (676)
MBG     3'   CUGUGUGUUUUUGUUCUCUA    (46)
             |||||||||||| || | |
        5' UGGACACACAAAAAAGAUGAAG    (585)
```

Family Paramyxoviridae

Genus *Henipavirus*

```
HV      3'   UGGCUUGUUCCCCUUUAUAC    (55)
             |||||||||||| ||| |
        5'   ACCGAACAAGGGUAAAGAGA    (33)
NV      3'   UGGUUUGUUCCCUCUUAUAC    (55)
             ||||||||||||  || | |
        5'   ACCGAACAAGGGUAAAGAAG    (33)
```

Table 1 (continued)

```
Genus Respirovirus
SeV     3'  UGGUUUGUUCUCUUGUUUGA   (55)
            ||||||||||||  ||||
        5'  ACCAGACAAGAGUUUAAGAG   (57)
HPIV1   3'  UGGUUUGUUCUCCUUUUUGA   (55)
            ||||||||||||   |||
        5'  ACCAGACAAGAGUUUAAGAA   (57)
HPIV3   3'  UGGUUUGUUCUCUUCUCUGA   (55)
            |||||||||||||| |  |
        5'  ACCAAACAAGAGAAAAACUC   (44)
Genus Rubulavirus
SV5     3'  UGGUUCCCCUUUUACUUCAC   (55)
            |||||||||||||  |||
        5'  ACCAAGGGGAAAACCAAGAU   (31)
MuV     3'  UGGUUCCCCUUUUACUUCUA   (55)
            ||||||||||||| | | |
        5'  ACCAAGGGGAGAAAGUAAAA   (24)
HPIV2   3'  UGGUUCCCCUCUUAGUCUAC   (70)
            |||||||||| |||||
        5'  ACCAAGGGGAAAAUCAAUAU   (21)
NDV     3'  UGGUUUGUCUCUUAGGCACU   (55)
            |||||||| ||| || ||||
        5'  ACCAAACAAAGAUUUGGUGA  (114)
Genus Morbillivirus
MeV     3'  UGGUUUGUUUCAACCCAUUC   (55)
            ||||||||||| |||| |
        5'  ACCAGACAAAGCUGGGAAUA   (40)
CDV     3'  UGGUCUGUUUCAACCGAUUC   (55)
            ||||||||||| ||| || |
        5'  ACCAGACAAAGCUGGGUAUG   (41)
RPV     3'  UGGUCUGUUUCGACCCAUUC   (55)
            |||||||||||||||| |
        5'  ACCAGACAAAGCUGGGGAUA   (40)
TPMV    3'  UGGUCUUUUCCCACCAGUGC   (55)
            |||||||||||   | ||||
        5'  ACCAGAAAAGGUGAGCUAUG  (590)
```

Table 1 (continued)

<table>
<tr><td colspan="4">Subfamily Pneumovirinae</td></tr>
<tr><td>HRSV</td><td>3′</td><td>UGCGCUUUUUUACGCAUGUU</td><td>(44)</td></tr>
<tr><td></td><td></td><td>||| ||||||| || | ||</td><td></td></tr>
<tr><td></td><td>5′</td><td>ACGAGAAAAAAAGUCUCAAA</td><td>(154)</td></tr>
<tr><td>BRSV</td><td>3′</td><td>UGCGCUUUUUUACGCAUAUU</td><td>(45)</td></tr>
<tr><td></td><td></td><td>||| ||||||| || | ||</td><td></td></tr>
<tr><td></td><td>5′</td><td>ACGAGAAAAAAAGUAUCAAA</td><td>(160)</td></tr>
</table>

The sequence of the first 20 nt of the genomic 3′ leader region are shown, together with the corresponding terminal 20 nt of the trailer region to indicate the extent of complementarity. The numbers in parentheses refer to the total number of nucleotides in the leader and trailer regions. The Ebola (EBO) virus and Marburg (MBG) virus alignments have been shifted to indicate the location of either 1 or 2 noncomplementary nucleotides at the genome ends.
VSV, vesicular stomatitis virus (Indiana strain); SVCV, spring viremia of carp virus; RV, rabies virus; SYNV, Sonchus yellow net virus; VHSV, viral hemorrhagic septicemia virus; BEFV, bovine ephemeral fever virus; HV, Hendra virus; NV, Nipah virus; SeV, Sendai virus; HPIV1, human parainfluenza virus 1; HPIV3, human parainfluenza virus 3; SV5, simian virus 5; MuV, mumps virus; HPIV2, human parainfluenza virus 2; NDV, Newcastle disease virus; CDV, canine distemper virus; RPV, rinderpest virus; TPMV, Tupaia paramyxovirus; HRSV, human respiratory syncytial virus; BRSV, bovine respiratory syncytial virus.

1.1.2
Conservation of Gene-Start, Gene-End, and Intergenic Sequences

At the beginning and end of each gene are sequences that are highly conserved in some families and semiconserved in others. In addition, between the genes there are intergenic regions that range from two to three highly conserved nucleotides at each junction in the rhabdoviruses and rubulaviruses to from 1 to 56 nt with no, or little, conservation in the pneumoviruses. These sequences modulate the activity of the polymerase during transcription, but these signals are ignored during the replication process. This review summarizes recent information on the functional determination of the requirements for each of these three sequence elements in the processes of transcription initiation and mRNA 5′ end modification and in control of transcript 3′ end polyadenylation and termination, including the finding that polyadenylation is integral to the termination process.

1.2 Overview of the Intracellular Replication Cycle

To initiate infection, NNS RNA viruses deliver into the cytoplasm an RNP complex of their genomic RNA encapsidated by the nucleocapsid protein and associated with the RdRP (Szilagyi and Uryvayev 1973). With the exception of the *Bornaviridae*, replication of all animal NNS RNA viruses occurs entirely within the cytoplasm of the cell; the steps that occur are depicted in Fig. 2. Before viral protein synthesis, the input RdRP copies the genomic template to provide a short uncapped leader RNA (Le+), and 5–10 capped and polyadenylated mRNAs that encode each of the viral proteins. After translation of these mRNAs into the viral proteins, genome replication can begin. During replication the RdRP initiates at the extreme 3′ end of the genome, ignores all the regulatory signals for production of discrete monocistronic mRNAs, and instead synthesizes a full-length complementary antigenome. Like the full-length genomic RNA, these antigenomic replicative intermediates are totally encapsidated by viral nucleocapsid protein. In turn, the antigenome RNA functions as template for synthesis of a short minus-sense leader RNA (Le−) and also for synthesis of full-length progeny genomes. These genomes can be used for synthesis of further viral mRNAs, can act as

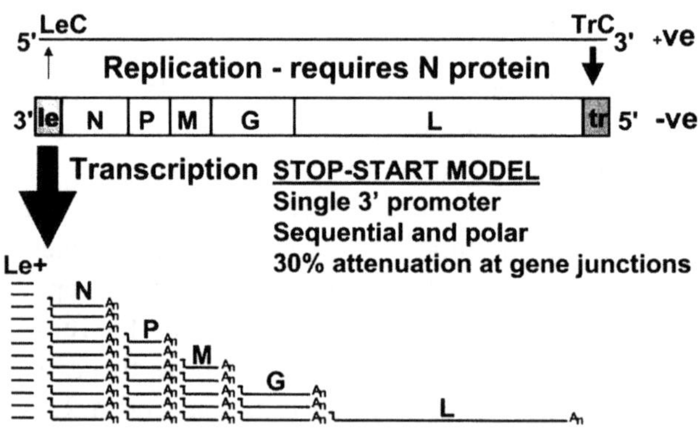

Fig. 2 Overview of RNA synthesis in NNS RNA virus infected cells. As a model, the vesicular stomatitis virus genome is shown 3′-5′, depicting the 50-nt leader region (*Le*) and the genes encoding the nucleocapsid protein (*N*), phosphoprotein (*P*), matrix protein (*M*), attachment glycoprotein (*G*), and large polymerase subunit (*L*) followed by the 59-nt trailer region (*Tr*). The steps of viral RNA synthesis are described in detail in the text

templates for further replication, or can be assembled into infectious particles. How the distinct RNA synthetic events are regulated is poorly understood, although the availability of a pool of soluble nucleocapsid protein is essential for encapsidation of the nascent genomic and antigenomic RNAs during RNA replication (Patton et al. 1984; Peluso and Moyer 1988; Howard and Wertz 1989).

1.3 Transcription

1.3.1 NNS RNA Virus Transcription Is Sequential

Transcription of the genes of NNS RNA viruses is obligatorily sequential. This was directly demonstrated for VSV, HRSV, NDV, RV and SeV with the technique of ultraviolet (UV) transcriptional mapping (Abraham and Banerjee 1976; Ball and White 1976; Ball 1977; Glazier et al. 1977; Flamand and Delagneau 1978; Collins et al. 1980; Dickens et al. 1984). A major effect of low-dose UV radiation of RNA is the formation of covalent dimers between adjacent uracil residues. These dimers act as blocks to progression of polymerase during RNA synthesis. For NNS RNA viruses it was reasoned that if each gene was transcribed independently, their sensitivity to UV radiation would be proportional to their physical size, provided that the UV target sites were randomly distributed throughout the genome. However, when the UV sensitivity of the viral genes was measured, it was found that the only gene with a target size corresponding to its physical size was the first gene on the genome. The UV target size of each subsequent gene was larger and corresponded precisely to the sum of its physical size and the size of each gene that preceded it in the viral gene order. These experiments showed that transcription in NNS RNA viruses initiates at a single point on the genome and that each mRNA is synthesized in an obligatory order, reflecting the position of each gene with respect to this single entry site.

1.3.2 NNS RNA Virus Transcription Is Polar

Measurements of the abundance of viral mRNAs in infected cells demonstrated that they were not produced in equimolar amounts; rather their abundance decreased with distance from the single 3′ polymerase entry site (Villarreal et al. 1976). For the prototypic rhabdovirus, VSV, it was shown that the intracellular half-lives of the viral mRNAs were indistinguishable from one another (Pennica et al. 1979), which implied that the differing amounts of each mRNA reflected differential synthesis rather than turnover. Kinetic measurements of the appearance of the 5′ and 3′ ends of VSV mRNAs indicated that transcription of a gene occurred at a rate of 3.1–4.3 nt s^{-1}, but significant pauses of 2.5–5.7 min occurred between detection of the 3′ end of the upstream transcript and the 5′ end of the downstream transcript (Iverson and Rose 1981). These experiments suggested that termination and polyadenylation of the upstream mRNA, and/or capping and methylation of the downstream mRNA, occurred at a slower rate than RNA synthesis. In addition, these measurements showed that there was little intragenic termination during mRNA synthesis and thus suggested that transcriptional attenuation was localized to the gene junctions (Iverson and Rose 1981).

1.3.3
Unusual Chemistry of mRNA Capping

The 5′ cap structure on cellular mRNAs is formed by a series of reactions, whereby an RNA triphosphatase trims the γ phosphate from a 5′pppNpNpN nascent RNA chain to provide a 5′ diphosphate substrate for a guanylyltransferase (GT). The GT catalyzes the formation of a 5′ppp5′ linkage between GMP and the nascent RNA chain through a covalent GT:GMP intermediate. The resultant cap 0 structure acts as template for the guanine-N7-methyltransferase and ribose 2′-*O*-methyltransferase, to ultimately yield the capped and methylated mRNA m7G(5′)pppNmpNpNpNp. For NNS RNA viruses, the cap structure of the mRNAs is formed by a mechanism that appears unique. For VSV (Abraham et al. 1975), HRSV (Barik 1993) and *Spring Vicemia of Carp Virus* (SVCV) (Gupta and Roy, 1980), the α and β phosphates of the 5′Gppp5′NpNpN triphosphate bridge were shown to be derived from a GDP donor. This difference, combined with the cytoplasmic location of RNA synthesis, led to the suggestion that the L protein subunit of the RdRP possesses a GT activity, although direct evidence for this is lacking. After capping, the 5′ terminus of the nascent transcript is methylated by guanine-N7-methyltransferase and nucleoside-2′-*O*-methyltransferase (Rhodes et al. 1974; Moyer et al. 1975; Rhodes and Banerjee 1975;

Rose 1975; Keene and Lazzarini 1976; Moyer and Banerjee 1976; Testa and Banerjee 1977; Hercyk et al. 1988). The mechanistic details of these reactions are not understood.

1.3.4 Models of Transcription

Three models were proposed to explain the sequential and polar nature of mRNA synthesis (shown in Fig. 2) and the unusual chemistry of mRNA capping.

1. Multiple entry site model:

   Proposed that the RdRP gained access to the genome at each gene-start site. To be compatible with the UV transcriptional mapping data, transcription of the upstream gene would be required before transcription of the downstream gene.

2. Cleavage model:

   Proposed that the RdRP initiated at a single 3′ entry site to produce a transcript that is cleaved into the leader RNA and the discrete mRNAs. This model was attractive because the unusual chemistry of capping is consistent with cleavage of a precursor.

3. Stop-start model:

   Proposed that the RdRP initiates all synthesis at position 1 of the genome, yielding a leader RNA and the viral mRNAs. In this model access of polymerase to downstream genes was entirely dependent on termination of synthesis of the upstream gene (hence, stop-start). This model is supported by much of the available data and has become the widely accepted model for NNS RNA virus transcription.

1.4 RNA Replication

Replication to produce the antigenomic and genomic RNAs of NNS RNA viruses remains poorly understood. The products of replication are completely encapsidated with the viral nucleocapsid protein. Replication is asymmetric, typically producing a 5- to 10-fold molar excess of genomic over antigenomic RNAs (Kiley and Wagner 1972; Schincariol and Howatson 1972; Soria et al. 1974; Wertz 1978; Simonsen et al. 1979). However, for the lyssavirus, rabies, genomic RNAs have been reported in a 50-fold molar excess of antigenomic RNAs (Finke and Conzelmann 1997). This asymmetry of replication suggested that the promoter for synthesis of full-length genomes was stronger than that for full-length antigenomes. Consistent with this, defective-interfering particles that exhibited the highest replication efficiency had replaced the promoter for antigenome synthesis with a copy of the promoter for genome synthesis.

The established model for regulation of transcription versus replication in NNS RNA viruses has been that the RdRP initiated all synthesis at position 1 of the genome (Emerson 1982). During leader synthesis a critical regulatory decision was made that dictated whether the RdRP terminated Le+ and initiated mRNA synthesis at the first gene-start site or, alternatively, read through the leader first gene junction, ignored all the transcriptional regulatory signals, and instead synthesized a full-length antigenomic replication product. Replication requires ongoing protein synthesis to provide a source of soluble N protein necessary for encapsidation of the nascent RNA strand. This led to the suggestion that RdRP activity was regulated by the availability of N protein to encapsidate the nascent leader RNA, such that when N protein concentration was low, the RdRP transcribed Le+ and the viral mRNAs, but when N protein concentration was high, termination of Le+ synthesis was suppressed and instead the RdRP replicated the genome (Blumberg et al. 1981, 1983). Thus the intracellular concentration of N protein was postulated to act as a regulator switching RdRP activity from transcription to RNA replication. Consistent with this leader RNA can be found encapsidated with N protein (Blumberg and Kolakofsky 1981). Subsequent work demonstrated that N protein was absolutely essential for RNA replication and is maintained in a soluble form by the P protein (Patton et al. 1984; Peluso and Moyer 1988; Howard and Wertz, 1989). This requirement for N protein in RNA replication is indisputable, but recent work indicates that other factors may regulate polymerase activity by controlling the site at which RNA synthesis is initiated (Whelan and Wertz 2002), and these are described in Sect. 3.

2 The Genomic Termini as Promoters

2.1 Introduction

The arrangement of the genomes of NNS RNA viruses and the products of RNA synthesis are shown in Figs. 1 and 2. Sequencing NNS RNA virus genomes identified conserved elements present at the termini (Table 1 and Sect. 1.1.1), suggesting that these regions were required for a common function of the two promoters. Reverse genetics systems have allowed a functional analysis of the role of the termini as promoters. The promoters of representative members of each genus in the family *Paramyxoviridae* have been examined, along with the lyssavirus, rabies virus and the vesiculovirus, VSV. For each virus, a major obstacle to dissecting the role of the termini as promoters was their multifunctional nature. The termini contain sequences required for encapsidation of the nascent RNA strand, binding of the RdRP, leader synthesis, mRNA synthesis, replication, and in some cases assembly and budding of particles. Consequently, caution must be applied when interpreting the effect of mutations engineered into the termini. For example, studies that rely exclusively on the translation of a reporter gene provide indirect information on the abundance of a capped, methylated, and polyadenylated messenger RNA, but provide little information on defects in encapsidation, polymerase binding, leader synthesis, mRNA initiation, mRNA processing, mRNA termination, or RNA replication. In the following section we attempt to summarize the work of several laboratories performed over the last decade. The majority of these studies were performed with VSV, HRSV, SeV, and SV5, and a comparison of these experiments highlights striking differences between viruses that are evolutionarily closely related.

2.2 Promoter Mapping in the *Rhabdoviridae*

One of the earliest experiments that delimited promoter function showed that the termini of a 5′ copy-back defective-interfering particle of VSV (DI-T) were necessary and sufficient for encapsidation, polymerase binding, replication, assembly, and budding of infectious particles

(Pattnaik et al. 1992). With synthetic VSV nucleocapsids in a reconstituted in vitro system, it was shown that the 3′ terminal 22 nt of the genome functioned as a promoter for transcription (Smallwood and Moyer 1993). However, in this case the polymerase ignored the presumed signals for synthesis of discrete leader RNA and initiation of the N mRNA. The importance of the 3′ leader region (Le) as a promoter for synthesis of a capped and polyadenylated mRNA was shown by comparing the template properties of subgenomic replicons that contained the wild-type genomic termini with those in which the 3′ Le had been replaced with increasing amounts of the corresponding sequences of the complement of the genomic trailer (TrC) (Wertz et al. 1994). Consistent with the abundance of transcription and replication products synthesized in infected cells, the wild-type termini directed abundant mRNA synthesis and comparatively low levels of replication. Incremental exchanges of Le sequence for TrC increased replication and ultimately decreased mRNA synthesis. Similarly when the Tr was engineered to be the perfect complement of Le for 47 nt, RNA synthesis was switched from predominantly transcription to almost exclusive replication (Wertz et al. 1994; Whelan and Wertz 1999b). This demonstration that changes at the opposite end of the viral genome have a dramatic effect on polymerase activity indicated that the ends of the genomic RNAs influence one another. The simplest interpretation of these data was that the ends of the genome interact with each other either directly or indirectly and that this interaction favors replication at the expense of transcription. This led to the suggestion that the increased terminal complementarity associated with the 3′ TrC-Tr 5′ arrangement of the termini (and found for the majority of VSV DI particles) offered a replicative advantage over the 3′ Le-Tr 5′ arrangement of the wild-type termini (Wertz et al. 1994). Other studies suggested that this complementarity between the termini did not influence replication efficiency, but rather that nt 31–45 of Tr functioned as a specific enhancer that favored replication of DI genomes (Li and Pattnaik 1997). However, this suggestion of a replication enhancer sequence could not account for the levels of replication observed for the 3′CB arrangement of the termini, because this replication enhancer element had been replaced with sequences of Le (Wertz et al. 1994; Whelan and Wertz 1999b). Further studies on the role of complementarity between the genomic termini of VSV suggested that its effects were confined to the natural terminal sequences and were most operative at the extreme genome ends (Whelan and Wertz 1999a, b). Metabolic labeling of RNAs synthesized by a recombinant virus in which the proposed replication enhancer sequence of Tr was replaced by the corresponding sequences of Le showed that this sequence did not significantly alter RNA synthesis in cell culture (Whelan and Wertz 1999b).

Deletions engineered throughout the leader and trailer regions of VSV genomic analogs and DI RNAs demonstrated that the terminal 14–18 nt contained a minimal signal for encapsidation and replication of RNA, although replication was significantly reduced (Li and Pattnaik 1999; Whelan and Wertz 1999a, b). Surprisingly, the previously defined binding site for the P subunit of the RdRP at nt 16–30 of Le (Keene et al. 1981) was not essential for RNA synthesis (Li and Pattnaik 1999; Whelan and Wertz 1999b). Some deletions to the termini, such as nt 47–50 of Le, decreased transcription and increased the levels of RNA replication, but this did not appear to be a general rule for VSV because other changes reduced both transcription and replication (Whelan and Wertz 1999b).

Comparison of the sequence of the two promoter regions of VSV (Indiana) shows nucleotide identity at 25 of 46 positions with 15 of the terminal 18 nt being identical (Table 1). To examine the role of nucleotides from positions 18–50 of the leader region, the regions that contained these nucleotide differences were exchanged between Le and TrC (Whelan and Wertz 1999b). These exchanges included compensatory changes in the trailer region to maintain the wild-type level of complementarity between the termini. These data showed that Le contains a specific signal for mRNA synthesis that could not be supplied by the corresponding sequences from TrC or by heterologous sequences (Whelan and Wertz 1999b).

For the lyssavirus rabies, the termini were also shown to contain essential promoter sequences (Conzelmann and Schnell 1994). A 3′CB rabies virus was generated, which was ambisense, assembled both positive and negative strands into infectious particles, and displayed a pronounced susceptibility to interference induced by simple internal deletion DI particles (Finke and Conzelmann 1997, 1999). Attempts to generate a 3′CB recombinant VSV were unsuccessful, and this was attributed to the low levels of transcription of such VSV templates as well as the inability of 3′CB genomes to assemble into infectious particles because they lack the signal provided by the 5′ trailer for assembly of replicated RNA into infectious particles (Whelan and Wertz 1999a, b).

2.3
Promoter Mapping in the *Paramyxoviridae*

Promoter mapping studies performed with representative members of both subfamilies of *Paramyxoviridae* revealed surprising differences from each other as well as from those obtained with VSV.

2.3.1
The *Paramyxovirinae*

A major difference that sets the *Paramyxovirinae* apart is a marked preference to maintain their genome lengths as an exact multiple of 6 nt. This was first described for Sendai virus, in which insertion of 1–7 nt at five different sites in the genome of a DI particle was tolerated only when the overall genome length was precisely divisible by 6 (Calain and Roux 1993). This "rule of six" reflects an apparent requirement for a single molecule of nucleocapsid protein (NP) to bind exactly 6 nt (Egelman et al. 1989), although precisely how this influences the ability of the polymerase to replicate the genome remains the subject of intense investigation. The rule of six applies with differing degrees of rigidity to the other *Paramyxovirinae* (Sidhu et al. 1995; Durbin et al. 1997; Murphy and Parks 1997; Clarke et al. 2000; Peeters et al. 2000; Skiadopoulos et al. 2003). By contrast, the 15,222-nt genome of the pneumovirus HRSV is an exact multiple of six, but studies on subgenomic replicons showed that there was no inherent disadvantage to genomes that do not conform to this integer rule (Samal and Collins 1996). Such an integer rule does not appear to apply to VSV.

A major difference between viruses that conform to this rule of six is the extended nature of their replication promoters. For example, an internal sequence element more than 72 nt away from the end of the genome was first described for SeV (Pelet et al. 1996). Comparison of active and inactive genomic and antigenomic promoters led to the identification of a three times repeated motif 3′-YNNRNN-5′ (Y=C/U, N=A/C/G/U, and R=A/G) present at nt 79–96 of the SeV genomic and antigenomic RNAs. With respect to the hexamer phasing of NP subunits relative to the genomic RNA, these residues corresponded precisely to hexamers 14–16 from the 3′ end of the template. In elegant experiments performed by Roux and colleagues the RNA sequence at each of these individual hexamers was randomized by mutagenesis of a cDNA clone of a

SeV DI particle. RNAs recovered from this cDNA clone were allowed to replicate, and the sequence through these hexamers was determined. These experiments identified a strong selection for the sequence 3′ CNNNNN 5′ in the genomic template at each hexamer position (Tapparel et al. 1998).

Comparison of the terminal sequences of *Rubulaviruses* proved equally as informative identifying two conserved regions (CR) that spanned nt 1–19 (CRI) and 73–90 (CRII). For SV5, it was shown that the spacing of these two elements was critical for efficient replication (Murphy et al. 1998). Mutational analysis of CRII demonstrated that the critical residues were distinct from those identified for SeV. For SV5, hexamers 13–15 have the common sequence 3′-NNNNGC-5′, and the most critical were the conserved positions of hexamers 14 and 15 (Murphy and Parks 1999). How appropriate spacing of an internal element relative to the genome end affects promoter function is not known. However, cryo-EM analysis suggested that the SeV nucleocapsid exists as a left-handed helix with 13 nucleocapsid protein subunits per turn (Egelman et al. 1989). This spacing would position the internal *cis*-acting element directly above the genomic terminus on the same face of the helix, and it raises the possibility that this acts as part of a critical recognition element for a component of the RdRP (Pelet et al. 1996; Murphy et al. 1998; Tapparel et al. 1998). Alternatively, this sequence element may function during encapsidation of the nascent RNA strand. Assembly of NP onto the nascent strand is intimately coupled with replication, and the completion of the first turn of the helix may represent a critical juncture at which the pitch of the RNP is determined. In other recent experiments with SV5, a third region in the termini (residues 51–66) was identified as essential for efficient RNA replication (Keller et al. 2001).

2.3.2 The *Pneumovirinae*

For HRSV, initial promoter mapping experiments indicated that 53 nt from the genomic 3′ terminus and 52 nt from the genomic 5′ terminus are necessary for encapsidation, mRNA synthesis, replication, and assembly of particles (Collins et al. 1991). These experiments showed that, in contrast to VSV, the genomic leader could be replaced by the complement of the trailer region to function as a promoter for mRNA synthesis (Collins et al. 1991). A similar arrangement of the termini was shown to enhance levels of replication compared with the wild-type termini (Yu et al. 1995). Significant homology is present between the first 26 nt of the termini of HRSV, BRSV, ovine RS, and avian pneumoviruses, with 10 of the first 11 positions being highly conserved (Table 1). This homology permits cross-recognition of these termini as promoters (Yunus et al. 1999; Marriott et al. 2001).

Detailed characterization of the promoter regions of HRSV (Fearns et al. 2000; Peeples and Collins 2000; Fearns et al. 2002) showed that nucleotide substitutions at positions 2, 3, 6, and 7 relative to the 5′ end of the genomic trailer severely downregulated synthesis of full-length genomes but did not affect mRNA synthesis or the production of full-length antigenomes and thus defined these positions as critical components of the promoter for genome synthesis (Peeples and Collins 2000). Substitution of the HRSV leader sequence with increasing amounts of the complement of the trailer region confirmed that TrC could function as a promoter of transcription provided that the spacing of the gene-start sequence from the 3′ end of the genome was maintained. These studies also suggested that complementarity between the termini did not affect RNA synthesis (Fearns et al. 2000). Mutagenesis of the first 26 nt of the leader region of RS virus demonstrated that of these nucleotides, 3C, 5C, 8U, 9U, 10U, and 11U were individually important for both mRNA synthesis and replication (Fearns et al. 2002). Additional nucleotides (1U, 2G, 6U, and 7U) were individually important for replication but not mRNA synthesis. At position 4 the presence of a G increased transcription and decreased replication, implying that the natural assignments of 4G in the leader region and 4U in the TrC are optimal for promoting transcription and replication, respectively (Fearns et al. 2002). These studies suggest that, like VSV, HRSV has promoter elements that affect both the processes of transcription and replication as well as separate elements that affect mRNA synthesis and replication.

3 Initiation of RNA Synthesis

3.1 Introduction

The established model for regulation of gene expression in NNS RNA viruses has been that the polymerase initiates all RNA synthesis at position 1 of the genome (Emerson 1982), and during the synthesis of the leader RNA a critical regulatory decision is made to either terminate Le+ and initiate mRNA synthesis at the first gene-start site or, alternatively, to read through the leader first gene junction to synthesize a full-length antigenomic replication product. The availability of N protein was proposed to regulate these two distinct RdRP activities (Sect. 1.4). The genomic replication products are completely encapsidated with the viral nucleocapsid protein, and the requirement for N to drive this process is indisputable; however, as described below, the processes of transcription and replication are also controlled by RdRP initiation at separate sites on the genome.

Experimental evidence in support of the single entry and initiation site model of RNA synthesis was provided by using purified VSV templates and RdRP under partial reaction conditions in vitro (Emerson 1982). Polymerase and template were separately purified from infectious virions and mixed together in the presence of ATP and CTP. The sequence of the genomic 3′ terminus is 3′-UGC, whereas the N gene-start of VSV is 3′-UUGUC-5′. In the presence of only ATP and CTP, polymerase that initiated at the 3′ end of the genome would synthesize the dinucleotide 5′-pppApC, whereas polymerase that initiated internally at the N gene-start site would synthesize the tetranucleotide 5′-pppApApCpA. Under such reaction conditions the predominant product was 5′-pppApC, implying that polymerase could only initiate at the 3′ end of the genome. The tetranucleotide 5′-pppApApCpA that corresponded to the N gene-start was only observed in a two-step reaction. In this reaction, polymerase was first incubated in the presence of all four NTPs, the reaction was then stopped, NTPs were removed, and synthesis was allowed to proceed in the presence of ATP and CTP. Under these conditions, the major product was the dinucleotide 5′-pppApC, but the tetranucleotide 5′-pppApApCpA was also detected, indicating that the RdRP also initiated at the N gene-start (Emerson 1982). The interpretation of this result was that preincubation with all four NTPs allowed synthesis of the leader RNA, which positioned RdRP at the N gene-start site where it could initiate when ATP and CTP were supplied. However, the relationships between the dinucleotide 5′-pppApC and Le+ and the tetranucleotide 5′-pppApApCpA and the N mRNA are not unequivocal. Sequences that could template synthesis of these products are present multiple times throughout the genome, and other short products of unknown origin such as 5′-pppGpC have been characterized in VSV transcription reactions in vitro (Chanda and Banerjee 1981b, a).

Because VSV transcription is both polar and sequential, one consequence of a single 3′ polymerase entry site is that upstream genes are transcribed in molar excess of downstream genes. Consequently, if polymerase initiates at nucleotide 1 of the genome, the leader RNA should always be synthesized in molar excess over the N mRNA. However a mutant VSV (polR1), which has a single amino acid change in the template-associated N protein (Perrault et al. 1983), produces an excess of N mRNA over leader RNA in vitro (Chuang and Perrault 1997). This finding is inconsistent with the single-entry site model and instead suggested that polymerase can enter the genome directly at the first gene-start site during transcription (Perrault et al. 1983; Chuang and Perrault 1997). Determining where the RdRP initiates transcription is crucial in understanding regulation of gene expression in NNS RNA viruses. The single-entry site model for RNA synthesis proposes that a regulatory event occurs during Le+ synthesis that determines whether polymerase terminates leader synthesis and transcribes the viral mRNAs or, alternately, replicates the genome to produce a full-length complementary antigenome. The two-polymerase entry site model posits that transcription and replication are regulated through the use of separate entry sites for the viral RdRP.

3.2 Transcription Initiates at the First Gene-Start in Infected Cells

For VSV the question of whether RdRP initiates transcription directly at the first gene-start site or by the prior transcription of a leader RNA was recently reexamined with the technique of UV mapping (Whelan and Wertz 2002). Low-dose UV radiation induces the formation of covalent dimers between adjacent uracil residues that act to block progression of the RdRP during RNA synthesis. By examining the effects of UV radiation on viral protein synthesis the order of the genes on the VSV genome was originally determined (Ball and White 1976). However, this approach could not be used to distinguish whether the RdRP that synthesized the N mRNA had transcribed the leader region, because the 50-nt leader region does not significantly contribute to the UV sensitivity of the 1,336-nt N gene. The development of reverse genetic systems for VSV allowed application of UV mapping to determine whether the RdRP initiates transcription at the first gene-start site or through the obligatorily sequential transcription of leader RNA. Recombinant viruses were generated that contained an inserted (I) 60- or 108-nt gene between the leader region and the N gene (Whelan et al. 2000). In addition, the sequence of the leader region was modified to either increase or decrease the number of adjacent uracil residues, such that the sensitivity of the leader RNA to UV radiation would be altered in a predictable manner (Whelan and Wertz, 2002). The rationale was that if transcription initiated at the 3′ end of the genome, these sequence alterations to the leader region should affect the UV sensitivity of the I-60 or I-108 mRNA. Conversely, if polymerase initiated transcription at the first gene-start site, these changes would not affect the UV sensitivity of the I-mRNA. On the basis of the maximum number of covalent uracil dimers that can form after exposure to UV radiation, the changes to the sequence of the leader region were predicted to alter the UV sensitivity of the leader RNA over a 2.7-fold range, and this corresponded well with the measured 3-fold alteration. However, these changes did not significantly affect the UV sensitivity of the I-60 or I-108 mRNA synthesized in infected cells. In addition, the relative sensitivity of the 60- and 108-nt mRNAs were compared and found to be independent of the leader region. These data indicated that transcription initiated directly at the first gene-start site in infected cells and not through the prior transcription of a leader RNA (Whelan and Wertz 2002).

These experiments provided strong support for the two-entry site model for VSV RNA synthesis (Chuang and Perrault 1997). In this model, the RdRP is proposed to initiate synthesis of the leader RNA and the antigenomic replication product by entry at position 1 of the genome; however, for mRNA synthesis, polymerase initiates by entry directly at the first gene-start site (Fig. 3). An alternative possibility is that the RdRP always enters the genome at the 3′ end, but for transcription scans through the leader region, in a UV-insensitive manner, to initiate mRNA synthesis at the first gene-start. What regulates the RdRP initiation site choice is unknown, but possibilities include a modification of the polymerase, the template, or both. To date, there is no direct evidence for different forms of polymerase in infected cells, but mutations in L protein have been identified that affect transcription and replication in separate ways (Perlman and Huang 1973; Hunt et al. 1976; Wertz 1978). Moreover, the phosphorylation status of the P protein can affect whether polymerase transcribes or replicates the genome (Pattnaik et al. 1997).

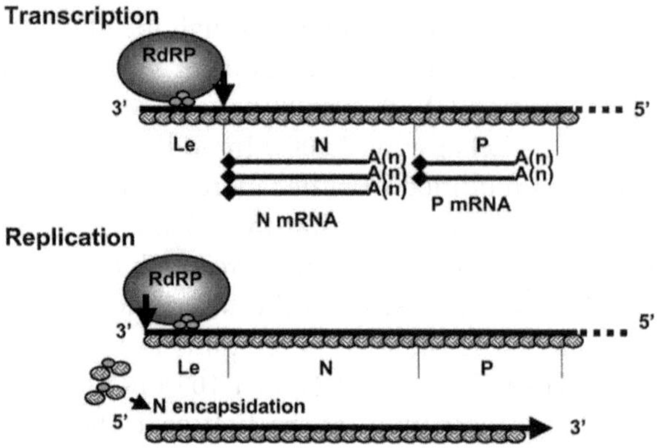

Fig. 3 Model for RNA synthesis in VSV-infected cells. A schematic of the 3′ end of the VSV genome is shown, depicting the leader region and the N and P genes. During mRNA synthesis, the RdRP, a complex of L (*large oval*) and a trimer of P (*small ovals*) binds to specific sequences in the leader region and initiates synthesis directly at the first gene-start site. During replication, the RdRP initiates at the 3′ end of the genome. Initiation at the 3′ end of the genome provides a leader RNA (not shown) and the full-length antigenome. Replication requires ongoing protein synthesis to supply N protein for encapsidation of the nascent RNA. N protein (*hatched oval*) is kept in a soluble form by interaction with P, in a 2:1 complex. *Solid arrows*, position of initiation of RNA synthesis

3.3 *Cis*-Acting Signals for Transcription Initiation

3.3.1 The Leader Region

The difficulties associated with accurately measuring Le+ synthesis mean that few studies have examined effects of changes within the leader region on synthesis of mRNA and Le+. The demonstration that stop-start sequential transcription does not apply at the VSV leader-N gene junction suggests that the effects of mutations in the leader region on mRNA synthesis are a direct consequence of alterations to a promoter sequence, rather than an indirect effect caused by alterations to the synthesis of the leader RNA. As described above, specific sequence elements within the leader region form part of the optimal promoter for the initiation of mRNA synthesis. Extensive analyses performed with VSV,

HRSV, and SeV were summarized in the preceding section and will not be repeated here.

For VSV the leader-N gene junction comprises four nucleotides 3′-GAAA-5′ that have no counterpart in either Le+ or the N mRNA. Deletion of the leader-N gene junction prevented mRNA synthesis from a subgenomic replicon that encodes a single gene. Point mutations engineered throughout the leader-N gene junction had only a minor effect on mRNA synthesis, with the exception that the sequence GAAU reduced mRNA levels to approximately 30% of wild type and shifted the site at which polymerase initiated mRNA synthesis to the −1 position (Whelan and Wertz, unpublished observations).

3.3.2
Role of the Gene-Start Sequence

For each NNS RNA virus, the beginning of each gene is marked by a conserved sequence of 10 nt. The role of these conserved nucleotides in mRNA synthesis has been well characterized with systematic analyses performed for VSV and HRSV. The gene-start sequences of some NNS RNA viruses are functionally equivalent in signaling mRNA synthesis. However, this does not hold true for all the viruses, and a striking example of this is provided by Sendai virus (discussed below).

For VSV the first 10 nt of each gene are highly conserved, having the sequence: 3′-UUGUCnnUAG-5′. Mutational analysis of this conserved sequence element identified a consensus sequence of 3′-UyGnnnnnnn-5′ required for efficient gene expression (Stillman and Whitt 1997). These studies were extended to examine the precise role of the first 3 nt in transcription. In a novel approach, the genome of VSV was separated onto two segments, the first comprising 3′ Le-N-P-M-L-Tr 5′ and the second the G gene together with the reporter genes green fluorescent protein (GFP) and chloramphenicol acetyl transferase (CAT), 3′ Le-G-GFP-CAT-Tr 5′ (Stillman and Whitt 1999). Alterations were made to the first 3 nt of the CAT gene, and the effects of these changes on mRNA synthesis were examined in infected cells and in vitro. In this system, alteration of the first U residue to a C, or alteration of the second U residue to an A, C, or G allowed synthesis of CAT transcripts in vitro, indicating that initiation per se was not prevented (Stillman and Whitt 1999). However, in infected cells these transcripts were not readily detected, suggesting that they were perhaps unstable. Further analysis of the transcripts produced in vitro indicated that the majority were truncated, ranging in size from 40 to 200 nt (Stillman and Whitt 1999). Immunoprecipitation of cDNA-RNA hybrids with an antibody that recognizes 2,2,7-trimethylguanosine indicated that the prematurely terminated transcripts lacked a methylated cap structure. These observations led to the formulation of a model in which polymerase processivity was linked to correct 5′ processing. This model suggested that the 5′ end of the nascent mRNA strand remains associated with the RdRP until correctly processed. Capping and methylation of the transcript would permit the release of the 5′ end from the RdRP, signaling to polymerase to proceed along the template. Failure of these modifications to occur would lead to the stalling of polymerase, premature termination of transcription, and the subsequent degradation of the RNA (Stillman and Whitt 1999).

Of the 10 genes of HRSV, the first 9 have the sequence 3′-CCCCGUUUA(U/C)-5′ at each gene-start site; the only departure from this is the sequence of the L gene-start 3′-CCCUGUUUUA-5′. Mutagenesis of the HRSV gene-start signal showed that positions 1, 3, 6, 7, and in particular 9 were most sensitive to alteration, whereas position 5 was relatively insensitive (Kuo et al. 1996b; Kuo et al. 1997). Surprisingly, the L gene start functioned equivalently despite having a nonconserved U residue at position 9. The negative effect of the U at position 9 appeared to be overcome by the presence of an A at position 10 of the L gene-start. Analysis of the transcripts synthesized from templates in which position 1 of the gene-start was altered from a C to a A, G, or U showed that the nucleotide incorporated into the mRNA was not always the templated nucleotide, but rather was the wild-type G residue approximately one-third of the time. The reason for this "pseudo-templated" initiation was not determined, but it was suggested to arise from either preloading of polymerase with the parental initiating nucleotide or slippage of polymerase that initiated at position 2, 3, or 4 of the gene-start (Kuo et al. 1997).

In contrast to the apparent equivalence for the gene-starts of HRSV, for other NNS RNA viruses subtle differences between the gene-starts appear highly significant. For SeV the P, M, and HN gene-starts are 3′-UCCCACUUUC-5′, whereas the N start is 3′-UCCCAgUUUC-5′, the F start is 3′-UCCCuaUUUC-5′, and the L start is 3′-UCCCCACUUaC-5′. To examine whether these gene-starts were functionally equivalent for transcription, the reporter gene firefly luciferase (LUC) was inserted between the N and P genes of SeV and the start sequence was altered to reflect each of the four possible sequences (Kato et al. 1999). Surprisingly, the F gene-start directed a lower level of transcription than each of the other three. The biological significance of this was examined by modifying the authentic F gene-start of wild-type SeV to that of the P, M, and HN genes. The resultant recombinant expressed significantly more F protein, as well as the two downstream genes HN and L, and replicated with faster kinetics in cultured cells and embryonated eggs. This recombinant virus was also more virulent, having a 20-fold lower LD_{50} for mice than wild type. Sequence analysis of multiple SeV genomes indicates that the gene-start-mediated downregulation of F is favored, implying that this offers a selective advantage for the virus in nature (Kato et al. 1999). These findings emphasize the fact that within the conserved *cis*-acting signals the nonconserved positions may be highly relevant to the biology of the virus in its natural environment.

3.3.3 Role of the Gene-End Sequence in Sequential Transcription

A central tenet of the stop-start model of sequential transcription is that termination of transcription at a gene-end sequence is essential for initiation of transcription at a subsequent gene-start site, and this is discussed in detail in Sect. 4. However, recent work with VSV has uncovered a previously unrecognized role of the gene-end sequence in transcription of a downstream mRNA. To examine the role of a gene-end sequence on transcription of the downstream mRNA, the VSV gene-end sequence was duplicated such that the upstream copy was used for termination and polyadenylation of the first mRNA. This approach allowed examination of the role of the second copy of the gene-end sequence on transcription of the downstream gene independent of its role in polyadenylation and termination. Surprisingly, the U7 tract of this second gene-end sequence was necessary for optimal transcription of the downstream gene. Altering the sequence or changing the length of this U7 tract significantly decreased transcription of the downstream gene. These experiments show that in the context of a wild-type gene junction, the U7 tract of the gene-end sequence is multifunctional, providing signals for termination and polyadenylation of the upstream mRNA and also signals that are essential for efficient transcription of the downstream gene (Hinzman et al. 2002).

4
Signals Involved in Termination of Transcription

4.1
Introduction

The concept that transcription of a downstream gene cannot occur without termination of the gene directly upstream is central to the generally accepted stop-start model of NNS RNA virus transcription. Expression of NNS RNA virus genes can therefore be regulated by not only *cis*-acting signals that affect transcription initiation, but also those that act at the level of transcription termination of an individual gene. Importantly, however, these two regulatory strategies are fundamentally different. Reduced transcription caused by an altered initiation signal globally reduces expression of all genes downstream of the alteration. In contrast, reduced transcription caused by changes at a termination signal reduces expression of the gene immediately downstream of the change, but be-

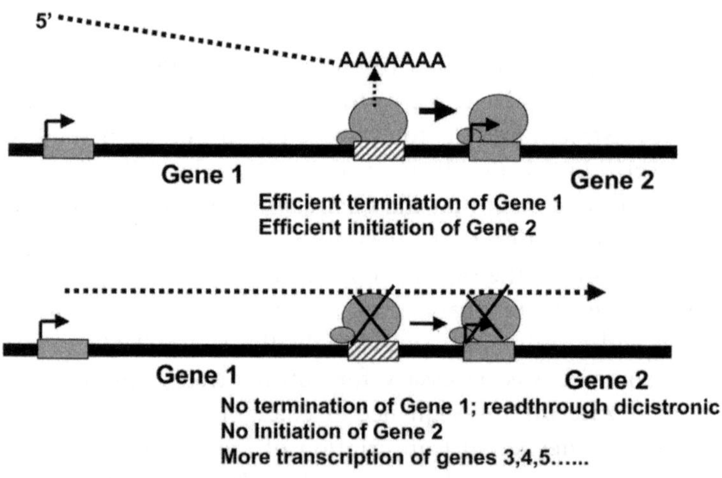

Fig. 4 Termination is a mechanism of gene regulation within the confines of a single promoter. A viral genome is depicted as a *solid black line* with gene-start and stop sequences indicated by the *gray* and *hatched boxes*, respectively. During mRNA synthesis (*upper panel*) efficient termination at a gene-end 1 leads to efficient initiation at gene-start 2. If termination fails to occur (*bottom panel*), expression of gene 2 is suppressed, but this leads to an increase in transcription of each subsequent gene, by bypassing an attenuation step

cause an attenuation step is by-passed, expression of genes located downstream will be increased (Fig. 4). Alteration of termination ability therefore provides the potential for great flexibility in regulation of gene expression.

4.2 The Signals Involved in Transcription Termination

Transcription termination in NNS RNA viruses is controlled by sequence elements that are located within gene junction regions that form the borders between upstream and downstream genes. These gene junctions are linear sequences that comprise the gene-end of an upstream gene, the gene-start of a downstream gene, and the intergenic sequence that separates these two elements. During the process of termination, sequences within the gene junction signal the RdRP to synthesize a 100- to 300-nt-long poly (A) tail at the 3′ end of the nascent transcript and then allow the modified transcript to be released or terminated, thus allowing the RdRP to proceed to transcribe the gene downstream. For VSV, mutagenic analysis of the gene junction has shown that transcription termination can be signaled by a linear sequence comprising the 11-nt-long gene-end sequence and the first nucleotide of the intergenic sequence (Barr et al. 1997a, b; Stillman and Whitt 1997; Hwang et al. 1998; Hinzman et al. 2002). This finding that the downstream gene-start sequence is not required for termination signaling correlates with the observation that the VSV L gene-end is not followed by a gene-start sequence, yet is able to efficiently signal termination of the L mRNA (Schubert and Lazzarini 1981).

Whether the termination signals of other NNS RNA viruses also comprise just the gene-end and the first position of the intergenic sequence, like that of VSV, remains to be tested. Recent evidence suggests that this is not the case for HPIV-1, because nonconserved sequences within the downstream gene of a gene junction have been shown to affect the ability of the termination signal immediately upstream to be recognized by the transcribing RdRP (Bousse et al. 2002).

4.3
The NNS RNA Virus Gene-End Sequence

Nucleotide sequence analysis of many viruses from the order *Mononegavirales* allows the identification and comparison of their gene-end and intergenic sequences, and many of these are listed in Table 2.

Indicative of their common functions in mRNA 3′ end poly (A) tail formation, the gene-end sequences share common features, the most striking of which is the presence of a U-tract between 4 and 8 nt in length. All available evidence indicates that this short run of U residues acts as the template for synthesis of the NNS RNA virus mRNA 3′ poly (A) tail through the process of reiterative transcription. This process involves correctly templated RNA synthesis followed by realignment of the template and nascent strand relative to the RdRP polymerization site. Cycles of these steps allow the nucleotides of the U-tract to be repeatedly copied and thus generate a poly (A) tail many times the length of the U-tract template. Comparison of gene-end sequences of numerous NNS RNA viruses reveals both strong conservation of a C residue directly upstream of the U tract and also the presence of nucleotides upstream of this position that are A/U rich in composition (Table 2). Examination of the 4 nt directly upstream of the U tract reveals that there is frequent preference for the sequence 3′-AUUC-5′. As with the U-tract, the degree of sequence conservation in this upstream region of the gene-end implies a common mode of signaling. Possible functions of these conserved sequences elements are discussed below.

4.3.1
Conservation of the NNS RNA Virus Gene-End Sequence

Some members of the *Mononegavirales* order (Table 2) have sequences at the end of each gene that are all alike, whereas others have little conservation outside of the presence of the ubiquitous U-tract. At one extreme, the gene-end sequences of individual members of the *Rhabdoviridae* family show a remarkably high level of sequence consistency, with most members of this family possessing gene-ends having uniform lengths of U-tract in addition to identical upstream sequences. An example of this is VSV, in which each gene-end sequence is invariant. A similar level of sequence consistency is also found for Ebola and Marburg viruses of the *Filoviridae* family. In contrast, within the diverse

Table 2 Comparison of consensus gene end and intergenic sequences of members of the order *Mononegavirales*[a]

Virus	Gene end	Intergenic	Accession no.
Family *Rhabdoviridae*			
VSIV	AUAC**UUUUUUU**	gA	NC001560
RV	AC**UUUUUUU**	Variable	NC001542
SYNV	AUUC**UUUUUUU**	GG	NC001615
SVCV	AUAC**UUUUUUU**	GA	NC002803
Sigma	GUAC**UUUUUUU**	Variable, 2 overlaps	N/A
LNYV	CUAAAUUC**UUUU**	Variable	N/A
VHSV	UCUAUC**UUUUUUU**	ND	NC000855
IHNV	UCUgUC**UUUUUUU**	ND	NC001652
HIRRV	UCUAUC**UUUUUUU**	ND	N/A
BEFV	aUAC**UUUUUUU**	ND, 1 overlap	NC002526
ARFV	aUAC**UUUUUUU**	ND, 1 overlap	N/A
Family *Filoviridae*			
EBO	UAAUUC**UUUUUu**	Variable, 3 overlaps	NC004161
MBG	UAAUUC**UUUUU**	Variable, 1 overlap	NC001608
Family *Paramyxoviridae*			
HV	UAAuUc**UUUUU**	GAA	NC001906
NV	AAuUn**UUUUU**	GAA	NC002728
Genus *Respirovirus*			
SeV	UnAUUC**UUUUU**	GaA	NC001552
HPIV-1	UUAUUC**UUUUU**	GcA	NC003461
HPIV-3	Uuauun**UUUUUu**	GAA	NC001796
Genus *Rubulavirus*			
SV5	aAaauUC**UUUUuu**	Variable	AF052755
SV41	AAaUUC**UUUUUu**	Variable	X64275
MuV	aauUC**UUUUUUu**	Variable	AB000388
HPIV-2	uUAaAUUC**UUUUUuu**	Variable	NC003443
NDV	AAUC**UUUUUUu**	Variable	AF077761
Genus *Morbillivirus*			
MeV	auun**UUUUuu**	GaA	AF016162
CDV	aaun**UUUUuuu**	GaA	AF014953
RPV	aaun**UUUUuu**	GaA	Z30697
TPMV	uAaUc**UUUUUu**	GAA	AF079780

Table 2 (continued)

Virus			Gene end	Intergenic	Accession no.
Subfamily *Pneumovirinae*					
HRSV	aUCAAU	3–4 nt	**UUUUuuu**	Variable, 1 overlap	AF013254
BRSV	Uca	2–4 nt	**UUUUuuu**	Variable, 1 overlap	AF092942
PVM	AUcAAUu	1–2 nt	**UUUUuuu**	Variable	N/A

[a] Invariant nucleotides are shown as upper case letters, and lower case letters indicate the most frequent nucleotide at positions showing variation between two possible nucleotide choices. Nucleotides having three or more choices are shown as n (nonconserved). The minimum U-tract lengths for each virus are shown in upper case, and variation in length is shown with lower case characters. The nucleotide lengths of the nonconserved *Pneumovirinae* central region are shown. Accession numbers for each virus are shown, where available.
VSIV, vesicular stomatitis Indiana virus; RV, rabies virus; SYNV, Sonchus yellow net virus; SVCV, spring viremia of carp virus; sigma, sigma rhabdovirus; LNYV, lettuce necrosis yellows virus; VHSV, viral hemorrhagic septicemia virus; IHNV, infectious hematopoietic necrosis virus; HIRRV, hirame rhabdovirus; BEFV, bovine ephemeral fever virus; ARV, Adelaide river virus; EBO, Ebola virus; MBG, Marburg virus; HV, Hendra virus; NV, Nipah virus; SeV, Sendai virus; HPIV-1 and 3, human parainfluenza viruses 1 and 3; SV5 and 41, simian viruses 5 and 41; MuV, mumps virus; HPIV-2, human parainfluenza virus 2; NDV, Newcastle disease virus; MeV, measles virus; CDV, canine distemper virus; RPV, rinderpest virus; TPMV, Tupaia paramyxovirus; HRSV, human respiratory syncytial virus; BRSV bovine respiratory syncytial virus; PVM, pneumonia virus of mice

Paramyxoviridae family, there is a broad range of sequence variation. The greatest conservation is found in two members of the *respirovirus* genus, namely Sendai virus (SeV) and human parainfluenza virus 1 (HPIV-1), whereas the greatest diversity is found in members of the *morbillivirus* genus. In fact, the low level of gene-end sequence conservation exhibited by several *morbillivirus* members, such as measles virus (MV), raises the question of how the gene-end sequences are distinguished from similar sequences that lie within coding regions of the MV genome. Perhaps the highly conserved intergenic regions play a greater role in termination in the *morbilliviruses.*

Lying midway between the variable *morbillivirus* gene-end sequences and the largely invariant *rhabdovirus* gene-ends are those of the *pneumoviruses.* The gene-end sequences of *pneumoviruses* possess both a highly conserved A/U-rich upstream element and a characteristic

U-tract, but these are separated by a short region that is variable in both sequence and length. Whether the unusual layout of the *pneumovirus* gene-end sequence is a consequence of fundamental differences in the mechanism of mRNA 3′ end formation is currently unknown.

4.4
The NNS RNA Virus Intergenic Sequence

The intergenic regions of members of the order *Mononegavirales* display a wide range of levels of sequence conservation. Many members of the *Rhabdoviridae* family exhibit highly conserved intergenic sequences in terms of both sequence composition and length, as do the respiroviruses and also the newly emergent NV and HeV. In each of these cases, the intergenic sequences are short and show a preference for the trinucleotide 3′-GAA-5′ or the related dinucleotide 3′-GA-5′. Surprisingly, the *morbilliviruses*, which display little sequence conservation within their gene-ends, show a high degree of conservation within the intergenic sequences, and again these viruses show a preference for the sequence 3′-GAA-5′. In contrast, the pneumovirus and rubulavirus intergenic regions show little or no sequence conservation, with the exception of a common A residue as the last nucleotide, and further display great heterogeneity in terms of sequence length.

4.5
Functional Analysis of the NNS RNA Virus Termination Signal

The ability to engineer the genomes of NNS RNA viruses through reverse genetics has allowed investigation of the roles of gene junction sequences in control of gene expression. The following section discusses the functional analysis of the gene junction regions, drawing in particular from studies involving the highly conserved gene-end sequence of the *rhabdovirus* VSV and also the less well-conserved gene-ends of the *rubulavirus* SV5 and the *pneumovirus* HRSV.

4.5.1 The VSV Termination Sequence Is Highly Conserved and Uniformly Effective

For VSV, the transcription termination signal is contained within the gene-end sequence and at least the first nucleotide of the intergenic region, with the second nucleotide of the intergenic and also the entire gene-start sequence exerting little influence on termination signaling ability (Barr et al. 1997a,b; Stillman and Whitt 1997; Hwang et al. 1998). For VSV, and in fact for many members of the *Rhabdoviridae* family, the gene-end and intergenic sequence elements are either invariant or display little divergence from a consensus sequence. Analysis of RNA synthesis during a VSV infection has shown that termination of mRNA synthesis for all genes examined is uniform (Iverson and Rose 1981), and this finding is consistent with the level of sequence conservation within this signal sequence. Furthermore, VSV readthrough RNAs that arise when termination signals are ignored are present in extremely low abundance during a VSV infection, confirming that the VSV termination signal is highly effective (Iverson and Rose 1981). For VSV, it appears that consistent and high termination signaling ability has been selected, suggesting that it is advantageous to the life cycle of this virus.

Functional Analysis of the VSV Termination Sequence. For ease of description, the VSV termination signal may be divided into the upstream tetranucleotide, the U-tract, and the intergenic sequence. The tetranucleotide is strictly conserved as 3′-AUAC-5′, and all nucleotide changes reduce the termination signaling ability of this sequence (Barr et al. 1997a; Hwang et al. 1998). However, the degree to which termination decreases varies according to both the position and identity of the nucleotide change. In general, the effect of nucleotide changes to 3′ positions of the 3′-AUAC-5′ sequence are less damaging to termination signaling than changes to those nucleotides more 5′. Remarkably, any changes made to the 5′ C residue abolish termination signaling. Intriguingly, when changes are made to the three upstream conserved nucleotides (3′-AUA-5′) termination signaling is most severely reduced when the replacing nucleotide is a C residue.

As with the conserved tetranucleotide sequence, the VSV U-tract is also strictly conserved, having a consistent length of seven U residues. Shortening the U-tract by a single nucleotide abolishes termination, leading to exclusive synthesis of readthrough RNAs (Barr et al. 1997a; Hwang et al. 1998). Similarly, alteration of any of the seven U residues to non-U nucleotides to interrupt the contiguous U-tract also abolishes termination (Barr et al. 1997a). Increasing the length of the U tract to U14 has little effect on the termination ability; however, it severely reduces the ability of the altered gene junction to signal downstream mRNA initiation (Barr et al. 1997a). This finding suggested that U-tract length may be limited because of a functional requirement for the process of initiation rather than its role in termination. More recently, additional functions of the VSV U7 tract in the process of initiation have been established (Hinzman et al. 2002), and these are discussed in Sect. 3.3.3.

The intergenic sequence is the only region within the VSV termination signal that does not exhibit total conservation. The P/M intergenic sequence 3′-CA-5′ deviates from the consensus 3′-GA-5′ at a single position. However, this difference does not seem to be significant for viral replication either in cultured BHK cells or animals (Ball et al. 1999; Flanagan et al. 2000). By altering both positions of the intergenic dinucleotide 3′-GA-5′ to all 15 other possible permutations, it was revealed that this sequence provides a signal not only for termination of upstream mRNA synthesis but also for initiation of downstream mRNA synthesis (Barr et al. 1997b; Stillman and Whitt 1997, 1998). Alteration of the first intergenic nucleotide was shown to primarily affect termination, whereas changes at the second position only affected downstream initiation. Studies that have more radically altered the intergenic sequence in terms of not only sequence composition but also sequence length have confirmed these findings (Barr et al. 1997b; Stillman and Whitt 1997, 1998; Hinzman et al. 2002). On the basis of the relative termination and initiation activities displayed by the various altered intergenic sequences, the role of the VSV intergenic region in transcription was suggested to be primarily in positioning a non-U residue between the U7 tract and the downstream gene-start sequence that begins 3′-UUGUC-5′ (Barr et al. 1997b).

In support of this notion, analysis of the variable rabies virus (RV) intergenic sequence, which shows variation in both sequence and length, has revealed that each of the four RV intergenic sequences signals approximately equal transcription termination, but they result in widely different levels of initiation from the downstream gene (Finke et al. 2000). The attenuation effect was found to correlate closely with intergenic sequence length, with the 24-nt-long G/L gene intergenic sequence signaling only 11% downstream mRNA synthesis compared with the 2-nt-long N/P intergenic sequence.

4.5.2
The Termination Sequences of HRSV and SV5 Exhibit Variable Sequence Conservation and Termination Signaling Ability

In contrast to VSV, the gene-ends and intergenic regions of both HRSV and SV5 exhibit considerable variation. Despite this, the gene-ends of both viruses conform to the previously described format, having both a conserved U-tract and an upstream sequence that is A/U rich in nucleotide composition. HRSV is different, however, in that these two sequence groups appear to be separated by a third set of nucleotides that are poorly conserved in both sequence and length.

Analysis of RNA synthesis in both HRSV- (Collins and Wertz 1983) and SV5-infected cells (Rassa and Parks, 1998) has shown that a substantial proportion (approximately 10%) of the viral transcripts exist as polycistronic readthrough RNAs, indicating that the termination signal at certain gene-ends is often ignored. Furthermore, for both of these viruses, the relative abundances of the various readthrough RNAs are not consistent. For HRSV, readthrough RNA synthesis is most abundant for the NS1/NS2 gene junction (Collins and Wertz 1983) whereas readthrough RNAs at the SH/G gene junction are not observed (Collins and Wertz 1983). For SV5, synthesis of readthrough RNAs containing the M and F gene sequences predominate (Rassa and Parks 1998). The variation within the gene-end sequences of these viruses allows a rational explanation for the observed differences in readthrough RNA abundances, and consequently the role of conserved nucleotides within these termination signals have been the subject of close scrutiny.

Defining the HRSV Termination Sequence: Involvement of the Intergenic Region. HRSV transcription termination was first assessed with a reporter gene assay in which each gene-end was separated from its accompanying intergenic sequence. In this context, all gene-end sequences signaled approximately equal levels of termination, with the sole exception of the NS1/NS2 gene-end, which was found to constitute the least effective signal of all (Kuo et al. 1996a). By positioning the highly variable HRSV intergenic sequences between consensus gene-end and gene-start sequences, initial studies also showed that the role played by the intergenic sequence in signaling termination was, at most, minor (Kuo et al. 1996a). However, analysis of termination signaling activity of each gene-end sequence when paired with its corresponding natural intergenic and gene-start sequences indicated that in this context the various gene-ends showed substantially different termination abilities (Hardy et al. 1999), which correlated with previous observations made in HRSV-infected cells (Collins and Wertz, 1983). This finding suggested that the HRSV termination signal is likely complex and involves interplay between the gene-end and intergenic sequences. In support of these findings, recent mutagenic analysis of the HRSV gene junction indicates that the intergenic region does contribute to termination signaling ability under circumstances in which the U-tract length is altered. Analysis of the HRSV F/M2 gene-end sequence revealed that at least the first nucleotide of the intergenic sequence makes a considerable contribution to the termination signal at this junction (Sutherland et al. 2001). In a separate study, analysis of the M/SH gene junction showed that the first intergenic nucleotide affected termination signaling, but only when the U-tract of the M/SH gene-end was shortened from U6 to U4 (Harmon and Wertz 2002). These findings suggest that the intergenic sequence is a contributing component of the HRSV termination signal when the gene-end sequence has a U4 tract, and because four HRSV gene ends have a U4 tract the intergenic sequence likely plays an important role in regulation of HRSV gene expression.

In addition to heterogeneous sequence composition, the HRSV intergenic sequences also exhibit extreme diversity in terms of sequence length, ranging from 1 to 56 nt. Insertion of related intergenic sequences of up to 160 nt at an artificial M/G gene junction within a cDNA-derived infectious HRSV variant conferred no detectable difference in steady-state levels of viral mRNA synthesis and only modest differences in plaque size and single step growth kinetics (Bukreyev et al. 2000). These findings indicated that when permissible intergenic sequences are used, their length plays little part in HRSV transcription.

Functional Analysis of the HRSV Gene-End Sequence. The HRSV gene-end sequence comprises the 3′-UCAAU-5′ conserved upstream sequence, an A/U-rich nonconserved central region, and a U-tract between 4 and 8 residues in length. To assess their role in signaling termination, these sequences have been subjected to extensive mutagenic analysis (Harmon et al. 2001; Sutherland et al. 2001; Harmon and Wertz 2002). In general, the sensitivity of nucleotide changes within these signal sequences mirrored the level of conservation seen in the gene-end sequences of the wild-type HRSV, with the most highly conserved nucleotides being sensitive to all types of nucleotide change. Of all 10 HRSV gene ends, only 1 possesses a deviation from the 3′-UCAAU-5′, which is the NS2 gene-end that has an A to U change at the final A residue of this sequence. This nucleotide change was found to considerably reduce termination signaling (Harmon et al. 2001), consistent with the low termination ability of the intact NS2 gene junction (Hardy et al. 1999). The variable central region was not equally tolerant of nucleotide changes, indicating that it does not function solely as a spacer region (Harmon et al. 2001; Sutherland et al. 2001). In general, nucleotide changes that maintained an A/U-rich sequence composition were tolerated with only slight reduction in termination signaling ability.

The observation that the HRSV gene-ends contain U-tracts of varying lengths suggests that the requirement for a specific length of U-tract is not rigid. This is in contrast to the case of VSV described above and has been confirmed by mutagenic analysis (Harmon et al. 2001; Sutherland et al. 2001). However, although the length of the U-tract appears not to be stringent, the relative position it occupies within the gene-end sequence has been shown to be critical. Increasing or decreasing the length of the central region such that the U-tract is either brought upstream or moved downstream by as little as one nucleotide position has been shown to severely reduce termination signaling (Harmon et al. 2001; Sutherland et al. 2001).

Poor Termination at the SV5 M/F Gene Junction. As with HRSV, the *rubulavirus*, SV5 generates abundant readthrough RNAs, the most populous being the M/F readthrough transcript (Rassa and Parks 1998). Correlating with this finding, the M gene-end sequence differs from that of the SV5 overall consensus sequence, most strikingly in that it is the only SV5 gene-end to possess a short U4-tract. However, mutational analysis of the SV5 gene-end sequence showed that the unusually short U-tract was not responsible for poor termination at the M gene-end (Rassa and Parks 1998). Rather, the poor termination ability of this gene-end sequence was found to be caused by a single nucleotide deviation from the SV5 consensus sequence within the conserved upstream A/U-rich sequence. Further mutagenic analysis of the SV5 M/F gene junction demonstrated that the intergenic region contributed to the SV5 termination signal, although this effect appeared to be specific for the M/F gene junction. For both SV5 and HRSV, the presence of a U4-tract at the gene-end allows sequences within the intergenic region to influence the RdRP during the process of transcription termination (Rassa and Parks 1998, 1999; Sutherland et al. 2001; Harmon and Wertz 2002). The question of whether the intergenic sequences of these viruses influence termination ability in other gene-end sequence contexts is a complex issue that is currently unresolved.

In conclusion, VSV possesses gene-end and intergenic sequences that are essentially invariant, and as a result their termination abilities are indistinguishable. In contrast, the gene-end and intergenic sequences of HRSV and SV5 display much greater sequence diversity and consequently exhibit widely different termination signaling abilities. Why these two different expression strategies have been adopted is currently not clear but is likely due to a requirement for differential expression of certain genes during the viral life cycle in response to various selective pressures.

4.6 Mechanism of Termination

As discussed above, the functional role played by many of the conserved nucleotides of the gene-end have been described in detail. However, precisely how these nucleotides modify the activity of the transcribing polymerase and cause transcriptional termination is still poorly understood.

The only mechanistically based studies to date have been for VSV, and these studies have centered around the ability of the gene-end sequence to signal reiterative transcription to generate the 3′ mRNA poly(A) tail. The observation that mutations that either shortened or interrupted the U-tract failed to signal termination and at the same time failed to support reiterative transcription provided strong evidence that reiterative transcription with the subsequent generation of 3′ poly (A) sequences were essential steps in the termination process (Barr et al. 1997a). The possibility that the process of reiterative transcription alone was sufficient to signal termination was ruled out by replacing the template U7-tract with an A7-tract. Although this sequence was able to signal reiterative transcription to generate a poly(U) sequences, it was unable to signal termination (Barr et al. 1997a). This indicated that either the template U7-tract or the subsequently generated poly(A) tail contained an essential termination signal.

The idea that reiterative transcription is an essential step in the termination process was further supported by the finding that a mutation of the conserved upstream 3′-AUAC-5′ tetranucleotide that prevented reiterative transcription also prevented termination (Barr et al. 1997a). Further mutational analysis of this conserved tetranucleotide also indicated that reiterative transcription was enhanced when this sequence was A/U rich in nucleotide composition but reduced when it was G/C rich (Barr and Wertz 2001). This finding implied that weak base pairing between the template and nascent strand within this region was likely an important requirement for allowing reiterative transcription to occur. Based on the probable role of reiterative transcription in the termination process of all NNS RNA viruses, this finding offered experimental evidence for why sequences upstream of the U-tract of most of these viruses are A/U rich.

The finding that NNS RNA virus gene-end sequences could function as transcription terminators in a variety of heterologous sequence contexts suggested that a termination signal consists entirely of a linear sequence element. However, a recent study has revealed that termination depends on more than just recognition of this linear sequence (Whelan et al. 2000). These experiments showed that the VSV RdRP is unable to terminate transcription at a wild-type gene-end sequence within a transcriptional unit less than 51 nt in length. This observation suggests that the ability of an RdRP to terminate transcription depends on a modification to an actively transcribing RdRP, or to the nascent mRNA itself during an early stage of the elongation process. These findings, in combination with previous studies that observed altered termination signaling as a result of defective 5′ mRNA cap methylation (Rose et al. 1977) and the demonstration that polymerase processivity was altered by changes to the gene-start sequence (Stillman and Whitt 1999), led to the suggestion that the process of termination may be mechanistically linked to that of initiation.

4.7
Cotranscriptional Editing in the Subfamily *Paramyxovirinae*

NNS RNA viruses share the ability to polyadenylate their mRNAs, and all available evidence indicates that this activity is performed by the viral RdRP in response to specific template signals located within the gene-ends of each respective virus. As described above, polyadenylation is thought to occur by reiterative transcription, in which repeated cycles of template and nascent strand realignment followed by correctly templated RNA synthesis lead to the generation of long stretches of poly(A) from a relatively short U-tract template.

With just one notable exception (HPIV-1), all members of the subfamily *Paramyxovirinae* of NNS RNA viruses perform an activity that appears related to polyadenylation known as editing or pseudo-templated transcription, which results in the genetically programmed insertion of one or more G residues at a defined $3'$-U_nC_n-$5'$ site within their respective P genes. This process was first described for the *rubulavirus* SV5 (Thomas et al. 1988). The specific nucleotide insertions generate three distinct mRNA species that cause the unique ORF located upstream of the editing site to be fused to each of the three ORFs positioned downstream of the site. For members of the *respirovirus* and *morbillivirus* genera (and also *Newcastle Disease virus,* NDV) the unedited P gene mRNA encodes the P protein and the frequently observed insertion of a single nucleotide accesses the alternate downstream V ORF (or D ORF in HPIV-3). In contrast, for members of the *rubulavirus* genus, an unedited mRNA allows access to the V ORF and the predominant insertion of 2 nt generates an mRNA that accordingly encodes the P protein. Because the downstream P protein ORF acts as an essential RdRP cofactor and the V protein ORF acts as an antagonist of the innate immune response, control of P gene editing and its effects on the relative expression of these P gene ORFs is likely to be vital to the viral life cycle.

Cotranscriptional editing has been best studied for the prototypic *respirovirus*, SeV, and by extensive mutagenesis of the $3'$-U_6C_3-$5'$ editing site and the surrounding nucleotides, the *cis*-acting sequences that control the editing process have been defined (Park and Krystal 1992; Hausmann et al. 1999a; Hausmann et al. 1999b). The frequency at which different numbers of G residues are inserted is controlled by both nucleotides within the conserved U_nC_n motif and also six nucleotides upstream of the U_n-tract, with the two nucleotides most proximal being of greatest influence (Hausmann et al. 1999b). Interestingly, specific alterations within the conserved $3'$-U_6C_3-$5'$ sequence switched the editing activity from one of precisely controlled G insertion to uncontrolled A insertion much like the process of polyadenylation and thus provide evidence that the signals that control these two RdRP activities may be related.

In an attempt to understand how the sequence of an editing site determines the pattern of nucleotide addition, a competitive kinetic model has been proposed (Hausmann et al. 1999b). This model suggests that the transcribing RdRP first stalls at the editing site, perhaps resulting from the low stability of the template-nascent strand hybrid in the vicinity of the editing sequence. The stalled RdRP can then enter one or more cycles of polymerase and nascent strand realignment followed by correctly templated transcription that continue until hybrid realignment is no longer energetically favored. Importantly, the nucleotide sequence of the editing site governs the number of cycles taken for this state to be reached and consequently determines the number of nucleotide insertions that occur.

All viruses that edit their P gene mRNAs also conform to the "rule of six", (described in Sect. 2.2), and the recent demonstration that the hexameric phase of the editing site can alter the pattern of G insertions led to the suggestion that these attributes are functionally linked and furthermore have likely coevolved (Kolakofsky et al. 1998). The ability of hexamer phase to affect RdRP stuttering activity led to the suggestion that the N protein may provide additional information that allows the *Paramyxovirinae* RdRP to distinguish editing sites from those that signal the functionally related activity of polyadenylation (Iseni et al. 2002).

4.8
Reduced Expression of L and F Genes

Within the framework of the stop-start model of transcription, attenuation at gene junctions results in reduced expression of genes located 5′ proximally compared with those positioned closer to the 3′ end. Without exception, the L genes of all NNS RNA viruses are located in the 5′ proximal location and consequently are expressed with low abundance because of their position. However, many viruses within this group have been shown to further reduce L gene expression through the presentation of *cis*-acting signals that result in the specific decrease of L gene mRNA synthesis.

For the *Rhabdoviridae* family members RV and VSV (New Jersey), low L gene expression has been shown to be a result of sequences within the intergenic regions that precede the L gene start (Stillman and Whitt

1998; Finke et al. 2000). For RV, the primary attenuating effect was thought to be due to the length of the intergenic region, whereas for VSV (New Jersey), the specific sequence of the intergenic was also considered to be important for the observed attenuation effect. In contrast, for VSV (Indiana), the observed low L gene expression is known to be controlled by sequences outside of the conserved gene junction sequence. Analysis of RNAs generated from rearranged VSV (Indiana) viruses suggest that sequences downstream from the L gene start site are involved in signaling low L gene expression (Wertz et al. 1998; Ball et al. 1999; Wertz et al. 2002).

Reduced L gene expression is also a feature of the *pneumovirus* member HRSV (Collins and Wertz 1983), and this effect is thought to be due to the unusual positioning of the L gene-start sequence upstream of the M2 gene-end sequence (Collins et al. 1987). The gene-overlap configuration implies that HRSV RdRP molecules that terminate M2 mRNA synthesis are able to migrate upstream and initiate transcription at the L gene-start, and this explanation is supported by experimental evidence (Fearns and Collins 1999). However, to transcribe full-length L mRNAs, the transcribing RdRP must ignore the M2 gene-end signal located just downstream of the L gene-start signal. Despite these apparent obstacles, attenuation at the M2/L gene junction is approximately only double that of other HRSV gene junctions (Fearns and Collins 1999). The low level of attenuation that results from the overlapping arrangement may be due to the ability of RdRPs that prematurely terminate L mRNA synthesis being able to repeatedly migrate upstream and reinitiate transcription at the L gene-start site. Studies with VSV demonstrate that a gene-end sequence must be positioned a minimal distance from a gene-start sequence for termination to occur (Whelan et al. 2000). Extrapolating from this study, it seems likely that the 68-nt overlapping arrangement of the RS M2/L genes reduces the efficiency of termination of polymerases that initiate at the L gene-start.

In addition to reduced L gene expression, several viruses have developed strategies to reduce expression of their respective F genes. However, in contrast to the case of reduced L gene expression described above, the common strategy for reduction of F protein expression is by modulating the termination ability of the upstream gene-end sequence, thus preventing RdRP access to the F gene-start site. Perhaps the most striking example of reduced F gene expression is that of the *rubulavirus* SV41 (Tsurudome et al. 1991). Transcription termination by the SV41

RdRP at the upstream M gene occurs so rarely that monocistronic F mRNAs are undetectable, and thus all F gene mRNA sequences are present within M/F readthrough RNAs. Reduced F gene expression is also a feature of the *respiroviruses*, HPIV-1 (Bousse et al. 1997) and the *rubulavirus*, SV5 (Rassa and Parks 1998). Recently it was proposed that high readthrough at the SV5 M/F gene junction serves as a mechanism to control viral transcription and growth by altering RdRP access to genes located more 5′ proximal, in particular by altering the ratio of the RdRP associated proteins, N, P, and L (Parks et al. 2001).

5 Importance of Gene Order and Transcriptional Control to the Viral Life Cycle

5.1 Introduction

NNS RNA virus genomes contain from 5 to 10 genes and, as described above, the order of the 4 genes common to all families is conserved. The N, P, and M genes are always located as a group at, or near, the genomic 3′ terminus in the invariant order 3′-N-P-M, and the L gene encoding the major component of the RdRP is always the most promoter distal gene. The genes encoding the various glycoproteins are located between these two groups and are found in a variety of orders. Because gene expression of NNS RNA viruses is controlled primarily at the level of transcription by the position of the genes relative to the single transcriptional promoter, this means that the promoter proximal gene is transcribed in greatest abundance and each successive gene is transcribed in progressively lower amounts because of attenuation at each gene junction.

How important is the observed conservation of gene order of the NNS viruses and the resultant impact it has on the transcriptional expression program to overall replicative fitness? This question was addressed directly once it was possible to engineer NNS RNA genomes at the cDNA level and recover infectious viruses. The importance of transcriptional control to the virus life cycle has been tested from several perspectives: (1) by manipulation of gene orders at the cDNA level and recovery of infectious viruses, (2) by addition of extra heterologous genes, and (3) by alteration of critical transcription signals.

5.2
Adding Genes to NNS RNA Virus Genomes

Heterologous genes have been inserted into the genomes of representatives of the genera of most of the NNS RNA viruses, at a variety of positions, with varying effects on replication and expression levels (Bukreyev et al. 1996; Schnell et al. 1996; Hasan et al. 1997; He et al. 1997; Spielhofer et al. 1998; Singh et al. 1999; Krishnamurthy et al. 2000; Walsh et al. 2000). The heterologous genes, which have been added to the genomes at the cDNA level by genetic engineering and then infectious virus recovered, have been stably maintained in the recovered viral genomes in all cases. However, the stability of *expression* of the added genes has varied and is discussed in detail below. The effects on viral replication of adding an additional gene have also varied considerably: from little or no effect on replication levels in the case of insertion of GFP into the rinderpest genome between the P and M genes (Walsh et al. 2000) to the lowering of replication by 100-fold in the case of insertion of the CAT gene into NDV between the HN and L genes (Krishnamurthy et al. 2000) or as observed in the case of addition of three genes to the genome of HPIV3 (Skiadopoulos et al. 2002).

5.2.1
Effects on Replication and Expression as a Function of Position of Insertion

There was no correlation between position of insertion and observed effects on replication that emerged from numerous reports described above until two different studies were reported, one with VSV and one with SeV, in which an additional gene was inserted into the genome at each of the gene junctions (Tokusumi et al. 2002; Wertz et al. 2002). These studies tested both the effect on replication of gene insertions at each junction and also the principle that gene expression is controlled by the position at which a gene is located relative to the single promoter. In the VSV study, engineered viruses were generated in which an additional, identical transcriptional unit with the conserved VSV transcriptional start and stop sequences was added to the genome at each of the viral gene junctions (Wertz et al. 2002). However, to avoid the possibility of feedback of a protein product affecting some step in transcription or replication, the added transcriptional unit in the VSV study contained no translational start site or major ORF. Significantly, because the added gene was not essential for viral replication there was no positive pressure for maintaining or expressing the gene. The levels of expression of the genes inserted at each gene junction, the relative replication levels of the viruses, and both genetic stability and stability of gene expression were examined. Direct analysis of transcripts confirmed that the level of transcription of the inserted "I" gene was determined by the position of the gene relative to the promoter. The most abundant relative synthesis was from between the first two VSV genes, N and P, with progressively lower amounts expressed from between successive gene pairs. The molar amounts of the I mRNA expressed from the N:P, P:M, M:G, and G:L junctions were 0.7, 0.5, 0.3, and 0.2 respectively, relative to a molar ratio of 1.00 for the first VSV gene, N. These data were consistent with attenuation of transcription of approximately 20%–30% across each gene junction, and they directly confirmed that gene expression is controlled by the position of a gene relative to the promoter.

The position at which the extra gene was inserted in the VSV genome, however, affected the replication potential of the viruses. Addition of a gene between the first two genes, N and P, reduced replication by over an order of magnitude, whereas addition of a gene at the other gene junctions had no effect on replication levels. All genes downstream of the inserted gene were found to have decreased levels of expression in all cases, because transcription of the extra gene introduced an additional transcriptional attenuation event.

In the SeV study, an additional reporter gene, secreted alkaline phosphatase (SEAP), was added at each gene junction and also between the L gene and trailer (Tokusumi et al. 2002). Expression of the inserted gene was measured by a fluorescent reporter assay rather than by direct mRNA analysis as in the VSV study. A progressive decrease in expression was observed also in this study as the position of the added gene was progressively farther from the 3′ genomic end. A 30-fold drop was seen in expression level between a gene placed at the 3′ most position versus one placed at the end of the L gene.

In the case of the SeV viruses with inserted genes, however, a progressive decrease in replication levels was observed that correlated with the position of the inserted gene: Generally, 3′ proximal insertions decreased replication levels more strongly than those farther from the promoter. The virus having the SEAP gene inserted upstream of the N gene was the slowest and lowest in replication, whereas the virus having the SEAP gene following the L gene replicated nearly the same as the wild-type virus. In all of the SeV constructs a unique *Not*I restriction site was introduced at the intergenic regions, but the "rule of six" was maintained (Tokusumi et al. 2002).

5.2.2
Physical Stability of Inserted Genes Versus Stability of Their Expression

The physical stability in the genome of additional genes inserted into several NNS RNA virus genomes has been analyzed. On repeated passage, the added genes have been found to be stably maintained in the genome in all cases (Bukreyev et al. 1996; Mebatsion et al. 1996; Schnell et al. 1996; He et al. 1997; Bukreyev et al. 1999). This may be a reflection of the fact that homologous recombination has not been observed in the NNS RNA viruses, providing no easy means for exact removal or exchange of genes. Additionally, the *expression* of added genes, when located distally from the promoter (between the G or HN and L genes, for example), of several viruses has been found to be stable throughout multiple passages (Bukreyev et al. 1996; Mebatsion et al. 1996; Schnell et al. 1996; He et al. 1997; Bukreyev et al. 1999). However, when the expression of the added gene was investigated as a function of the *position of insertion* as in the case of VSVs having genes inserted at all of the gene junctions, position was found to be an important factor in stability of expression (Wertz et al. 2002). Expression of the added genes in VSV was stable at only three of the four positions. In the case of insertion between the N and P genes, where insertion resulted in a significant inhibition of replication, a virus population arose within two passages that had restored replication to wild-type levels (Wertz et al. 2002). In this population, expression of the additional gene as a monocistronic mRNA was suppressed by mutations at the end of the upstream (N) gene that abrogated transcriptional termination. Because transcription is obligatorily sequential, this prevented transcription of the inserted downstream gene as a monocistronic mRNA and resulted instead in polymerase reading through the gene junction to produce a bicistronic mRNA. This eliminated the additional attenuation step and restored expression of all downstream genes and viral replication to wild-type levels. These data reinforce previous data showing that appropriate relative molar ratios of the N and P proteins are important for genome replication of VSV (Howard and Wertz 1989). The presence of the additional gene between the N and P genes that introduced an additional transcriptional attenuation step presented a selective disadvantage to replication. These data also show that transcriptional termination is a key element in control of gene expression of the negative-strand RNA viruses and a means by which expression of individual genes may be regulated within the framework of a single transcriptional promoter. Furthermore, these results are directly relevant to the use of NNS viruses as vectors and vaccine delivery agents because they show that the level of expression of an added gene can be controlled by its insertion position, but that not all positions of insertion yield stable expression of the added gene.

5.2.3
Selective Pressure Identifies Sequences Critical to Transcription

The selective disadvantage of introducing an additional gene with its attendant additional attenuation step between the N and P genes of VSV, and thereby upsetting the molar ratios of the N and P proteins, provided an unexpected bonus. It provided selective pressure to eliminate expression of the added gene and its extra attenuation step. This occurred within two passages by accumulation of mutations that eliminated termination at the end of the upstream N gene, thereby preventing initiation of the downstream inserted gene as a monocistronic mRNA. This provided a means to identify sequence elements essential for transcriptional termination by natural selective pressure (Wertz et al. 2002).

The RdRP of RNA viruses including VSV have a high error rate (Steinhauer and Holland 1986; Steinhauer et al. 1989) and no proofreading mechanism (Steinhauer et al. 1992), and the viruses thus exist as a quasi-species (Eigen et al. 1981; Eigen and Biebricher 1988). Despite the fact that the engineered viruses were recovered from cDNA, a stringent cloning experience, by the time that they had been passaged they had been amplified many thousandfold and could be expected to exist as a quasispecies with a broad range of sequence options available for selection. Analysis of the gene junction sequences of the initial and passaged viruses are consistent with this. The original passage of the virus that contained a transcriptional unit inserted (I) between the N and P genes (NIP virus) had wild-type sequence at the N/I junction and expressed abundant monocistronic I mRNA, but after two passages, the expression of I mRNA was almost completely suppressed. Correlating with that suppression was the accumulation of sequence alterations at the N gene-end identified by analysis of individual clones of independent RT-PCR amplifications, consistent with a viral population representing viruses with N gene-end sequences selected that abrogated termination. Roughly 80% of the mutations represented shortening of the WT U7-tract to U6 or interrupting it with a heterologous nucleotide, whereas 10% altered the conserved AUAC gene sequence (Wertz et al. 2002). All these observed changes had been tested during previous site-directed mutagenesis of the gene junction sequence, and each had been shown to abrogate termination of transcription (Barr et al. 1997a). Thus the importance of the conserved 3′-AUACU7-5′ tract in termination has now been confirmed by two different approaches—one involving natural selection. These data also confirmed the obligatorily sequential nature of transcription and the requirement for termination of transcription of each upstream gene in order to initiate synthesis of each downstream gene.

5.3 Rearranging the Order of Genes in the Genome

There is strong conservation of the overall order of genes among all Mononegaviruses as described above. Because homologous recombination has not been observed in these viruses, there appears to be no easy natural opportunity for rearrangement of gene orders. As a result, the question of whether the existing orders are optimal arises. A method was described to rearrange the order of the genes of VSV in an infectious cDNA clone by using remote cutting restriction enzymes in such a manner that the genes could be translocated without introducing other sequence changes and most especially without introducing changes in critical gene junction sequences (Ball et al. 1999). This allowed rearrangement of gene orders so that one could test the importance of gene order and its mandate on transcriptional control in the viral life cycle.

In one instance, engineered VSVs were generated in which the position of the nucleocapsid gene, N, required in stoichiometric amounts to support genome replication was translocated from its normal 3′-most position to successively promoter distal positions (Wertz et al. 1998). The expression of the N gene was reduced in a stepwise manner as it was moved to successively promoter distal positions. As N gene expression decreased, so did overall viral genome expression, showing that the overall replication potential of the virus could be controlled by altering the expression levels of this one gene.

Moreover, the pathogenesis of the virus in both swine (a natural host) and in a mouse model was attenuated in a stepwise manner by the translocation of the N gene to successively promoter distal locations (Wertz et al. 1998; Flanagan et al. 2000; Flanagan et al. 2001). Furthermore, it was shown to be feasible to move genes involved in the establishment of protective immunity such as the G gene to promoter proximal positions to increase their level of expression and increase the level and speed of the immune response (Flanagan et al. 2000). These data showed that it was possible to use the principles of transcriptional control, in which position of a gene is a major determining factor in its expression level, to provide an approach for systematically altering the expression of key genes involved in replication and pathogenesis of these viruses. This approach allows one to maintain a wild-type complement of genes but to alter the expression of key genes involved in replication or in immune responses at will to either increase or decrease their expression as needed. Thus the approach of utilizing gene rearrangement to systematically alter gene expression of the negative-strand RNA viruses provides a new paradigm for development of live attenuated viral vaccines. This has been successfully demonstrated with live attenuated VSV viruses where a virus engineered to have the G gene in the first position and the N gene moved to the fourth position protectively immunized swine, one of the natural hosts of the virus, against wild-type VSV, without causing disease (Flanagan et al. 2001).

In other engineered VS viruses, the N and L genes were maintained at their 3′- and 5′-most positions, respectively, but the three central genes were rearranged to all possible positions (Ball et al. 1999). In all cases, the transcription level of a gene was shown to be a function of its position relative to the single promoter for transcription. Effects of these rearrangements on replication levels were less predictable, and in some cases viruses were generated that replicated at least as well as those having the wild-type gene order, when examined in single-step growth assays in cell culture or in replication in mouse brain. The ability to rearrange the highly conserved gene order of these viruses at the cDNA level and successfully recover infectious viruses suggests that the ancestral gene order may have been frozen by the inability of these viruses to carry out homologous recombination. Apparently as long as the expression levels of the N and L genes are not disturbed, these viruses can tolerate other genome arrangements that lead to alternate programs of transcriptional expression. Importantly, however, experiments examining the relative competitive fitness of these viruses have been initiated (I. Novella, personal communication), and to date none of the viruses having rearranged genomes is as fit as the wild-type gene order, suggesting that within the extant order these viruses have evolved the optimal sequence.

References

Abraham G, Banerjee AK (1976) Sequential transcription of the genes of vesicular stomatitis virus. Proc Natl Acad Sci USA 73:1504–1508

Abraham G, Rhodes DP, Banerjee AK (1975) The 5′ terminal structure of the methylated mRNA synthesized in vitro by vesicular stomatitis virus. Cell 5:51–58

Ball LA (1977) Transcriptional mapping of vesicular stomatitis virus in vivo. J Virol 21:411–414

Ball LA, Pringle CR, Flanagan B, Perepelitsa VP, Wertz GW (1999) Phenotypic consequences of rearranging the P, M, and G genes of vesicular stomatitis virus. J Virol 73:4705–4712

Ball LA, White CN (1976) Order of transcription of genes of vesicular stomatitis virus. Proc Natl Acad Sci USA 73:442–446

Baltimore D, Huang AS, Stampfer M (1970) Ribonucleic acid synthesis of vesicular stomatitis virus, II. An RNA polymerase in the virion. Proc Natl Acad Sci USA 66:572–576

Banerjee AK, Chattopadhyay D (1990) Structure and function of the RNA polymerase of vesicular stomatitis virus. Adv Virus Res 38:99–124

Barik S (1993) The structure of the 5′ terminal cap of the respiratory syncytial virus mRNA. J Gen Virol 74 (Pt 3):485–490

Barr JN, Wertz GW (2001) Polymerase slippage at vesicular stomatitis virus gene junctions to generate poly(A) is regulated by the upstream 3′-AUAC-5′ tetranucleotide: implications for the mechanism of transcription termination. J Virol 75:6901–6913

Barr JN, Whelan SP, Wertz GW (1997a) *cis*-Acting signals involved in termination of vesicular stomatitis virus mRNA synthesis include the conserved AUAC and the U7 signal for polyadenylation. J Virol 71:8718–8725

Barr JN, Whelan SP, Wertz GW (1997b) Role of the intergenic dinucleotide in vesicular stomatitis virus RNA transcription. J Virol 71:1794–1801

Blumberg BM, Giorgi C, Kolakofsky D (1983) N protein of vesicular stomatitis virus selectively encapsidates leader RNA in vitro. Cell 32:559–567

Blumberg BM, Kolakofsky D (1981) Intracellular vesicular stomatitis virus leader RNAs are found in nucleocapsid structures. J Virol 40:568–576

Blumberg BM, Leppert M, Kolakofsky D (1981) Interaction of VSV leader RNA and nucleocapsid protein may control VSV genome replication. Cell 23:837–845

Bousse T, Matrosovich T, Portner A, Kato A, Nagai Y, Takimoto T (2002) The long noncoding region of the human parainfluenza virus type 1 f gene contributes to the read-through transcription at the m-f gene junction. J Virol 76:8244–8251

Bousse T, Takimoto T, Murti KG, Portner A (1997) Elevated expression of the human parainfluenza virus type 1 F gene downregulates HN expression. Virology 232:44–52

Bukreyev A, Camargo E, Collins PL (1996) Recovery of infectious respiratory syncytial virus expressing an additional, foreign gene. J Virol 70:6634–6641

Bukreyev A, Murphy BR, Collins PL (2000) Respiratory syncytial virus can tolerate an intergenic sequence of at least 160 nucleotides with little effect on transcription or replication in vitro and in vivo. J Virol 74:11017–11026

Bukreyev A, Whitehead SS, Bukreyeva N, Murphy BR, Collins PL (1999) Interferon gamma expressed by a recombinant respiratory syncytial virus attenuates virus replication in mice without compromising immunogenicity. Proc Natl Acad Sci USA 96:2367–2372

Calain P, Roux L (1993) The rule of six, a basic feature for efficient replication of Sendai virus defective interfering RNA. J Virol 67:4822–4830

Chanda PK, Banerjee AK (1981a) Identification of promoter-proximal oligonucleotides and a unique dinucleotide, pppGpC, from in vitro transcription products of vesicular stomatitis virus. J Virol 39:93–103

Chanda PK, Banerjee AK (1981b) Purified vesicular stomatitis virus contains an enzyme activity that synthesizes cytidylyl (5′-3′) guanosine 5′-triphosphate in vitro. J Biol Chem 256:11393–11396

Chuang JL, Perrault J (1997) Initiation of vesicular stomatitis virus mutant polR1 transcription internally at the N gene in vitro. J Virol 71:1466–1475

Clarke DK, Sidhu MS, Johnson JE, Udem SA (2000) Rescue of mumps virus from cDNA. J Virol 74:4831–4838

Collins PL, Hightower LE, Ball LA (1980) Transcriptional map for Newcastle disease virus. J Virol 35:682–693

Collins PL, Hill MG, Camargo E, Grosfeld H, Chanock RM, Murphy BR (1995) Production of infectious human respiratory syncytial virus from cloned cDNA confirms an essential role for the transcription elongation factor from the 5′ proximal open reading frame of the M2 mRNA in gene expression and provides a capability for vaccine development. Proc Natl Acad Sci USA 92:11563–11567

Collins PL, Mink MA, Stec DS (1991) Rescue of synthetic analogs of respiratory syncytial virus genomic RNA and effect of truncations and mutations on the expression of a foreign reporter gene. Proc Natl Acad Sci USA 88:9663–9667

Collins PL, Olmsted RA, Spriggs MK, Johnson PR, Buckler-White AJ (1987) Gene overlap and site-specific attenuation of transcription of the viral polymerase L gene of human respiratory syncytial virus. Proc Natl Acad Sci USA 84:5134–5138

Collins PL, Wertz GW (1983) cDNA cloning and transcriptional mapping of nine polyadenylylated RNAs encoded by the genome of human respiratory syncytial virus. Proc Natl Acad Sci USA 80:3208–3212

Conzelmann KK (1998) Nonsegmented negative-strand RNA viruses: genetics and manipulation of viral genomes. Annu Rev Genet 32:123–162

Conzelmann KK, Schnell M (1994) Rescue of synthetic genomic RNA analogs of rabies virus by plasmid-encoded proteins. J Virol 68:713–719

De BP, Banerjee AK (1993) Rescue of synthetic analogs of genome RNA of human parainfluenza virus type 3) Virology 196:344–348

Dickens LE, Collins PL, Wertz GW (1984) Transcriptional mapping of human respiratory syncytial virus. J Virol 52:364–369

Dimock K, Collins PL (1993) Rescue of synthetic analogs of genomic RNA and replicative-intermediate RNA of human parainfluenza virus type 3) J Virol 67:2772–2778

Durbin AP, Siew JW, Murphy BR, Collins PL (1997) Minimum protein requirements for transcription and RNA replication of a minigenome of human parainfluenza virus type 3 and evaluation of the rule of six. Virology 234:74–83

Egelman EH, Wu SS, Amrein M, Portner A, Murti G (1989) The Sendai virus nucleocapsid exists in at least four different helical states. J Virol 63:2233–2243

Eigen M, Biebricher CK (1988) Sequence space and quasispecies distribution. In: Domingo E, Holland J, P. A, eds. RNA genetics. Boca Raton, FL: CRC Press. pp 211–245

Eigen M, Gardiner W, Schuster P, Winkler-Oswatitsch R (1981) The origin of genetic information. Sci Am 244:88–92, 96, et passim

Emerson SU (1976) Vesicular stomatitis virus: structure and function of virion components. Curr Top Microbiol Immunol 73:1–34

Emerson SU (1982) Reconstitution studies detect a single polymerase entry site on the vesicular stomatitis virus genome. Cell 31:635–642

Emerson SU, Wagner RR (1972) Dissociation and reconstitution of the transcriptase and template activities of vesicular stomatitis B and T virions. J Virol 10:297–309

Emerson SU, Yu Y (1975) Both NS and L proteins are required for in vitro RNA synthesis by vesicular stomatitis virus. J Virol 15:1348–1356

Fearns R, Collins PL (1999) Model for polymerase access to the overlapped L gene of respiratory syncytial virus. J Virol 73:388–397

Fearns R, Collins PL, Peeples ME (2000) Functional analysis of the genomic and antigenomic promoters of human respiratory syncytial virus. J Virol 74:6006–6014

Fearns R, Peeples ME, Collins PL (2002) Mapping the transcription and replication promoters of respiratory syncytial virus. J Virol 76:1663–1672

Finke S, Conzelmann KK (1997) Ambisense gene expression from recombinant rabies virus: random packaging of positive- and negative-strand ribonucleoprotein complexes into rabies virions. J Virol 71:7281–7288

Finke S, Conzelmann KK (1999) Virus promoters determine interference by defective RNAs: selective amplification of mini-RNA vectors and rescue from cDNA by a 3′ copy-back ambisense rabies virus. J Virol 73:3818–3825

Finke S, Cox JH, Conzelmann KK (2000) Differential transcription attenuation of rabies virus genes by intergenic regions: generation of recombinant viruses overexpressing the polymerase gene. J Virol 74:7261–7269

Flamand A, Delagneau JF (1978) Transcriptional mapping of rabies virus in vivo. J Virol 28:518–523

Flanagan EB, Ball LA, Wertz GW (2000) Moving the glycoprotein gene of vesicular stomatitis virus to promoter-proximal positions accelerates and enhances the protective immune response. J Virol 74:7895–7902

Flanagan EB, Zamparo JM, Ball LA, Rodriguez LL, Wertz GW (2001) Rearrangement of the genes of vesicular stomatitis virus eliminates clinical disease in the natural host: new strategy for vaccine development. J Virol 75:6107–6114

Garcin D, Pelet T, Calain P, Roux L, Curran J, Kolakofsky D (1995) A highly recombinogenic system for the recovery of infectious Sendai paramyxovirus from cDNA: generation of a novel copy-back nondefective interfering virus. EMBO J 14:6087–6094

Glazier K, Raghow R, Kingsbury DW (1977) Regulation of Sendai virus transcription: evidence for a single promoter in vivo. J Virol 21:863–871

Gupta KC, Roy P (1980) Alternate capping mechanisms for transcription of Spring Viremia of Carp Virus: Evidence for independent mRNA initiation. J Virol 33:292–303

Hardy RW, Harmon SB, Wertz GW (1999) Diverse gene junctions of respiratory syncytial virus modulate the efficiency of transcription termination and respond differently to M2-mediated antitermination. J Virol 73:170–176

Harmon SB, Megaw AG, Wertz GW (2001) RNA sequences involved in transcriptional termination of respiratory syncytial virus. J Virol 75:36–44

Harmon SB, Wertz GW (2002) Transcriptional termination modulated by nucleotides outside the characterized gene end sequence of respiratory syncytial virus. Virology 300:304–315

Hasan MK, Kato A, Shioda T, Sakai Y, Yu D, Nagai Y (1997) Creation of an infectious recombinant Sendai virus expressing the firefly luciferase gene from the 3′ proximal first locus. J Gen Virol 78 (Pt 11):2813–2820

Hausmann S, Garcin D, Delenda C, Kolakofsky D (1999a) The versatility of paramyxovirus RNA polymerase stuttering. J Virol 73:5568–5576

Hausmann S, Garcin D, Morel AS, Kolakofsky D (1999b) Two nucleotides immediately upstream of the essential A6G3 slippery sequence modulate the pattern of G insertions during Sendai virus mRNA editing. J Virol 73:343–351

He B, Paterson RG, Ward CD, Lamb RA (1997) Recovery of infectious SV5 from cloned DNA and expression of a foreign gene. Virology 237:249–260

Hercyk N, Horikami SM, Moyer SA (1988) The vesicular stomatitis virus L protein possesses the mRNA methyltransferase activities. Virology 163:222–225

Hinzman EE, Barr JN, Wertz GW (2002) Identification of an upstream sequence element required for vesicular stomatitis virus mRNA transcription. J Virol 76:7632–7641

Howard M, Wertz G (1989) Vesicular stomatitis virus RNA replication: a role for the NS protein. J Gen Virol 70 (Pt 10):2683–2694

Hunt DM, Emerson SU, Wagner RR (1976) RNA-temperature-sensitive mutants of vesicular stomatitis virus: L-protein thermosensitivity accounts for transcriptase restriction of group I mutants. J Virol 18:596–603

Hwang LN, Englund N, Pattnaik AK (1998) Polyadenylation of vesicular stomatitis virus mRNA dictates efficient transcription termination at the intercistronic gene junctions. J Virol 72:1805–1813

Iseni F, Baudin F, Garcin D, Marq JB, Ruigrok RW, Kolakofsky D (2002) Chemical modification of nucleotide bases and mRNA editing depend on hexamer or nucleoprotein phase in Sendai virus nucleocapsids. Rna 8:1056–1067

Iverson LE, Rose JK (1981) Localized attenuation and discontinuous synthesis during vesicular stomatitis virus transcription. Cell 23:477–484

Kato A, Kiyotani K, Hasan MK, Shioda T, Sakai Y, Yoshida T, Nagai Y (1999) Sendai virus gene start signals are not equivalent in reinitiation capacity: moderation at the fusion protein gene. J Virol 73:9237–9246

Keene JD, Lazzarini RA (1976) A comparison of the extents of methylation of vesicular stomatitis virus messenger RNA. Virology 69:364–367

Keene JD, Thornton BJ, Emerson SU (1981) Sequence-specific contacts between the RNA polymerase of vesicular stomatitis virus and the leader RNA gene. Proc Natl Acad Sci USA 78:6191–6195

Keller MA, Murphy SK, Parks GD (2001) RNA replication from the simian virus 5 antigenomic promoter requires three sequence-dependent elements separated by sequence-independent spacer regions. J Virol 75:3993–3998

Kiley MP, Wagner RR (1972) Ribonucleic acid species of intracellular nucleocapsids and released virions of vesicular stomatitis virus. J Virol 10:244–255

Kolakofsky D, Pelet T, Garcin D, Hausmann S, Curran J, Roux L (1998) Paramyxovirus RNA synthesis and the requirement for hexamer genome length: the rule of six revisited. J Virol 72:891–899

Krishnamurthy S, Huang Z, Samal SK (2000) Recovery of a virulent strain of newcastle disease virus from cloned cDNA: expression of a foreign gene results in growth retardation and attenuation. Virology 278:168–182

Kuo L, Fearns R, Collins PL (1996a) The structurally diverse intergenic regions of respiratory syncytial virus do not modulate sequential transcription by a dicistronic minigenome. J Virol 70:6143–6150

Kuo L, Fearns R, Collins PL (1997) Analysis of the gene start and gene end signals of human respiratory syncytial virus: quasi-templated initiation at position 1 of the encoded mRNA. J Virol 71:4944–4953

Kuo L, Grosfeld H, Cristina J, Hill MG, Collins PL (1996b) Effects of mutations in the gene-start and gene-end sequence motifs on transcription of monocistronic and dicistronic minigenomes of respiratory syncytial virus. J Virol 70:6892–6901

Lamb RA, Kolakofsky D (2001) Paramyxoviridae: The viruses and their replication. In: Knipe D, Howley PM, eds. Fields Virology: Lippincott Williams and Wilkins. pp 1305–1340

Lawson ND, Stillman EA, Whitt MA, Rose JK (1995) Recombinant vesicular stomatitis viruses from DNA. Proc Natl Acad Sci USA 92:4477–4481

Li T, Pattnaik AK (1997) Replication signals in the genome of vesicular stomatitis virus and its defective interfering particles: identification of a sequence element that enhances DI RNA replication. Virology 232:248–259

Li T, Pattnaik AK (1999) Overlapping signals for transcription and replication at the 3′ terminus of the vesicular stomatitis virus genome. J Virol 73:444–452

Marriott AC, Smith JM, Easton AJ (2001) Fidelity of leader and trailer sequence usage by the respiratory syncytial virus and avian pneumovirus replication complexes. J Virol 75:6265–6272

Mebatsion T, Schnell MJ, Cox JH, Finke S, Conzelmann KK (1996) Highly stable expression of a foreign gene from rabies virus vectors. Proc Natl Acad Sci USA 93:7310–7314

Moyer SA, Abraham G, Adler R, Banerjee AK (1975) Methylated and blocked 5′′ termini in vesicular stomatitis virus in vivo mRNAs. Cell 5:59–67

Moyer SA, Banerjee AK (1976) In vivo methylation of vesicular stomatitis virus and its host-cell messenger RNA species. Virology 70:339–351

Murphy SK, Ito Y, Parks GD (1998) A functional antigenomic promoter for the paramyxovirus simian virus 5 requires proper spacing between an essential internal segment and the 3′ terminus. J Virol 72:10–19

Murphy SK, Parks GD (1997) Genome nucleotide lengths that are divisible by six are not essential but enhance replication of defective interfering RNAs of the paramyxovirus simian virus 5) Virology 232:145–157

Murphy SK, Parks GD (1999) RNA replication for the paramyxovirus simian virus 5 requires an internal repeated (CGNNNN) sequence motif. J Virol 73:805–809

Park KH, Huang T, Correia FF, Krystal M (1991) Rescue of a foreign gene by Sendai virus. Proc Natl Acad Sci USA 88:5537–5541

Park KH, Krystal M (1992) In vivo model for pseudo-templated transcription in Sendai virus. J Virol 66:7033–7039

Parks GD, Ward KR, Rassa JC (2001) Increased readthrough transcription across the simian virus 5 M-F gene junction leads to growth defects and a global inhibition of viral mRNA synthesis. J Virol 75:2213–2223

Pattnaik AK, Ball LA, LeGrone AW, Wertz GW (1992) Infectious defective interfering particles of VSV from transcripts of a cDNA clone. Cell 69:1011–1020

Pattnaik AK, Hwang L, Li T, Englund N, Mathur M, Das T, Banerjee AK (1997) Phosphorylation within the amino-terminal acidic domain I of the phosphoprotein of vesicular stomatitis virus is required for transcription but not for replication. J Virol 71:8167–8175

Patton JT, Davis NL, Wertz GW (1984) N protein alone satisfies the requirement for protein synthesis during RNA replication of vesicular stomatitis virus. J Virol 49:303–309

Peeples ME, Collins PL (2000) Mutations in the 5′ trailer region of a respiratory syncytial virus minigenome which limit RNA replication to one step. J Virol 74:146–155

Peeters BP, Gruijthuijsen YK, de Leeuw OS, Gielkens AL (2000) Genome replication of Newcastle disease virus: involvement of the rule-of-six. Arch Virol 145:1829–1845

Pelet T, Delenda C, Gubbay O, Garcin D, Kolakofsky D (1996) Partial characterization of a Sendai virus replication promoter and the rule of six. Virology 224:405–414

Peluso RW, Moyer SA (1988) Viral proteins required for the in vitro replication of vesicular stomatitis virus defective interfering particle genome RNA. Virology 162:369–376

Pennica D, Lynch KR, Cohen PS, Ennis HL (1979) Decay of vesicular stomatitis virus mRNAs in vivo. Virology 94:484–487

Perlman SM, Huang AS (1973) RNA synthesis of vesicular stomatitis virus. V. Interactions between transcription and replication. J Virol 12:1395–1400

Perrault J, Clinton GM, McClure MA (1983) RNP template of vesicular stomatitis virus regulates transcription and replication functions. Cell 35:175–185

Radecke F, Spielhofer P, Schneider H, Kaelin K, Huber M, Dotsch C, Christiansen G, Billeter MA (1995) Rescue of measles viruses from cloned DNA. EMBO J 14:5773–5784

Rassa JC, Parks GD (1998) Molecular basis for naturally occurring elevated read-through transcription across the M-F junction of the paramyxovirus SV5) Virology 247:274–286

Rassa JC, Parks GD (1999) Highly diverse intergenic regions of the paramyxovirus simian virus 5 cooperate with the gene end U tract in viral transcription termination and can influence reinitiation at a downstream gene. J Virol 73:3904–3912

Rhodes DP, Banerjee AK (1975) 5′-Terminal sequence of vesicular stomatitis virus mRNA's synthesized in vitro. J Virol 17:33–42

Rhodes DP, Moyer SA, Banerjee AK (1974) In vitro synthesis of methylated messenger RNA by the virion-associated RNA polymerase of vesicular stomatitis virus. Cell 3:327–333

Rose JK (1975) Heterogeneous 5′-terminal structures occur on vesicular stomatitis virus mRNAs. J Biol Chem 250:8098–8104

Rose JK, Lodish HF, Brock ML (1977) Giant heterogeneous polyadenylic acid on vesicular stomatitis virus mRNA synthesized in vitro in the presence of S-adenosylhomocysteine. J Virol 21:683–693

Samal SK, Collins PL (1996) RNA replication by a respiratory syncytial virus RNA analog does not obey the rule of six and retains a nonviral trinucleotide extension at the leader end. J Virol 70:5075–5082

Schincariol AL, Howatson AF (1972) Replication of vesicular stomatitis virus. II. Separation and characterization of virus-specific RNA species. Virology 49:766–783

Schlender J, Bossert B, Buchholz U, Conzelmann KK (2000) Bovine respiratory syncytial virus nonstructural proteins NS1 and NS2 cooperatively antagonize alpha/beta interferon-induced antiviral response. J Virol 74:8234–8242

Schnell MJ, Buonocore L, Whitt MA, Rose JK (1996) The minimal conserved transcription stop-start signal promotes stable expression of a foreign gene in vesicular stomatitis virus. J Virol 70:2318–2323

Schnell MJ, Mebatsion T, Conzelmann KK (1994) Infectious rabies viruses from cloned cDNA. EMBO J 13:4195–4203

Schubert M, Lazzarini RA (1981) In vivo transcription of the 5′-terminal extracistronic region of vesicular stomatitis virus RNA. J Virol 38:256–262

Sidhu MS, Chan J, Kaelin K, Spielhofer P, Radecke F, Schneider H, Masurekar M, Dowling PC, Billeter MA, Udem SA (1995) Rescue of synthetic measles virus minireplicons: measles genomic termini direct efficient expression and propagation of a reporter gene. Virology 208:800–807

Simonsen CC, Batt-Humphries S, Summers DF (1979) RNA synthesis of vesicular stomatitis virus-infected cells: in vivo regulation of replication. J Virol 31:124–132

Singh M, Cattaneo R, Billeter MA (1999) A recombinant measles virus expressing hepatitis B virus surface antigen induces humoral immune responses in genetically modified mice. J Virol 73:4823–4828

Skiadopoulos MH, Surman SR, Riggs JM, Orvell C, Collins PL, Murphy BR (2002) Evaluation of the replication and immunogenicity of recombinant human parainfluenza virus type 3 vectors expressing up to three foreign glycoproteins. Virology 297:136–152

Skiadopoulos MH, Vogel L, Riggs JM, Surman SR, Collins PL, Murphy BR (2003) The genome length of human parainfluenza virus type 2 follows the rule of six,

and recombinant viruses recovered from non-polyhexameric-length antigenomic cDNAs contain a biased distribution of correcting mutations. J Virol 77:270–279
Smallwood S, Moyer SA (1993) Promoter analysis of the vesicular stomatitis virus RNA polymerase. Virology 192:254–263
Soria M, Little SP, Huang AS (1974) Characterization of vesicular stomatitis virus nucleocapsids. I. Complementary 40 S RNA molecules in nucleocapsids. Virology 61:270–280
Spielhofer P, Bachi T, Fehr T, Christiansen G, Cattaneo R, Kaelin K, Billeter MA, Naim HY (1998) Chimeric measles viruses with a foreign envelope. J Virol 72:2150–2159
Steinhauer DA, de la Torre JC, Holland JJ (1989) High nucleotide substitution error frequencies in clonal pools of vesicular stomatitis virus. J Virol 63:2063–2071
Steinhauer DA, Domingo E, Holland JJ (1992) Lack of evidence for proofreading mechanisms associated with an RNA virus polymerase. Gene 122:281–288
Steinhauer DA, Holland JJ (1986) Direct method for quantitation of extreme polymerase error frequencies at selected single base sites in viral RNA. J Virol 57:219–228
Stillman EA, Whitt MA (1997) Mutational analyses of the intergenic dinucleotide and the transcriptional start sequence of vesicular stomatitis virus (VSV) define sequences required for efficient termination and initiation of VSV transcripts. J Virol 71:2127–2137
Stillman EA, Whitt MA (1998) The length and sequence composition of vesicular stomatitis virus intergenic regions affect mRNA levels and the site of transcript initiation. J Virol 72:5565–5572
Stillman EA, Whitt MA (1999) Transcript initiation and 5′-end modifications are separable events during vesicular stomatitis virus transcription. J Virol 73:7199–7209
Sutherland KA, Collins PL, Peeples ME (2001) Synergistic effects of gene-end signal mutations and the M2-1 protein on transcription termination by respiratory syncytial virus. Virology 288:295–307
Szilagyi JF, Uryvayev L (1973) Isolation of an infectious ribonucleoprotein from vesicular stomatitis virus containing an active RNA transcriptase. J Virol 11:279–286
Tapparel C, Maurice D, Roux L (1998) The activity of Sendai virus genomic and antigenomic promoters requires a second element past the leader template regions: a motif (GNNNNN)3 is essential for replication. J Virol 72:3117–3128
Testa D, Banerjee AK (1977) Two methyltransferase activities in the purified virions of vesicular stomatitis virus. J Virol 24:786–793
Thomas SM, Lamb RA, Paterson RG (1988) Two mRNAs that differ by two nontemplated nucleotides encode the amino coterminal proteins P and V of the paramyxovirus SV5) Cell 54:891–902
Tidona CA, Kurz HW, Gelderblom HR, Darai G (1999) Isolation and molecular characterization of a novel cytopathogenic paramyxovirus from tree shrews. Virology 258:425–434
Tokusumi T, Iida A, Hirata T, Kato A, Nagai Y, Hasegawa M (2002) Recombinant Sendai viruses expressing different levels of a foreign reporter gene. Virus Res 86:33–38

Tsurudome M, Bando H, Kawano M, Matsumura H, Komada H, Nishio M, Ito Y (1991) Transcripts of simian virus 41 (SV41) matrix gene are exclusively dicistronic with the fusion gene which is also transcribed as a monocistron. Virology 184:93–100

Villarreal LP, Breindl M, Holland JJ (1976) Determination of molar ratios of vesicular stomatitis virus induced RNA species in BHK21 cells. Biochemistry 15:1663–1667

Walsh EP, Baron MD, Anderson J, Barrett T (2000) Development of a genetically marked recombinant rinderpest vaccine expressing green fluorescent protein. J Gen Virol 81:709–718

Wang LF, Yu M, Hansson E, Pritchard LI, Shiell B, Michalski WP, Eaton BT (2000) The exceptionally large genome of Hendra virus: support for creation of a new genus within the family Paramyxoviridae. J Virol 74:9972–9979

Wertz GW (1978) Isolation of possible replicative intermediate structures from vesicular stomatitis virus-infected cells. Virology 85:271–285

Wertz GW, Moudy R, Ball LA (2002) Adding genes to the RNA genome of vesicular stomatitis virus: positional effects on stability of expression. J Virol 76:7642–7650

Wertz GW, Perepelitsa VP, Ball LA (1998) Gene rearrangement attenuates expression and lethality of a nonsegmented negative strand RNA virus. Proc Natl Acad Sci USA 95:3501–3506

Wertz GW, Whelan S, LeGrone A, Ball LA (1994) Extent of terminal complementarity modulates the balance between transcription and replication of vesicular stomatitis virus RNA. Proc Natl Acad Sci USA 91:8587–8591

Whelan SP, Ball LA, Barr JN, Wertz GT (1995) Efficient recovery of infectious vesicular stomatitis virus entirely from cDNA clones. Proc Natl Acad Sci USA 92:8388–8392

Whelan SP, Barr JN, Wertz GW (2000) Identification of a minimal size requirement for termination of vesicular stomatitis virus mRNA: implications for the mechanism of transcription. J Virol 74:8268–8276

Whelan SP, Wertz GW (1999a) The 5′ terminal trailer region of vesicular stomatitis virus contains a position-dependent *cis*-acting signal for assembly of RNA into infectious particles. J Virol 73:307–315

Whelan SP, Wertz GW (1999b) Regulation of RNA synthesis by the genomic termini of vesicular stomatitis virus: identification of distinct sequences essential for transcription but not replication. J Virol 73:297–306

Whelan SP, Wertz GW (2002) Transcription and replication initiate at separate sites on the vesicular stomatitis virus genome. Proc Natl Acad Sci USA 99:9178–9183

Yu Q, Hardy RW, Wertz GW (1995) Functional cDNA clones of the human respiratory syncytial (RS) virus N, P, and L proteins support replication of RS virus genomic RNA analogs and define minimal *trans*-acting requirements for RNA replication. J Virol 69:2412–2419

Yunus AS, Krishnamurthy S, Pastey MK, Huang Z, Khattar SK, Collins PL, Samal SK (1999) Rescue of a bovine respiratory syncytial virus genomic RNA analog by bovine, human and ovine respiratory syncytial viruses confirms the "functional integrity" and "cross-recognition" of BRSV *cis*-acting elements by HRSV and ORSV. Arch Virol 144:1977–1990

Orthomyxovirus Replication, Transcription, and Polyadenylation

G. Neumann[1] · G. G. Brownlee[2] · E. Fodor[2] · Y. Kawaoka[1, 3]

[1] Department of Pathobiological Sciences, School of Veterinary Medicine, University of Wisconsin-Madison, 2015 Linden Drive, Madison, WI 53706, USA
E-mail: kawaokay@svm.vetmed.wisc.edu

[2] Sir William Dunn School of Pathology, University of Oxford, South Parks Road, Oxford, OX1 3RE, UK

[3] Department of Microbiology and Immunology, Institute of Medical Science, University of Tokyo, 108-9639, Tokyo, Japan

Abstract Efficient in vitro and in vivo systems are now in place to study the role of viral proteins in replication and/or transcription, the regulation of these processes, polyadenylation of viral mRNAs, the viral promoter structures, or the significance of noncoding regions for virus replication. In this chapter, we review the status of current knowledge of the orthomyxovirus RNA synthesis.

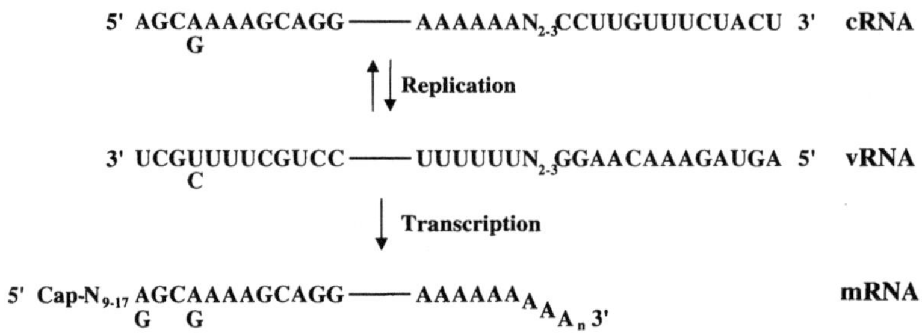

Fig. 1 Schematic diagram of vRNA, cRNA, and mRNA synthesized during viral transcription and replication in infected cells. The conserved 13 and 12 nucleotides at the 5′ and 3′ ends of vRNA, respectively, are shown. The U_6 poly(A) signal near the 5′ end of vRNA, separated by 2 or 3 segment-specific residues from the conserved 13 5′-terminal residues, is also shown. The natural variation occurring at position 4 at the 3′ end of vRNA (either U or C) is indicated. The cap structure and the 9–17 heterologous nucleotides (Cap-N_{9-17}) at the 5′ end of mRNA are derived from host cell pre-mRNAs

1 Overview of Influenza Virus Replication and Transcription

The minimal influenza virus replication complex consists of three polymerase subunits, PB2, PB1, and PA (encoded by the three largest segments of the influenza viral genome), the negative-sense viral RNA (vRNA), and the nucleoprotein NP (encoded by the fifth segment), which encapsidates the vRNAs. The resulting complex is referred to as the viral ribonucleoprotein (vRNP) complex. Individual vRNP complexes are formed for each of the eight influenza A viral RNA segments. During transcription, the polymerase complex transcribes the negative-sense vRNAs into positive-sense mRNAs that are capped and polyadenylated (Fig. 1). During replication a full-length copy of the vRNA, termed complementary RNA (cRNA), is synthesized and subsequently serves as a template for the synthesis of progeny viral RNA.

1.1 Transcription

Initiation of mRNA synthesis requires 5′-cap structures that are generated by cleavage of capped cellular mRNAs (Fig. 2A) (Caton and Robertson 1980; Beaton and Krug 1981; Krug 1981; Plotch et al. 1981).

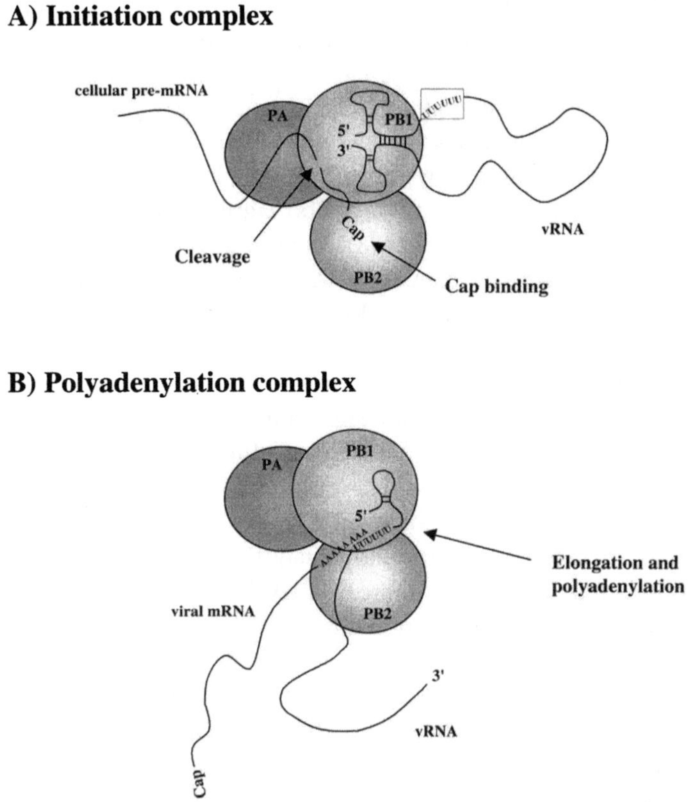

Fig. 2A, B Schematic representation of the initiation and polyadenylation complexes. **A** Initiation of transcription requires 5′-cap structures that are generated by cleavage of capped cellular mRNAs. Cap binding is performed by PB2, followed by capped RNA cleavage by PB1. Cap binding and cleavage activities are dependent on the interaction of the polymerase complex with vRNA. Association of vRNA with PB1 is shown in a corkscrew conformation. **B** Transcription proceeds until the polymerase complex reaches the polyadenylation signal, which is formed by a stretch of uridine residues located 15–22 nucleotides from the 3′ end of the vRNA. Polyadenylation of viral mRNA occurs by reiterative copying of the uridine sequence (Poon et al. 1999, 2000)

Biochemical data indicated that the PB2 subunit recognizes and binds to the cap structure of cellular mRNAs (Ulmanen et al. 1981; Blaas et al. 1982; Nakagawa et al. 1995; Honda et al. 1999; Li et al. 2001). Subsequently, the endonuclease activity, which has been located in the PB1 subunit (Li et al. 2001), cleaves mRNAs 9–17 nucleotides from their 5′

ends, preferentially after a purine residue (Beaton and Krug 1981; Krug 1981; Plotch et al. 1981). The polymerase complex acquires cap binding and endonuclease activities in the presence of an influenza viral RNA (Fig. 2A) (Hagen et al. 1994; Cianci et al. 1995; Brownlee et al. 1995; Li et al. 1998). Initially it was thought that the 5′ end of vRNA was sufficient to stimulate capped mRNA binding by the polymerase complex, whereas both the 5′ and 3′ ends of vRNA were required for significant endonuclease activity (Cianci et al. 1995; Li et al. 1998). However, recent work (Rao et al. 2003) suggests that when a capped RNA substrate contains a CA cleavage site, the 5′ terminal sequence of vRNA alone is sufficient for endonuclease activation. The viral polymerase often utilizes capped CA-containing RNA fragments as primers for viral mRNA synthesis in infected cells (Beaton and Krug 1981; Shaw and Lamb 1984). Although an earlier model suggested that the polymerase first binds to the 5′ end of vRNA and only then does binding to the 3′ end occur (Lamb and Krug 2001), this model has been recently challenged by Lee et al. (2003), who reported that a polymerase bound simultaneously to a preannealed duplex of the 5′ and 3′ ends had greatly increased levels of primer binding and endonuclease activities compared with a sequentially assembled complex.

Short model 5′ terminal cRNAs also stimulated cap binding but did not efficiently activate the viral endonuclease (Cianci et al. 1995; Honda et al. 2001). However, somewhat surprisingly, endonuclease cleavage of a capped CA-containing RNA substrate could be efficiently activated by a 5′ cRNA alone, although the activation of the CA-specific endonuclease by 5′ cRNA did not lead to capped RNA-primed transcription of cRNA because 3′ cRNA does not apparently function as a template for such synthesis (Rao et al. 2003).

Transcription is initiated by the addition of a "G" residue onto the 3′ end of the capped primer, directed by the penultimate "C" of vRNAs (Beaton and Krug 1981; Kawakami et al. 1983). Initiation with a "'C," directed by the "G" in the third position of vRNA, has also been reported (Fodor et al. 1995). This addition is catalyzed by the PB1 protein, which contains nucleotide binding sites (Romanos and Hay 1984; Asano et al. 1995; Asano and Ishihama 1997), as well as four consensus motifs for nucleic acid polymerases (Argos 1988; Poch et al. 1989; Biswas and Nayak 1994; Ishihama and Barbier 1994). Alteration of the conserved SDD sequence in motif C (amino acid position 443–446) rendered the polymerase complex nonfunctional (Biswas and Nayak 1994), demonstrating that these amino acids are crucial for the biological activity of the polymerase complex. Transcription proceeds to a uridine stretch (U_{5-7}), located 15–22 nucleotides before the 5′ end of the vRNA (Li and Palese 1994). Stuttering at the uridine stretch or its reiterative copying results in the addition of polyA tails to nascent mRNA transcripts (Fig. 2B; see below for details).

1.2 Replication

Replication is primer-independent and, initially, results in the synthesis of full-length cRNA copies of the viral genomic RNA. Early studies suggested that replication requires de novo protein synthesis, and it was proposed that there must be a switch from mRNA to cRNA synthesis. The mechanism(s) that triggers the switch from primer-dependent mRNA synthesis to primer-independent cRNA synthesis is not understood. The production of full-length cRNA molecules is reported to require soluble NP protein (Honda et al. 1988; Shapiro and Krug 1988). However, short replicative intermediates can be synthesized in vitro in the absence of NP (Honda et al. 1988; Lee et al. 2002), whereas the addition of NP results in the generation of full-length RNA molecules (Honda et al. 1988). Further indication for a role of NP in replication came from studies with a temperature-sensitive (*ts*) NP mutant (Shapiro and Krug 1988), in which cRNA, but not mRNA, synthesis in vitro was decreased at the nonpermissive temperature. One possible scenario is that NP binds to the nascent RNA transcript, thus preventing the uridine stretch near the 5′ end of the vRNA from slipping backwards; consequently, reiterative copying (hence polyadenylation) would be blocked in the presence of NP. Alternatively, NP might play a purely structural role, by binding to cRNA, together with the viral polymerase complex, and thus preventing the degradation of newly synthesized cRNA molecules by cellular nucleases. Indeed, specific interactions between NP and the PB1 and PB2 subunits of the viral polymerase have been demonstrated (Biswas et al. 1998; Medcalf et al. 1999; Mena et al. 1999).

During the second step of replication, full-length antigenomic cRNA serves as a template for progeny vRNA synthesis. vRNA synthesis, like cRNA synthesis, is primer-independent, requires NP protein, and results in the generation of full-length viral transcripts (reviewed in Lamb and Krug 2001).

2 The RNA Polymerase Complex

Both transcription and replication are catalyzed by the viral RNA-dependent RNA polymerase complex, which is composed of three subunits, PB1, PB2, and PA proteins. The exact nature of the polymerase complexes involved in the synthesis of the three different RNA species, namely, mRNA, cRNA, and vRNA, is not known. A recent report showed that recombinant dimeric complexes of PB1-PB2 and PB1-PA have distinct transcriptase and replicase activities, respectively (Honda et al. 2002). However, other reports concluded that for efficient RNA synthesis all three subunits are required and they suggested that a trimeric complex consisting of one of each of the three subunits is involved in all three RNA synthetic reactions (Brownlee and Sharps 2002; Fodor et al. 2002; Martín-Benito et al. 2001; Perales and Ortín 1997). The distinct activities of the polymerase complex, e.g., endonuclease cleavage of capped RNA primers, transcription, polyadenylation, or replication, could be modulated by its interaction with vRNA, cRNA, or host factors.

2.1 The PB1 Protein

The PB1 subunit of the RNA polymerase complex plays a central role in the catalytic activities of the viral polymerase. It contains the conserved motifs characteristic of RNA-dependent RNA polymerase (Argos 1988; Biswas and Nayak 1994). It binds specifically to the terminal sequences of vRNA and cRNA and, dependent on its interaction with RNA, performs capped RNA endonuclease activities (Cianci et al. 1995; González and Ortín 1999a, b; Li et al. 1998, 2001; Rao et al. 2003). Li et al. (2001) identified three acidic amino acids (E508, E519, D522), essential for endonuclease cleavage, that could form an active site similar to that of other enzymes that cleave polynucleotides to produce 3′-OH ends.

2.2 The PB2 Protein

The PB2 subunit plays a crucial role in mRNA synthesis by being responsible for recognition and binding of host mRNAs. The cap-binding domain was localized to amino acids 533–564 in a cross-linking study

(Li et al. 2001), but an earlier study proposed that, in addition, the region 242–252 of PB2 was also involved (Honda et al. 1999). More recently, Fechter et al. (2003) proposed that two aromatic amino acids, F363 and F404, form an "aromatic sandwich" that is involved in cap binding analogous to the evolutionarily unrelated cap-binding proteins VP39, eIF4E, and CBP20. PB2 also seems to play an important role in replication, because single amino acid mutations in the N-terminal region of PB2 specifically affected replication but not transcription (Gastaminza et al. 2003).

2.3 The PA Protein

Although the PA subunit is essential for the biological activity of the polymerase complex, its function in influenza virus transcription, replication, and polyadenylation is less well established than either PB1 or PB2. Expression of PA reduces the half-lives of coexpressed proteins, suggesting that it has proteolytic activity or the ability to induce proteolytic activity of a cellular protein (Sanz-Ezquerro et al. 1995, 1996). Studies by Perales et al. (2000) suggested a correlation between this function and the biological activity of the polymerase complex, although Naffakh et al. (2001) contradicted this conclusion. Hara et al. (2001) found proteolytic activity in purified PA and identified serine at position 624 as the active site of a serine protease. However, because Sanz-Ezquerro et al. (1996) mapped the PA proteolytic activity to the N-terminal third of the protein, the proteolytic activities observed by the two groups must differ from one other. In addition, the serine protease activity of PA, identified by Hara et al. (2001) is not critical for polymerase activity or viral growth in cell culture (Fodor et al. 2002). Thus, the biological significance of the proteolytic activity associated with PA for influenza viral replication remains obscure.

New information on the role of the PA subunit in transcription and replication has emerged from a detailed mutagenic analysis of the C-terminal half of the molecule. A histidine at position 510 that is conserved in influenza A, B, C, and Thogotovirus, has been implicated in endonuclease cleavage of capped RNA primers (Fodor et al. 2002). Mutation of another conserved amino acid, arginine at position 638, led to the formation of defective RNAs, implying that PA could act as an elongation factor during RNA synthesis (Fodor et al. 2003).

The recent reports of the direct roles of PB2 in replication and of PA in transcription (Gastaminza et al. 2003; Fodor et al. 2002, 2003), in addition to their previously recognized roles in transcription and replication, support the idea that the RNA polymerase complex is a trimeric complex with residues from all three subunits contributing to specific functions. Indeed, a low-resolution three-dimensional structural model of a recombinant influenza virus RNP particle generated by electron microscopy indicates a rather compact structure for the polymerase complex (Martín-Benito et al. 2001).

3
Systems Used to Study the Influenza A Virus Life Cycle

Efforts to study the influenza virus life cycle posed numerous technical challenges. Systems had to be established to examine viral replication both in vitro and in vivo, and methods had to be developed to introduce mutations into viral RNAs as well as NP and polymerase proteins and to monitor the effects of these mutations on the viral life cycle. The in vitro reconstitution of functional RNP complexes was achieved with use of purified NP and polymerase proteins derived from virions (Honda et al. 1987, 1988) or from micrococcal nuclease-treated viral cores (Seong and Brownlee 1992a; Seong et al. 1992). Parvin et al. (1989) first used synthetic viral RNA templates for RNP reconstruction, thus permitting mutations to be introduced into the viral RNA template.

Huang et al. (1990) devised a helper virus-independent system to study influenza A virus replication in eukaryotic cells. RNP complexes were assembled in vitro from purified polymerase complexes and a virus-like RNA encoding the reporter gene chloramphenicol-acetyltransferase (CAT). Cells were then transfected with the in vitro-assembled RNP complexes and infected with recombinant vaccinia virus expressing the NP and polymerase proteins. CAT expression demonstrated the in vivo formation of functional RNP complexes that replicated and transcribed the virus-like RNA. Alternatively, the influenza virus proteins can be provided from SV40 recombinant viruses (de la Luna et al. 1993) or from cell lines stably expressing the polymerase and NP proteins (Nakamura et al. 1991; Kimura et al. 1992). A transcription system based on the cellular enzyme RNA polymerase I allowed the in vivo synthesis of influenza virus transcripts (Zobel et al. 1993; Neumann et al. 1994). Subsequently, this approach led to the establishment of a plasmid-based minireplicon system for influenza A virus (Pleschka et al. 1996) and, eventually, to the establishment of plasmid-based rescue systems for generating fully recombinant influenza A (Fodor et al. 1999; Neumann et al. 1999), influenza B (Hoffmann et al. 2002, Jackson et al. 2002), and Thogotovirus (Wagner et al. 2001).

4
The Influenza A Virus RNA Promoter

The 13 5′ terminal and 12 3′ terminal nucleotides of the eight influenza A virus segments are highly conserved among all strains (Fig. 1). Studies with short synthetic viral RNAs or mutated vRNAs suggested that the first 12–14 nucleotides at the 3′ end of the vRNAs (Parvin et al. 1989; Yamanaka et al. 1991; Seong and Brownlee 1992a, b; Piccone et al. 1993) and the first 11–13 nucleotides at the 3′ end of the cRNA (Li and Palese 1992; Seong and Brownlee 1992a, b) constitute the core promoter elements. Early in vitro experiments suggested that the 3′ end of the vRNA is sufficient to transcribe short synthetic templates (Parvin et al. 1989). However, subsequent biochemical studies revealed that the polymerase complex interacts with both the 3′ and 5′ ends of vRNAs (Fodor et al. 1993, 1994; Tiley et al. 1994; Klumpp et al. 1997). Mutational analysis of the promoter elements additionally demonstrated that the influenza A virus vRNA and cRNA promoters encompass both the 5′ and 3′ ends of the vRNA or cRNA, respectively (Fodor et al. 1994; Neumann and Hobom 1995; Pritlove et al. 1995; Flick et al. 1996).

The significance of nucleotides at the 5′ and 3′ ends of the vRNA and cRNA for influenza virus replication and transcription has been addressed in vitro (Li and Palese 1992; Piccone et al. 1993; Fodor et al. 1994, 1995; Pritlove et al. 1995) and in vivo (Li and Palese 1992; Piccone et al. 1993; Neumann et al. 1994; Neumann and Hobom 1995; Flick et al. 1996; Kim et al. 1997). In general, the effects of mutations as judged by in vivo expression of a reporter gene are more pronounced than in the in vitro transcription of a model template (Li and Palese 1992; Piccone et al. 1993). This is because only a single step (e.g., the synthesis of cRNA from vRNA) is monitored in in vitro experiments, whereas multiple steps including replication, transcription, polyadenylation, and translation are involved in the expression of a reporter gene construct in the in vivo systems. Thus, discrepancies between in vitro and in vivo studies likely reflect the fact that nucleotides within the promoter region contribute to more than one function.

4.1 The vRNA Promoter

Sequence alignment of the conserved vRNA termini suggested a partially double-stranded structure consisting of two discrete base-paired regions (Fig. 3A). Region I encompasses nucleotides 1–9 at the 3′ and 5′ ends of the vRNA, whereas region II is composed of nucleotides 10–15 and 11–16 at the 3′ and 5′ ends of the vRNA, respectively (Fodor et al. 1995; Flick et al. 1996). These regions are likely connected by a flexible joint formed by the unpaired adenosine residue at position 10 in the 5′ end of the vRNA (Neumann and Hobom 1995; Flick et al. 1996). Experimental evidence from electron microscopic studies and nuclease S1 protection assays supported the existence of an interaction between the 5′ and 3′ vRNA termini and led to the proposal of the so-called "panhandle model" (Hsu et al. 1987). The panhandle model (Fig. 3A) predicted a partially double-stranded structure encompassing the influenza virus core promoter. The results of structural studies performed with naked RNA or RNA bound to NP, using chemical or enzymatic probing or nuclear magnetic resonance (NMR), are consistent with the existence of a panhandle structure (Bae et al. 2001; Baudin et al. 1994; Cheong et al. 1996; Klumpp et al. 1997).

Mutagenic analyses of the vRNA termini resulted in the proposal of alternative models to the panhandle structure. Mutations that disrupted base pairing in region II abolished promoter activity, which could be restored by introducing complementary mutations that reestablished base pairing (Fodor et al. 1994; Neumann and Hobom 1995; Flick et al. 1996; Kim et al. 1997). Because base pairing between the 5′ and 3′ vRNA termi-

Fig. 3A–C Influenza A virus promoter models. **A** Panhandle model, which proposes a partially double-stranded structure for the 5′ and 3′ terminal sequences of vRNA. **B** Corkscrew model for the vRNA promoter, predicting base pairing within the 5′ and 3′ ends of the viral RNAs. **C** Corkscrew model for the cRNA promoter. The Watson-Crick base pairs (indicated by *lines*) and wobble base pairs (indicated by *dots*) in the RNA structure are shown. Numbering of nucleotides starts from the 5′ end in the 5′-terminal sequence and from the 3′ end in the 3′-terminal sequence

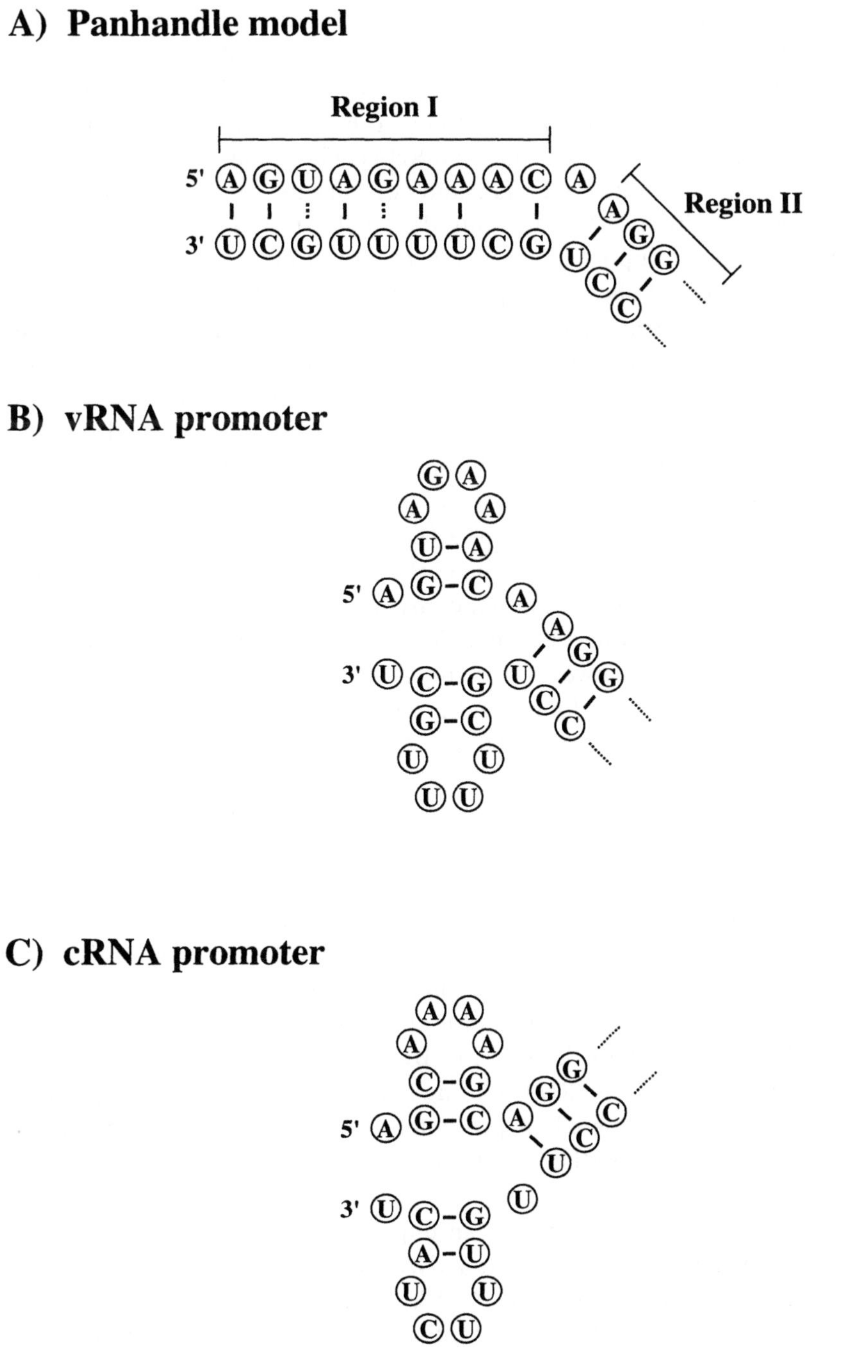
A) Panhandle model
Region I
Region II
5'
3'
B) vRNA promoter
5'
3'
C) cRNA promoter
5'
3'

ni was critical in region II but not region I, Fodor et al. (1994, 1995) suggested the "RNA-fork model," which predicts a double-stranded formation for the former region and a single-stranded conformation for the latter. However, the "corkscrew model" (Fig. 3B) predicted base pairing within the 5′ and 3′ ends of the viral RNAs (Flick et al. 1996). Accordingly, base pairs at positions 2 and 9 and at positions 3 and 8 form a hairpin loop with positions 4 to 7 forming the single-stranded loops. The corkscrew model is supported by several recent findings. Nucleotide replacements that would allow base pairing in a panhandle conformation, but destroy base pairing between the short predicted stems (formed by nucleotides at positions 2 and 9 and positions 3 and 8), abolished promoter activity (Flick and Hobom 1999). Moreover, the short stem at the 5′ end of the vRNA is critical for the synthesis of polyadenylated mRNA (Pritlove et al. 1999), whereas the short loops formed by nucleotides at positions 4–7 at both ends of the vRNA are crucial for the endonuclease activity of the polymerase complex (Leahy et al. 2001a, b). Furthermore, mutations that interfered with the corkscrew structure destabilized polymerase-vRNA promoter binding, consistent with the idea that the corkscrew structure was needed for efficient polymerase binding (Brownlee and Sharps 2002). However, for the cleavage of CA-containing capped RNA primers the stem-loop structure in the vRNA 3′ end does not appear to be critical (Rao et al. 2003).

4.2
The cRNA Promoter

The 5′ and 3′ termini of cRNA have the potential to form a panhandle-like structure similar, but not identical, to the vRNA panhandle. Pritlove et al. (1995) proposed that initiation of RNA synthesis from a 3′ cRNA template requires the presence of a cRNA 5′ end, and binding sites for both termini have been located in PB1 that are reported to be different from those involved in vRNA binding (González and Ortín 1999b). It has been reported that the cRNA promoter also functions in a "corkscrew" configuration, in a similar way to the vRNA promoter (Fig. 3C) (Azzeh et al. 2001).

The cRNA promoter contains an unpaired U residue at position 10 at the 3′ end (Fig. 3C), whereas the vRNA promoter is characterized by an unpaired A residue at position 10 of its 5′ end (Fig. 3B). These unpaired residues could affect the overall conformation of the influenza virus promoters and lead to the selective nuclear export of vRNA promoter-containing RNAs (Shapiro et al. 1987; Tchatalbachev et al. 2001).

5
Influenza A Virus Polyadenylation

mRNA transcription proceeds until the polymerase complex encounters a stretch of uridine residues adjacent to the promoter element at the 5′ end of the vRNA (Fig. 2B). Polyadenylation requires an uninterrupted stretch of five to seven uridine residues adjacent to the base-paired region II (see Fig. 3A) of the influenza virus promoter (16 nucleotides are the optimal distance between the 5′ end of the vRNA and the uridine stretch) (Luo et al. 1991; Li and Palese 1994; Poon et al. 1999; Zheng et al. 1999). The current model, originally proposed by Fodor et al. (1994) and Hagen et al. (1994), predicts stable binding of the polymerase complex to the 5′ end of the vRNA throughout mRNA synthesis, with the vRNA threaded through the polymerase complex (Fig. 2B). In this scenario, the polymerase complex bound to the promoter region would function as a physical barrier that causes reiterative copying at the neighboring uridine stretch. Evidence for this model has come from the introduction of mutations in the vRNA promoter region that weaken the interaction of the polymerase complex with the vRNA, hence affecting the efficiency of polyadenylation (Fodor et al. 1998; Poon et al. 1998; Pritlove et al. 1998). Studies performed both in vitro and in vivo confirmed that the uridine stretch near the 5′ end of vRNA acts directly as a template for poly(A) addition; replacement of the U sequence with an A sequence resulted in the synthesis of poly(U)-containing RNA transcripts (Poon et al. 1999, 2000).

6
Noncoding Regions

The noncoding regions that lie between the conserved promoter elements and the start or stop codon are of variable length and nucleotide composition, and their function is still unknown. Deletions, insertions, or mutations in these regions demonstrated that they are not absolutely required for viral replication or transcription (García-Sastre et al. 1994; Barclay and Palese 1995; Bergmann and Muster 1995, 1996; Zheng et al. 1996) but may affect these processes, possibly by interacting with the

NP protein and/or the polymerase complex. Furthermore, a sequence motif in the noncoding region of segment 7 stimulated the expression of the encoded protein, whereas transcription levels remained unaffected (Enami et al. 1994). This finding suggests that the noncoding regions contain signals that affect mRNA stability, or the initiation of translation, or both.

7
Influenza B and C Virus Promoters

The overall structures of influenza A, B, and C virus promoters are highly conserved, although some differences exist. For example, among influenza A viruses, 12 and 13 nucleotides are conserved at the 3′ and 5′ ends of the vRNAs (with the exception of residue 4 of the 3′ end of vRNA, see Fig. 1), respectively, whereas among influenza B viruses, 9 and 10 nucleotides are identical. Limited sequence data indicate that 11 nucleotides are conserved among the influenza C virus segments. The double-stranded promoter region II encompasses 4 or 5 base pairs in influenza C viruses, 4–8 base pairs in influenza A viruses, but 8–10 base pairs in influenza B viruses. The uridine stretch that functions as the polyadenylation signal varies from 5 to 7 or 5 to 6 uridine residues in influenza A or B viruses, respectively, but is restricted to 4 or 5 residues in influenza C virus. Like their influenza A virus counterparts, the influenza B and C virus promoters can be divided into two regions: Within region II, base pair formation is crucial (Lee and Seong 1996, 1998), whereas in region I, the nature of the nucleotide is critical (Lee and Seong 1998).

Minireplicon systems established for influenza B (Jambrina et al. 1997; Stevens and Barclay 1998; Crescenzo-Chaigne et al. 1999) and C (Crescenzo-Chaigne et al. 1999) viruses have proven that the three polymerase and NP proteins are necessary and sufficient for the replication of virus-like templates. Plasmid-based minireplicon systems and helper virus-dependent replication systems demonstrated that the influenza A, B, or C viral polymerase complexes transcribed heterologous templates, although with different efficiencies (Muster et al. 1991; Crescenzo-Chaigne et al. 1999). Mutational analysis of influenza A and C virus promoters revealed that nucleotides at position 5 and 6 at the 3′ and 5′ ends of the vRNA, respectively, contribute to subtype specificity (Crescenzo-Chaigne and van der Werf 2001). Additionally, the data demonstrated the importance of the base pair formed between nucleotides at positions

3 and 8 at the 5′ end of the vRNA, suggesting a "corkscrew" conformation for the influenza C virus promoter.

8
Thogotovirus Replication

Thogotovirus is a tick-transmitted member of the family *Orthomyxoviridae*. RNP complexes can be reconstituted in vitro (Leahy et al. 1997a, c; Gomez-Puertas et al. 2000) or in vivo (Weber et al. 1998; Wagner et al. 2000, 2001), thus allowing researchers to study replication and transcription of this virus. As with influenza viruses, the NP and polymerase proteins are necessary and sufficient for the amplification of a virus-like RNA (Weber et al. 1998). The initiation of mRNA synthesis relies on host cell-derived cap structures; however, only the cap structures, but no additional nucleotides, are cleaved from host cell mRNAs (Albo et al. 1996; Weber et al. 1996; Leahy et al. 1997a). The promoter consists of the terminal 14–15 nucleotides at the 3′ and 5′ ends of the vRNAs, both of which are required for promoter activity (Leahy et al. 1997a, c). In vitro assays demonstrated that a model vRNA promoter, but not its cRNA counterpart, stimulated the endonuclease activity of the polymerase complex (Leahy et al. 1998a). This might be due to the fact that the Thogotovirus vRNA and cRNA promoters adopt different conformation. The vRNA promoter is characterized by a stem-loop structure at its 5′ end (similar to that found in influenza A viruses) (Leahy et al. 1997b; Weber et al. 1997). A similar structure has not been identified at the 5′ end of the cRNA promoter (Leahy et al. 1997b, 1998b; Weber et al. 1997). Interestingly, introduction of a stem-loop structure into the 5′ end of the cRNA promoter stimulated the endonuclease activity of the polymerase complex (Leahy et al. 1998b), indicating that this structural feature is critical for the activation of endonuclease activity.

9
Concluding Remarks

Over the last decade, efficient in vitro and in vivo systems have been established to study the replication of influenza A and B viruses, as well as Thogotovirus, leading to elucidation of the viral promoters and identification of polyadenylation signals. Despite these advances, much remains unknown, including a detailed understanding of the roles of PB1, PB2, and PA in replication and transcription, the switch between transcription and replication, the mechanism of vRNA packaging (random versus selective), and the significance of the noncoding regions in virus amplification. It is probable that structural analysis of the influenza polymerase complex will ultimately be required to elucidate the subtleties of how the synthesis of mRNA, cRNA, and vRNA are regulated in the infected cell and how vRNA is differentially packaged into virions.

References

Albo C, Martin J, Portela A. (1996) The 5′ ends of Thogoto virus (*Orthomyxoviridae*) mRNAs are homogeneous in both length and sequence. J Virol 70:9013–9017

Argos P (1988) A sequence motif in many polymerases. Nucleic Acids Res 16:9909–9916

Asano Y, Mizumoto K, Maruyama T, Ishihama A. (1995) Photoaffinity labeling of influenza virus RNA polymerase PB1 subunit with 8-azido GTP. J Biochem 117:677–682

Asano Y, Ishihama A (1997) Identification of two nucleotide-binding domains on the PB1 subunit of influenza virus RNA polymerase. J Biochem (Tokyo) 122:627–634

Azzeh M, Flick R, Hobom G (2001) Functional analysis of the influenza A virus cRNA promoter and construction of an ambisense transcription system. Virology 289:400–410

Bae SH, Cheong HK, Lee JH, Cheong C, Kainosho M, Choi BS (2001) Structural features of an influenza virus promoter and their implications for viral RNA synthesis. Proc Natl Acad Sci USA 98:10602–10607

Barclay WS, Palese P (1995) Influenza B viruses with site-specific mutations introduced into the HA gene. J Virol 69, 1275–1279

Baudin F, Bach C, Cusack S, Ruigrok RWH (1994) Structure of influenza virus RNP. I. Influenza virus nucleoprotein melts secondary structure in panhandle RNA and exposes the bases to the solvent. EMBO J 13:3158–3165

Beaton AR, Krug RM (1981) Selected host cell capped RNA fragments prime influenza viral RNA transcription in vivo. Nucleic Acids Res 9:4423–4436

Bergmann M, Muster T (1995) The relative amount of an influenza A virus segment present in the viral particle is not affected by a reduction in replication of that segment. J Gen Virol 76:3211–3215

Bergmann M, Muster T (1996) Mutations in the nonconserved noncoding sequences of the influenza A virus segments affect viral vRNA formation. Virus Res 44:23–31

Biswas SK, Boutz PL, Nayak DP (1998) Influenza virus nucleoprotein interacts with influenza virus polymerase proteins. J Virol 72:5493–5501

Biswas SK, Nayak DP (1994) Mutational analysis of the conserved motifs of influenza A virus polymerase basic protein 1. J Virol 68:1819–1826

Blaas D, Patzelt E, Kuechler E (1982) Identification of the cap binding protein of influenza virus. Nucleic Acids Res 10:4803–4812

Brownlee GG, Fodor E, Pritlove DC, Gould KG, Dalluge JJ (1995) Solid phase synthesis of 5′-diphosphorylated oligoribonucleotides and their conversion to capped m^7Gppp-oligoribonucleotides for use as primers for influenza A virus RNA polymerase in vitro. Nucleic Acids Res 23: 2641–2647

Brownlee GG, Sharps JL (2002) The RNA polymerase of influenza A virus is stabilized by interaction with its viral RNA promoter. J Virol 76:7103–7113

Caton AJ, Robertson JS (1980) Structure of the host-derived sequences present at the 5′ ends of influenza virus mRNA. Nucleic Acids Res 8:2591–2603

Cheong HK, Cheong C, Choi BS (1996) Secondary structure of the panhandle RNA of influenza virus A studied by NMR spectroscopy. Nucleic Acids Res 24:4197–4201

Cianci C, Tiley L, Krystal M (1995) Differential activation of the influenza virus polymerase via template RNA binding. J Virol 69:3995–3999

Crescenzo-Chaigne B, Naffakh N, van der Werf S (1999) Comparative analysis of the ability of the polymerase complexes of influenza viruses type A, B and C to assemble into functional RNPs that allow expression and replication of heterotypic model RNA templates in vivo. Virology 265:342–353

Crescenzo-Chaigne B, van der Werf S (2001) Nucleotides at the extremities of the viral RNA of influenza C virus are involved in type-specific interactions with the polymerase complex. J Gen Virol 82:1075–1083

de la Luna S, Martin J, Portela A, Ortín J (1993) Influenza virus naked RNA can be expressed upon transfection into cells co-expressing the three subunits of the polymerase and the nucleoprotein from simian virus 40 recombinant viruses. J Gen Virol 74:535–539

Enami K, Sato TA, Nakada S, Enami M (1994) Influenza virus NS1 protein stimulates translation of the M1 protein. J Virol 68:1432–1437

Fechter P, Mingay L, Sharps J, Chambers A, Fodor E, Brownlee GG (2003) Two aromatic residues in the PB2 subunit of influenza A RNA polymerase are crucial for cap-binding. J Biol Chem 278:20381–20388

Flick R, Neumann G, Hoffmann E, Neumeier E, Hobom G (1996) Promoter elements in the influenza vRNA terminal structure. RNA 2:1046–1057

Flick R, Hobom G (1999) Interaction of influenza virus polymerase with viral RNA in the "corkscrew"' conformation. J Gen Virol 80:2565–2572

Fodor E, Crow M, Mingay LJ, Deng T, Sharps J, Fechter P, Brownlee GG (2002) A single amino acid mutation in the PA subunit of the influenza RNA polymerase inhibits endonucleolytic cleavage of capped RNAs. J Virol 76:8989–9001

Fodor E, Devenish L, Engelhardt OG, Palese P, Brownlee GG, García-Sastre A. (1999) Rescue of influenza A virus from recombinant DNA. J Virol 73:9679–9682

Fodor E, Mingay LJ, Crow M, Deng T, Brownlee GG (2003) A single amino acid mutation in the PA subunit of the influenza RNA polymerase promotes the generation of defective interfering RNAs. J Virol 77:5017–5020

Fodor E, Seong BL, Brownlee GG (1993) Photochemical cross-linking of influenza A polymerase to its virion RNA promoter defines a polymerase binding site at residues 9 to 12 of the promoter. J Gen Virol 74:1327–1333

Fodor E, Pritlove DC, Brownlee GG (1994) The influenza virus panhandle is involved in the initiation of transcription. J Virol 68:4092–4096

Fodor E, Pritlove DC, Brownlee GG (1995) Characterization of the RNA-fork model of virion RNA in the initiation of transcription in influenza A virus. J Virol 69:4012–4019

Fodor E, Palese P, Brownlee GG, García-Sastre A (1998) Attenuation of influenza A virus mRNA levels by promoter mutations. J Virol 72:6283–6290

García-Sastre A, Percy N, Barclay W, Palese P (1994) Introduction of foreign sequences into the genome of influenza A virus. Dev Biol Stand 82, 237–246

Gastaminza P, Perales B, Falcón AM, Ortín J (2003) Mutations in the N-terminal region of influenza virus PB2 protein affect virus RNA replication but not transcription. J Virol 77:5098–5108

Gomez-Puertas P, Leahy MB, Nuttall PA, Portela A (2000) Rescue of synthetic RNAs into thogoto and influenza A virus particles using core proteins purified from Thogoto virus. Virus Res 67:41–48

González S, Ortín J (1999a) Characterization of influenza virus PB1 protein binding to viral RNA: two separate regions of the protein contribute to the interaction domain. J Virol 73: 631–637

González S, Ortín J (1999b). Distinct regions of influenza virus PB1 polymerase subunit recognize vRNA and cRNA templates. EMBO J 18: 3767–3775

Hagen M, Chung TD, Butcher JA, Krystal M (1994) Recombinant influenza virus polymerase: requirement of both 5′ and 3′ viral ends for endonuclease activity. J Virol 68:1509–1515

Hara K, Shiota M, Kido H, Ohtsu Y, Kashiwagi T, Iwahashi J, Hamada N, Mizoue K, Tsumura N, Kato H, Toyoda T (2001) Influenza virus RNA polymerase PA subunit is a novel serine protease with Ser624 at the active site. Genes Cells 6:87–97

Hoffmann E, Mahmood K, Yang CF, Webster RG, Greenberg HB, Kemble G (2002) Rescue of influenza B virus from eight plasmids. Proc Natl Acad Sci USA 99:11411–11416

Honda A, Ueda K, Nagata K, Ishihama A (1987) Identification of the RNA polymerase-binding site on genome RNA of influenza virus. J Biochem (Tokyo) 102:1241–1249

Honda A, Ueda K, Nagata K, Ishihama A (1988) RNA polymerase of influenza virus: role of NP in RNA chain elongation. J Biochem (Tokyo) 104:1021–1026

Honda A, Mizumoto K, Ishihama A (1999) Two separate sequences of PB2 subunit constitute the RNA cap-binding site of influenza virus RNA polymerase. Genes Cells 4:475–485

Honda A, Endo A, Mizumoto K, Ishihama A (2001) Differential roles of viral RNA and cRNA in functional modulation of the influenza virus RNA polymerase. J Biol Chem 276:31179–31185

Honda A, Mizumoto K, Ishihama A (2002) Minimum molecular architectures for transcription and replication of the influenza virus. Proc Natl Acad Sci USA 99:13166–13171

Hsu MT, Parvin JD, Gupta S, Krystal M, Palese P (1987) Genomic RNAs of influenza viruses are held in a circular conformation in virions and in infected cells by a terminal panhandle. Proc Natl Acad Sci USA 84:8140–8144

Huang TS, Palese P, Krystal M (1990) Determination of influenza virus proteins required for genome replication. J Virol 64:5669–5673

Ishihama A, Barbier P (1994) Molecular anatomy of viral RNA-directed RNA polymerases. Arch Virol 134:235–258

Jackson D, Cadman A, Zürcher T, Barclay WS (2002) A reverse genetics approach for recovery of recombinant influenza B viruses entirely from cDNA. J Virol 76:11744–11747

Jambrina E, Barcena J, Uez O, Portela A (1997) The three subunits of the polymerase and the nucleoprotein of influenza B virus are the minimum set of viral proteins required for expression of a model RNA template. Virology 235:209–217

Kawakami K, Mizumoto K, Ishihama A (1983) RNA polymerase of influenza virus. IV. Catalytic properties of the capped RNA endonuclease associated with the RNA polymerase. Nucleic Acids Res 11:3637–3649

Kim HJ, Fodor E, Brownlee GG, Seong BL (1997) Mutational analysis of the RNA-fork model of the influenza A virus vRNA promoter in vivo. J Gen Virol 78:353–357

Kimura N, Nishida M, Nagata K, Ishihama A, Oda K, Nakada S (1992) Transcription of a recombinant influenza virus RNA in cells that can express the influenza virus RNA polymerase and nucleoprotein genes. J Gen Virol 73:1321–1328

Klumpp K, Ruigrok RW, Baudin F (1997) Roles of the influenza virus polymerase and nucleoprotein in forming a functional RNP structure. EMBO J 16:1248–1257

Krug RM (1981) Priming of influenza viral RNA transcription by capped heterologous RNAs. Curr Top Microbiol Immunol 93: 125–149

Lamb RA, Krug RM (2001) *Orthomyxoviridae*: The Viruses and Their Replication. In: D.M. Knipe, Howley, P. M., Griffin, D. E., Martin, M. A., Lamb, R. A., Roizman, B., Straus, S. E. (Ed), Fields Virology, pp. 1487–1532. Vol. 1. 2 vols. Lippincott Williams & Wilkins, Philadelphia

Leahy MB, Dessens JT, Nuttall PA (1997a) In vitro polymerase activity of Thogoto virus: evidence for a unique cap-snatching mechanism in a tick-borne orthomyxovirus. J Virol 71:8347–8351

Leahy MB, Dessens JT, Nuttall PA (1997b) Striking conformational similarities between the transcription promoters of Thogoto and influenza A viruses: evidence for intrastrand base pairing in the 5′ promoter arm. J Virol 71:8352–8356

Leahy MB, Dessens JT, Weber F, Kochs G, Nuttall PA (1997c) The fourth genus in the *Orthomyxoviridae*: sequence analyses of two Thogoto virus polymerase proteins and comparison with influenza viruses. Virus Res 50:215–224

Leahy MB, Dessens JT, Pritlove DC, Nuttall PA (1998a) An endonuclease switching mechanism in the virion RNA and cRNA promoters of Thogoto orthomyxovirus. J Virol 72:2305–2309

Leahy MB, Dessens JT, Pritlove DC, Nuttall PA (1998b) The Thogoto orthomyxovirus cRNA promoter functions as a panhandle but does not stimulate cap snatching in vitro. J Gen Virol 79:457–460

Leahy MB, Dobbyn HC, Brownlee GG (2001a) Hairpin loop structure in the 3′ arm of the influenza a virus virion RNA promoter is required for endonuclease activity. J Virol 75:7042–7049

Leahy MB, Pritlove DC, Poon LL, Brownlee GG (2001b) Mutagenic analysis of the 5′ arm of the influenza A virus virion RNA promoter defines the sequence requirements for endonuclease activity. J Virol 75:134–142

Lee MT, Bishop K, Medcalf L, Elton D, Digard P, Tiley L (2002) Definition of the minimal viral components required for the initiation of unprimed RNA synthesis by influenza virus RNA polymerase. Nucleic Acids Res 30:429–438
Lee MT, Klumpp K, Digard P, Tiley L (2003). Activation of influenza virus RNA polymerase by the 5′ and 3′ terminal duplex of genomic RNA. Nucleic Acids Res 31:1624–1632
Lee YS, Seong BL (1996) Mutational analysis of influenza B virus RNA transcription in vitro. J Virol 70:1232–1236
Lee YS, Seong BL (1998) Nucleotides in the panhandle structure of the influenza B virus virion RNA are involved in the specificity between influenza A and B viruses. J Gen Virol 79:673–681
Li ML, Ramirez BC, Krug RM (1998) RNA-dependent activation of primer RNA production by influenza virus polymerase: different regions of the same protein subunit constitute the two required RNA-binding sites. EMBO J 17:5844–5852
Li ML, Rao P, Krug RM (2001) The active sites of the influenza cap-dependent endonuclease are on different polymerase subunits. EMBO J 20:2078–2086
Li X, Palese P (1992) Mutational analysis of the promoter required for influenza virus virion RNA synthesis. J Virol 66:4331–4338
Li X, Palese P (1994) Characterization of the polyadenylation signal of influenza virus RNA. J Virol 68:1245–1249
Luo GX, Luytjes W, Enami M, Palese P (1991) The polyadenylation signal of influenza virus RNA involves a stretch of uridines followed by the RNA duplex of the panhandle structure. J Virol 65:2861–2867
Martín-Benito J, Area E, Ortega J, Llorca O, Valpuesta JM, Carrascosa JL, Ortín J (2001) Three-dimensional reconstruction of a recombinant influenza virus ribonucleoprotein particle. EMBO Rep 2:313–317
Medcalf L, Poole E, Elton D, Digard P (1999) Temperature-sensitive lesions in two influenza A viruses defective for replicative transcription disrupt RNA binding by the nucleoprotein. J Virol 73:7349–7356
Mena I, Jambrina E, Albo C, Perales B, Ortín J, Arrese M, Vallejo D, Portela A (1999) Mutational analysis of influenza A virus nucleoprotein: identification of mutations that affect RNA replication. J Virol 73:1186–1194
Muster T, Subbarao EK, Enami M, Murphy BR, Palese P (1991) An influenza A virus containing influenza B virus 5′ and 3′ noncoding regions on the neuraminidase gene is attenuated in mice. Proc Natl Acad Sci USA 88:5177–5181
Naffakh N, Massin P, van der Werf S (2001) The transcription/replication activity of the polymerase of influenza a viruses is not correlated with the level of proteolysis induced by the pa subunit. Virology 285:244–252
Nakagawa Y, Kimura N, Toyoda T, Mizumoto K, Ishihama A, Oda K, Nakada S (1995) The RNA polymerase PB2 subunit is not required for replication of the influenza virus genome but is involved in capped mRNA synthesis. J Virol 69:728–733
Nakamura Y, Oda K, Nakada S (1991) Growth complementation of influenza virus temperature-sensitive mutants in mouse cells which express the RNA polymerase and nucleoprotein genes. J Biochem (Tokyo) 110:395–401
Neumann G, Zobel A, Hobom G (1994) RNA polymerase I-mediated expression of influenza viral RNA molecules. Virology 202:477–479

Neumann G, Hobom G (1995) Mutational analysis of influenza virus promoter elements in vivo. J Gen Virol 76:1709–1717

Neumann G, Watanabe T, Ito H, Watanabe S, Goto H, Gao P, Hughes M, Perez DR, Donis R, Hoffmann E, Hobom G, Kawaoka Y. (1999) Generation of influenza A viruses entirely from cloned cDNAs. Proc Natl Acad Sci USA 96:9345–9350

Parvin JD, Palese P, Honda A, Ishihama A, Krystal M (1989) Promoter analysis of influenza virus RNA polymerase. J Virol 63:5142–5152

Perales B., Ortín J (1997) The influenza A virus PB2 polymerase subunit is required for the replication of viral RNA. J Virol 71:1381–1385

Perales B, Sanz-Ezquerro JJ, Gastaminza P, Ortega J, Santaren JF, Ortín J, Nieto A (2000) The replication activity of influenza virus polymerase is linked to the capacity of the PA subunit to induce proteolysis. J Virol 74:1307–1312

Piccone ME, Fernandez-Sesma A, Palese P (1993) Mutational analysis of the influenza virus vRNA promoter. Virus Res 28:99–112

Pleschka S, Jaskunas R, Engelhardt OG, Zürcher T, Palese P, García-Sastre A (1996) A plasmid-based reverse genetics system for influenza A virus. J Virol 70:4188–4192

Plotch SJ, Bouloy M, Ulmanen I, Krug RM (1981) A unique cap(m7GpppXm)-dependent influenza virion endonuclease cleaves capped RNAs to generate the primers that initiate viral RNA transcription. Cell 23:847–858

Poch O, Sauvaget I, Delarue M, Tordo N (1989) Identification of four conserved motifs among the RNA-dependent polymerase encoding elements. EMBO J 8:3867–3874

Poon LL, Pritlove DC, Sharps J, Brownlee GG (1998) The RNA polymerase of influenza virus, bound to the 5′ end of virion RNA, acts in cis to polyadenylate mRNA. J Virol 72:8214–8219

Poon LL, Pritlove DC, Fodor E, Brownlee GG (1999) Direct evidence that the poly(A) tail of influenza A virus mRNA is synthesized by reiterative copying of a U track in the virion RNA template. J Virol 73:3473–3476

Poon LL, Fodor E, Brownlee GG (2000) Polyuridylated mRNA synthesized by a recombinant influenza virus is defective in nuclear export. J Virol 74:418–427

Pritlove DC, Fodor E, Seong BL, Brownlee GG (1995) In vitro transcription and polymerase binding studies of the termini of influenza A virus cRNA: evidence for a cRNA panhandle. J Gen Virol 76:2205–2213

Pritlove DC, Poon LL, Fodor E, Sharps J, Brownlee GG (1998) Polyadenylation of influenza virus mRNA transcribed in vitro from model virion RNA templates: requirement for 5′ conserved sequences. J Virol 72:1280–1286

Pritlove DC, Poon LL, Devenish LJ, Leahy MB, Brownlee GG (1999) A hairpin loop at the 5′ end of influenza A virus virion RNA is required for synthesis of poly(A)+ mRNA in vitro. J Virol 73:2109–2114

Rao P, Yuan W, Krug RM (2003) Crucial role of CA cleavage sites in the cap-snatching mechanism for initiating viral mRNA synthesis. EMBO J 22:1188–1198

Romanos MA, Hay AJ (1984) Identification of the influenza virus transcriptase by affinity-labeling with pyridoxal 5′-phosphate. Virology 132:110–117

Sanz-Ezquerro JJ, de la Luna S, Ortín J, Nieto A (1995) Individual expression of influenza virus PA protein induces degradation of coexpressed proteins. J Virol 69:2420–2426

Sanz-Ezquerro JJ, Zürcher T, de la Luna S, Ortín J, Nieto A (1996) The amino-terminal one-third of the influenza virus PA protein is responsible for the induction of proteolysis. J Virol 70:1905–1911

Seong BL, Brownlee GG (1992a) A new method for reconstituting influenza polymerase and RNA in vitro: a study of the promoter elements for cRNA and vRNA synthesis in vitro and viral rescue in vivo. Virology 186:247–260

Seong BL, Brownlee GG (1992b) Nucleotides 9 to 11 of the influenza A virion RNA promoter are crucial for activity in vitro. J Gen Virol 73:3115–3124

Seong BL, Kobayashi M, Nagata K, Brownlee GG, Ishihama A (1992) Comparison of two reconstituted systems for in vitro transcription and replication of influenza virus. J Biochem 111:496–499

Shapiro GI, Gurney T, Jr., Krug RM (1987) Influenza virus gene expression: control mechanisms at early and late times of infection and nuclear-cytoplasmic transport of virus-specific RNAs. J Virol 61:764–773

Shapiro GI, Krug RM (1988) Influenza virus RNA replication in vitro: synthesis of viral template RNAs and virion RNAs in the absence of an added primer. J Virol 62:2285–2290

Shaw MW, Lamb RA (1984) A specific sub-set of host-cell mRNAs prime influenza virus mRNA synthesis. Virus Res 1:455–467

Stevens MP, Barclay WS (1998) The N-terminal extension of the influenza B virus nucleoprotein is not required for nuclear accumulation or the expression and replication of a model RNA. J Virol 72:5307–5312

Tchatalbachev S, Flick R, Hobom G (2001) The packaging signal of influenza viral RNA molecules. RNA 7, 979–989

Tiley LS, Hagen M, Matthews JT, Krystal M (1994) Sequence-specific binding of the influenza virus RNA polymerase to sequences located at the 5′ ends of the viral RNAs. J Virol 68:5108–5116

Ulmanen I, Broni BA, Krug RM (1981) Role of two of the influenza virus core P proteins in recognizing cap 1 structures (m7GpppNm) on RNAs and in initiating viral RNA transcription. Proc Natl Acad Sci USA 78:7355–7359

Wagner E, Engelhardt OG, Weber F, Haller O, Kochs G (2000) Formation of virus-like particles from cloned cDNAs of thogoto virus. J Gen Virol 81:2849–2853

Wagner E, Engelhardt OG, Gruber S, Haller O, Kochs G (2001) Rescue of recombinant Thogoto virus from cloned cDNA. J. Virol. 75:9282–9286

Weber F, Haller O, Kochs G (1996) Nucleoprotein viral RNA and mRNA of Thogoto virus: a novel "cap-stealing" mechanism in tick-borne orthomyxoviruses? J Virol 70:8361–8367

Weber F, Haller O, Kochs G (1997) Conserved vRNA end sequences of Thogoto-orthomyxovirus suggest a new panhandle structure. Arch Virol 142:1029–1033

Weber F, Jambrina E, Gonzalez S, Dessens JT, Leahy M, Kochs G, Portela A, Nuttall PA, Haller O, Ortín J, Zürcher T (1998) In vivo reconstitution of active Thogoto virus polymerase: assays for the compatibility with other orthomyxovirus core proteins and template RNAs. Virus Res 58:13–20

Yamanaka K, Ogasawara N, Yoshikawa H, Ishihama A, Nagata K (1991) In vivo analysis of the promoter structure of the influenza virus RNA genome using a transfection system with an engineered RNA. Proc Natl Acad Sci USA 88:5369–5373

Zheng H, Palese P, García-Sastre A (1996) Nonconserved nucleotides at the 3′ and 5′ ends of an influenza A virus RNA play an important role in viral RNA replication. Virology 217:242–251

Zheng H, Lee HA, Palese P, García-Sastre A (1999) Influenza A virus RNA polymerase has the ability to stutter at the polyadenylation site of a viral RNA template during RNA replication. J Virol 73:5240–5243

Zobel A, Neumann G, Hobom G (1993) RNA polymerase I catalysed transcription of insert viral cDNA. Nucleic Acids Res 21:3607–3614

Escaping from the Cell: Assembly and Budding of Negative-Strand RNA Viruses

A. P. Schmitt · R. A. Lamb

Department of Biochemistry, Molecular Biology, and Cell Biology,
Howard Hughes Medical Institute, Northwestern University, Evanston, IL, 60208-3500, USA
E-mail: ralamb@northwestern.edu

Abstract Negative-strand RNA virus particles are formed by a process that includes the assembly of viral components at the plasma membranes of infected cells and the subsequent release of particles by budding. Here, we review recent progress that has been made in understanding the mechanisms of negative-strand RNA virus assembly and budding. Important topics for discussion include the key role played by the viral matrix proteins in assembly of viruses and viruslike particles, as well as roles played by additional viral components such as the viral glycoproteins. Various interactions that contribute to virus assembly are discussed, including interactions between matrix proteins and membranes, interactions between matrix proteins and glycoproteins, interactions between matrix proteins and nucleocapsids, and interactions that lead to matrix protein self-assembly. Selection of specific sites on plasma membranes to be used for virus assembly and budding is described, including the asymmetric budding of some viruses in polarized epithelial cells and assembly of viral components in lipid raft microdomains. Evidence for the involvement of cellular proteins in the late stages of rhabdovirus and filovirus budding is discussed as well as the possible involvement of similar host factors in the late stages of budding of other negative-strand RNA viruses.

1
Introduction

Virus particles are vehicles designed to transfer genetic information from one cell to another. Once formed and released from infected cells, they protect viral genomes during transit and mediate entry of genomes into new target cells. Strategies for enclosing and protecting viral genomes vary widely. Some viruses employ icosahedral protein shells as genome containers and consequently have defined, uniform structures. Negative-strand RNA viruses lack icosahedral protein shells and typically are heterogeneous in size and shape. The RNA genomes of these viruses are associated with subunits of the viral nucleocapsid protein, which polymerizes to form helical complexes called nucleocapsids or ribonucleoproteins (RNPs). In the cases of the rhabdoviruses and paramyxoviruses the RNA is protected from RNAse digestion by the nucleocapsid protein, whereas for influenza virus the RNA is exposed such that it can be digested with RNAse. The nucleocapsid subunits are required for transcription and replication of the RNA genome. The RNPs are packaged into lipid envelopes derived from host cells, which are coated with viral matrix protein on the inner surface and enriched with viral spike glycoproteins. Thus the virus particle can be thought of as a lipid vesicle reinforced with matrix protein and having glycoprotein spikes and a cargo of encapsidated viral genome associated with a viral polymerase complex. In many cases, the RNPs of negative-strand RNA viruses are packaged into roughly spherical particles. An example is shown in Fig. 1a, in which a purified paramyxovirus particle has been visualized by electron microscopy. The rhabdoviruses vesicular stomatitis virus (VSV) and rabies virus give rise to bullet-shaped virus particles that are considerably more uniform than the pleiomorphic particles formed by the majority of other negative-strand RNA viruses (Fig. 1b).

Negative-strand RNA viruses, like other enveloped viruses, are formed by a budding process. Buds emerge from sites on the plasma membrane where viral components have assembled and then pinch off, resulting in the release of particles (Fig. 1c–e). A number of steps appear to be involved in the formation of infectious virus particles, including the transport of viral glycoproteins through the secretory pathway of the cell to reach the cell surface, the transport of soluble viral components such as assembled RNPs and matrix proteins to the plasma membrane of the cell, coordination between the glycoproteins and the soluble components in selection of a budding site, and interactions between viral components and the host machinery that allow bud formation and membrane fission. Here, we review progress that has been made toward understanding the mechanism of assembly and budding of negative-strand RNA viruses. Important questions for discussion include the following:

1. All viral structural components must be packaged together to generate a standard infectious particle, but which of these structural components actively contribute to the budding process? What is the minimum set of proteins required for budding to occur? Do glycoproteins play an important role in virus budding, and if so, in cases where viruses encode multiple glycoproteins, do the different glycoproteins have distinct roles?
2. Viral glycoproteins and soluble components must coalesce at discrete sites on the plasma membrane for budding, despite the fact that they arrived at the plasma membrane by separate mechanisms. How does coordination between these components occur in selection of a budding site? If matrix proteins act as bridges between glycoproteins and RNPs, what is the nature of these interactions? What happens if these interactions are disrupted?
3. In some cases, budding sites on the plasma membrane are chosen very specifically, for example, budding might occur preferentially from either the apical or the basolateral surface of polarized cells or

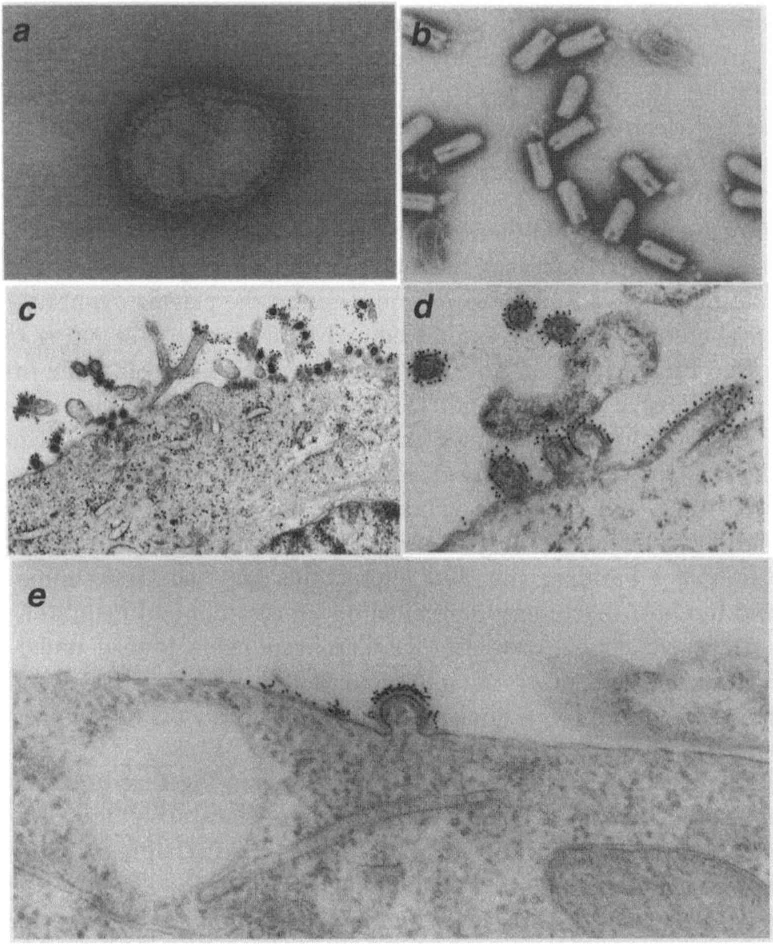

Fig. 1a–e Electron micrographs of purified virions and virions that are budding from cells. **a** Purified SV5 particle negatively stained with phosphotungstic acid. Spikes consisting of the glycoproteins HN and F can be observed at the periphery of the particle. **b** Purified VSV particles negatively stained with phosphotungstic acid. The particles have a characteristic bullet shape and are surrounded by a layer of spikes consisting of the glycoprotein G. **c** Thin section of an influenza virus-infected MDCK cell. Pleiomorphic influenza virions can be seen budding from the cell. Colloidal gold staining of HA is shown. **d**, **e** Thin section of SV5-infected CV-1 cells, with colloidal gold staining of HN. (Electron micrographs for panels **a**, **c**, **d**, and **e** courtesy of George Leser, Northwestern University. Electron micrograph for panel **b** courtesy of Michael Whitt, University of Tennessee-Memphis)

occur from specific membrane microdomains. How are these specific budding sites selected? Do the viral glycoproteins specify the budding site and recruit soluble components to the proper location on the plasma membrane, or is it the soluble components that recruit the glycoproteins?

4. The final stage of virus budding requires a membrane fission step that allows virus particles to be released. Do negative-strand RNA viruses recruit host machinery to assist in this process, as has been described recently for retroviruses? If so, are these interactions mediated by the same "L domain" sequences found in many retroviral Gag proteins, or do negative-strand RNA viruses have a distinct strategy for recruiting the host machinery involved in budding?

2
Critical Role for Viral Matrix Proteins in Virus Budding

For a long time it has been thought that the matrix proteins of negative-strand RNA viruses have important roles in virus assembly and budding. Matrix proteins are positioned in virions beneath the lipid envelope, so that they have the potential to contact both RNP cores and envelope glycoprotein cytoplasmic tails and are therefore likely to be the key organizers of virus assembly that induce separate viral components to concentrate together at defined budding sites on the plasma membranes of infected cells. The atomic structures of matrix proteins derived from three different viruses (influenza A virus, Ebola virus, and VSV) have been obtained by X-ray crystallographic methods, and each protein has a completely different three-dimensional structure that bears little resemblance to the other two (Dessen et al. 2000; Gaudier et al. 2002; Sha and Luo 1997). Despite this lack of structural similarity between different matrix proteins, all viral matrix proteins appear to be functionally similar and to have common biochemical properties that are consistent with their presumed role in organizing virus assembly. Matrix proteins bind to viral RNPs in vitro and are found stably attached to RNPs when purified from virions, they bind to lipid membranes both in vitro and in living cells, and they self-assemble into ordered structures as purified proteins in vitro and in virus-infected cells. These observations are all suggestive of a model in which matrix proteins assemble as layers beneath the plasma membranes of infected cells and induce other viral components to gather at these locations, from which virus budding can then occur. Confirmation that matrix proteins are in fact important for proper budding of virus has been obtained through experiments with temperature-sensitive viruses that have defective M genes, through direct manipulation of M genes in recombinant viruses by reverse genetics, and through reconstitution of budding directed by matrix proteins in transfected cells.

2.1 Studies Using Temperature-Sensitive Mutants

A role for matrix proteins in virus budding was defined initially through use of temperature-sensitive (ts) mutants. A variety of VSV mutants have been characterized that contain ts mutations in their M genes, and these viruses were found to be defective in assembly and budding of particles at the restrictive temperature (Knipe et al. 1977). In many cases, not only were particles released inefficiently, but the particles that were released were "empty" particles that lacked the normal complement of nucleocapsid protein and RNA and had a spherical or pleiomorphic morphology instead of the bullet-shaped morphology characteristic of wild-type (wt) VSV virions (Lyles et al. 1996; Schnitzer et al. 1979). Normal particle morphology could be restored by expressing the wt M protein *in trans* (Lyles et al. 1996). Some of the mutant M proteins appear to be misfolded at the restrictive temperature and form aggregates that are incapable of associating with viral nucleocapsids (Flood et al. 2000; McCreedy and Lyles 1989; Ono et al. 1987). Today, many of these older observations made with ts mutants can be explained more completely with information gained from the recently obtained crystal structure of the wt M protein, which predicts that these mutations would lower M protein stability (Gaudier et al. 2002).

A ts mutant of Sendai virus, Cl.151, has been isolated that is defective in a function required for virus assembly at the nonpermissive temperature because of mutations in its M gene (Kondo et al. 1993; Yoshida et al. 1979). The mutant M protein is unstable and fails to accumulate to normal levels at the nonpermissive temperature, suggesting that lack of M protein is preventing assembly of virus particles. Particle production was restored when this virus was used to infect cell lines that stably produce wt M protein. Furthermore, massive overexpression of the mutant M protein was also able to complement the virus (Kondo et al. 1993), suggesting that the budding defect was due to insufficient accumulation of M protein and not to a defect in the function of the mutant M protein. Thus Sendai virus budding seems to require a threshold level of M protein accumulation to proceed.

2.2
Studies Using Reverse Genetics

Subsequent studies to confirm the importance of viral matrix proteins in assembly and budding have relied on the use of reverse genetics techniques, which allow the generation of negative-strand RNA viruses entirely from plasmid DNA. A rabies virus mutant has been generated that lacks the entire M gene (Mebatsion et al. 1999). In the absence of the M gene and protein, the yield of infectious virus released into the media was reduced as much as 5×10^5-fold and the release of physical particles into the media was also severely impaired. The few particles that were released had rod-shaped or spherical morphologies instead of the normal bullet-shaped morphology. Thus there were severe defects in virus assembly and budding in the absence of rabies virus M protein.

A similar strategy was used to assess the importance of matrix proteins in the budding of paramyxoviruses. The M proteins of measles virus and Sendai virus had previously been implicated in budding based on analysis of viruses derived from persistent viral infections. Cells persistently infected with Sendai virus were found to express an unstable M protein, and lack of stable M protein correlated with a reduction in virus particle formation (Roux and Waldvogel 1982). A role for the measles virus M protein in budding was suggested based on analysis of viruses isolated from patients with subacute sclerosing panencephalitis (SSPE), a rare and lethal disease of the human central nervous system caused by persistent measles virus infection. SSPE viruses are defective for the production of progeny virus particles, yet they spread throughout the brains of affected patients by virus-induced cell-to-cell fusion, without significant formation of virions. Nucleotide sequence analysis has revealed extensive defects in the M genes of SSPE measles virus strains (Cattaneo et al. 1988; Wong et al. 1991). Thus a correlation has been established between mutations in the M gene, defective virion production, and lethal measles virus persistence. To test for the existence of a causal relationship, recombinant measles viruses with defective M genes were generated. In one case, measles virus completely lacking the M gene was rescued (Cathomen et al. 1998a), and in another case, measles virus containing an M gene derived from an SSPE measles virus strain was rescued (Patterson et al. 2001). Both of these viruses were shown to be severely defective in budding from the surfaces of infected cells. Furthermore, assembly of viral glycoproteins into concentrated patches at the surfaces of infected cells was defective in the M-deleted virus. Interestingly, this virus was observed to spread mainly by cell-to-cell fusion, which occurred more efficiently and extensively in cells infected with M-deleted virus than in cells infected with standard measles virus (Cathomen et al. 1998a). In mouse model systems, the M-deleted virus was found to spread more deeply into the brain than the standard virus and the SSPE M virus caused a chronic, progressive disease of the central nervous system (Cathomen et al. 1998a; Patterson et al. 2001). These results strongly implicate M protein function in measles virus assembly and in measles virus-associated SSPE.

2.3
Generation of Viruslike Particles

The importance of matrix proteins for virus budding has been shown not only by observing the effects on virus budding when M is removed or altered but also by reconstituting budding in cells that express M protein alone. This approach was first used to demonstrate budding of viruslike particles (VLPs) from insect cells expressing VSV M protein from a recombinant baculovirus (Li et al. 1993). M protein was released into the culture medium of the cells in association with lipid vesicles having irregular shapes and sizes. Subsequent experiments have confirmed that VSV M protein induces budding of lipid vesicles from mammalian cells when it is expressed from a recombinant vaccinia virus (Justice et al. 1995), from a recombinant Sendai virus (Sakaguchi et al. 1999), or with eukaryotic expression vectors (Harty et al. 1999; Lyles et al. 1996). Electron microscopic analysis of these particles reveals that they are seemingly "empty" and that they are surrounded by a thick membrane, perhaps consisting of a double layer of membranes (Li et al. 1993; Lyles et al. 1996; Sakaguchi et al. 1999). These characteristics are very similar to those observed for particles released from VSV ts mutants with altered M proteins that fail to associate with nucleocapsids (Lyles et al. 1996; Schnitzer et al. 1979), which led to the suggestion that this type of particle might be produced as a result of residual membrane binding activity of M protein in the absence of binding to nucleocapsids (Lyles et al. 1996).

Budding of VLPs has been observed from cells expressing a wide variety of viral matrix proteins in addition to VSV M protein. For example, expression of the Ebola virus matrix protein VP40 alone in cells results in efficient budding of VLPs (Bavari et al. 2002; Harty et al. 2000; Jasenosky et al. 2001; Noda et al. 2002; Timmins et al. 2001), and electron microscopy has shown that these particles are filamentous and resemble spikeless virions (Noda et al. 2002). Furthermore, expression of the influenza A virus matrix protein either from cDNA (Gomez-Puertas et al. 2000) or from a recombinant baculovirus (Latham and Galarza 2001) results in VLP budding, as does expression of matrix proteins from the paramyxoviruses human parainfluenza virus type 1 (hPIV1) (Coronel et al. 1999) and Sendai virus (Takimoto et al. 2001).

In many cases, additional viral components, when coexpressed with the viral matrix proteins, have been found to be incorporated into the VLPs. For example, influenza virus glycoproteins can be coexpressed with influenza virus matrix protein to achieve budding of particles that contain glycoprotein spikes and thus have a morphology that more closely resembles that of authentic virions (Gomez-Puertas et al. 2000; Latham and Galarza 2001). Similar observations have been made for Sendai virus matrix protein when coexpressed with Sendai virus glycoproteins (Takimoto et al. 2001) and for Ebola virus matrix protein when coexpressed with the Ebola virus glycoprotein (Bavari et al. 2002; Noda et al. 2002). Additionally, coexpression of hPIV1 matrix protein together with the hPIV1 nucleocapsid protein led to the formation of vesicles enclosing nucleocapsid-like structures (Coronel et al. 1999).

In some cases, coexpression of matrix proteins along with additional viral components can substantially increase the efficiency with which particles bud. For example, the simian virus 5 (SV5) matrix protein expressed by itself does not induce efficient budding of particles, but when this protein is coexpressed with nucleocapsid protein and a viral glycoprotein, budding of particles becomes very efficient, approaching the budding efficiency observed in virus-infected cells (Schmitt et al. 2002). VLP budding that is normally observed on expression of the Sendai virus matrix protein alone can be made more efficient by coexpressing the Sendai virus F glycoprotein (Takimoto et al. 2001), and budding normally induced by Ebola virus VP40 expression can be made more efficient by coexpressing the Ebola virus glycoprotein GP (Bavari et al. 2002).

Thus, although matrix proteins of many negative-strand RNA viruses are capable of directing budding of particles in the absence of other viral components, in many cases budding is made more efficient on inclusion of multiple viral proteins.

3
Role for Viral Glycoproteins in Promoting Efficient Virus Budding

3.1
Contributions Made by the Orthomyxovirus Glycoproteins

The orthomyxovirus influenza A virus encodes two spike glycoproteins, hemagglutinin (HA) and neuraminidase (NA). HA protein allows attachment of virus to sialic acid on the surfaces of target cells and also mediates membrane fusion during virus entry. NA protein has a receptor-destroying activity that removes sialic acid and prevents newly formed particles from aggregating at the surfaces of infected cells. Initial experiments to determine the importance of glycoproteins in influenza virus assembly relied on the use of temperature-sensitive virus mutants. One ts influenza virus was characterized in which HA protein fails to be transported to the cell surface at the restrictive temperature (Pattnaik et al. 1986). Noninfectious virions that lack HA spikes still formed from cells infected with this mutant virus at an efficiency similar to that found in wt virus-infected cells, suggesting that HA protein is not required for assembly and budding of influenza virus particles, at least in the presence of a normal NA protein. Similarly, influenza viruses have been characterized that have a ts mutation in NA protein (Palese et al. 1974) or have a defective NA gene such that the virus replicates only on addition of exogenous sialidase (Liu and Air 1993; Liu et al. 1995). Virions lacking NA protein were found to assemble and bud normally from cells and were similar morphologically to wt virions. However, once formed, these particles remained attached to the cell surface and to each other, forming large aggregates. Similar observations have also been made for budding of influenza virus-like particles from transfected cells. Expression of all the influenza virus structural components together in cells leads to efficient release of VLPs, and when NA expression is omitted, particles are still formed but fail to release normally and instead associate with the cell surface or form large aggregates (Gomez-Puertas et al. 2000; Gomez-Puertas et al. 1999; Mena et al. 1996).

Virion aggregation could be prevented and virus release could be restored by addition of exogenous bacterial sialidase to the culture medium (Liu and Air 1993; Liu et al. 1995; Palese et al. 1974). Thus the influenza virus NA protein does not appear to be required for normal assembly and budding of particles, but its enzymatic activity is required to prevent reattachment and/or self-attachment of progeny virions, which inhibits infection of new cells. Together, these results suggested either that influenza virus glycoproteins are not important for the assembly and budding process or that there is a redundant assembly function and either HA or NA alone may be sufficient. This possibility was addressed in subsequent experiments in which reverse genetics was used to isolate viruses that lack both the HA and NA protein cytoplasmic tails. Thus neither protein would be expected to be capable of making contacts with soluble viral components such as matrix protein during assembly. Influenza virus lacking both HA and NA protein cytoplasmic tails was found to be severely defective in assembly, as illustrated in Fig. 2. Particle production was reduced 10-fold compared with wt virus, and the particles that were formed had a greatly distorted morphology (Jin et al. 1997). These distended and irregularly shaped particles also exhibited 1 to 3 logs lower infectivity than wt virions and were defective in genome packaging such that they contained a much broader distribution in the number of packaged RNA segments (Zhang et al. 2000a). Interestingly, these severe defects were observed only when both glycoprotein cytoplasmic tails were deleted. Deletion of only the HA protein cytoplasmic tail led to virus that assembled similarly to wt virus and formed particles with the normal spherical morphology (Jin et al. 1994), and deletion of only the NA protein cytoplasmic tail led to defects in virus assembly and particle morphology that were much less severe than those observed in the double-mutant virus (Garcia-Sastre and Palese 1995; Mitnaul et al. 1996). These results suggest that the cytoplasmic domains of influenza virus HA and NA glycoproteins have redundant functions that allow efficient budding of spherically shaped virions.

3.2 Contributions Made by the Rhabdovirus Glycoproteins

Rhabdoviruses encode only a single spike glycoprotein, G, that functions for both attachment to target cells and membrane fusion to allow virus

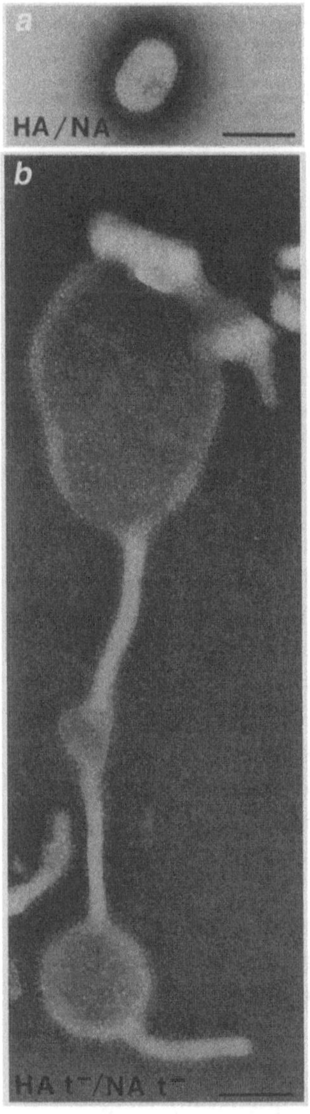

Fig. 2a, b Morphology of influenza virus with altered glycoprotein cytoplasmic tails. Purified virions were negatively stained with phosphotungstic acid and examined by electron microscopy. *HA/NA*, wt virus and *HA t⁻/NA t⁻*, virus lacking both of the glycoprotein cytoplasmic tails. *Bar*, 80 nm. (Electron micrographs courtesy of George Leser, Northwestern University)

entry. G protein contains a C-terminal cytoplasmic tail that has the potential to contact other viral components during virus assembly. Remarkably, G proteins of both VSV and rabies virus, when expressed by themselves, have been found to induce the formation of budding vesicles from cells, demonstrating that rhabdovirus G proteins have autonomous exocytosis activities that could potentially be important for the budding of virus (Mebatsion et al. 1996; Rolls et al. 1994). However, although it has become clear that G proteins do play a role in efficient budding of virus, it has also been established that virus budding can occur even in the absence of G protein. This conclusion was initially made on the basis of studies of a ts mutant of VSV (*ts*O45) in which it was found that low levels of noninfectious spikeless VSV particles could be produced at the restrictive temperature (Knipe et al. 1977; Schnitzer et al. 1979). However, it was later found that these released particles, although devoid of intact G protein, contained G protein membrane-spanning fragments that included the C-terminal cytoplasmic domain (Metsikko and Simons 1986), and so it could not be ruled out that the observed particle budding might have depended on this residual G "stump." This situation was clarified later through use of reverse genetics that allowed rescue of recombinant rabies virus and VSV that completely lack G genes (Mebatsion et al. 1996; Schnell et al. 1997). Budding of characteristic bullet-shaped rhabdovirus particles could still occur in cells infected with these viruses, formally showing that interactions between G protein and soluble viral components are not required for rhabdovirus budding and do not specify virion morphology. However, although budding of these recombinant viruses could be detected, it occurred with an efficiency that was reduced at least 30-fold compared with wt virus, suggesting that rhabdoviruses in fact require the G protein to make budding efficient. Initial efforts to localize regions within the G protein responsible for directing efficient rhabdovirus budding focused on the cytoplasmic tail, and it was found that deletion of the G protein cytoplasmic tail in recombinant rhabdoviruses led to poor budding (Mebatsion et al. 1996; Schnell et al. 1997), in agreement with a model in which specific contacts between the G protein cytoplasmic tail and the condensed M-RNP core drive budding. Indeed, biochemical evidence for glycoprotein-matrix protein interactions has been obtained (Lyles et al. 1992; see Sect. 4.2). However, subsequent experiments with VSV showed that the G protein cytoplasmic tail could be replaced with an unrelated sequence derived from human CD4 with very little effect on virus budding and that, furthermore, a revertant to the cytoplasmic tail-deleted virus could be selected and this revertant encoded a new 8-amino acid-long cytoplasmic tail having a sequence unrelated to the normal G protein cytoplasmic tail but capable of promoting normal VSV budding (Schnell et al. 1998). Thus, although efficient budding of VSV requires a G protein cytoplasmic tail, the specific amino acid sequence of the cytoplasmic tail is seemingly unimportant, contrary to what would be expected if efficient VSV budding were driven by specific interactions between the G protein cytoplasmic tail and soluble cores. It should be noted that these observations do not completely rule out the possibility of specific interactions between M protein and the G protein cytoplasmic tail and that "fitness" assays to rigorously compare replication efficiencies between wt VSV and VSV variants with altered G protein cytoplasmic tails have not yet been reported. Recent studies have identified an additional domain in the membrane-proximal stem region of the G protein ectodomain that is important for efficient budding of VSV (Robison and Whitt 2000). How this domain functions to promote budding is not yet understood. One possibility is that it facilitates concentration of G protein and curvature of the plasma membrane at the site of budding so that the inherent exocytosis activity of G protein can work in concert with budding activities directed by the matrix and nucleocapsid proteins to enable efficient particle release.

3.3
Contributions Made by the Paramyxovirus Glycoproteins

Paramyxoviruses encode two glycoproteins, the fusion protein F and an attachment protein, HN, H, or G. Early studies using a temperature-sensitive mutant of Sendai virus (*ts*271) demonstrated that the Sendai virus HN protein is dispensable for budding of virus particles. This mutant HN protein is synthesized efficiently in infected cells but is degraded during transport to the cell surface at the restrictive temperature, yet virions lacking HN are still released (Markwell et al. 1985; Portner et al. 1974, 1975; Stricker and Roux 1991; Tuffereau et al. 1985). The conclusion that HN is dispensable for Sendai virus budding was further confirmed with VLP budding systems. In one system, VLP budding was observed on coexpression of the Sendai virus proteins N, P, M, F, HN, and L along with a minigenome and budding was found to be unaffected on omission of HN protein expression (Leyrer et al. 1998). In a second system, VLP budding was observed on expression of M protein alone and the efficiency of budding was found to be stimulated on coexpression of F protein, but coexpression of HN protein had no effect on budding efficiency (Takimoto et al. 2001). These results suggested that of the two Sendai virus glycoproteins, perhaps only the F protein is important for efficient budding of virus. Consistent with this view, F protein was found to have an intrinsic exocytosis activity as it induced the release of particles bearing F spikes when expressed alone, whereas HN protein was not observed to have this ability (Takimoto et al. 2001). Recombinant Sendai viruses with altered glycoprotein cytoplasmic tails have been generated in an attempt to define more precisely the domains important for efficient budding. On truncation of the F protein cytoplasmic tail, budding of recombinant virus was poor, and a specific amino acid sequence spanning amino acids 538 to 550 of the F protein cytoplasmic tail was found to be important for efficient virus budding (Fouillot-Coriou and Roux 2000). Consistent with this finding, release of particles on expression of F protein alone was found to depend on the amino acid sequence TYTLE comprising amino acids 542 to 546 of the F protein cytoplasmic tail (Takimoto 2001). This TYTLE sequence is also present in the hPIV1 F protein cytoplasmic tail, which otherwise bears little sequence resemblance to the Sendai virus F protein cytoplasmic tail. Truncation of the HN protein cytoplasmic tail in a recombinant Sendai virus also was found to result in poor budding (Fouillot-Coriou and Roux 2000), which was surprising in light of previous data showing that the entire HN protein is dispensable for virus budding. However, other recombinant viruses were also characterized harboring altered HN proteins and these viruses budded normally even when the mutant HN proteins failed to be incorporated efficiently into virions. Thus it appears that the absence of HN protein is not necessarily detrimental for Sendai virus production and release, unlike severe truncation of its cytoplasmic tail, perhaps because the cytoplasmic tail-deleted HN protein is inhibitory for budding (Fouillot-Coriou and Roux 2000).

Similar studies have been carried out with another paramyxovirus, SV5, to define further the roles of the viral glycoproteins in paramyxovirus budding. Here recombinant viruses were made that harbored HN proteins with truncated cytoplasmic tails, and budding was found to be inefficient on HN protein cytoplasmic tail deletion (Schmitt et al. 1999). Furthermore, these viruses were found to be defective in assembly of different viral components together at the plasma membranes of infected cells as judged by confocal microscopy. The same approach was used to define the role of the SV5 F protein cytoplasmic tail in virus budding, and it was found that recombinant viruses with deleted F protein cytoplasmic tails replicated in tissue culture and were released from infected cells with efficiency similar to that of wt virus (Waning et al. 2002). This result was surprising in light of work with Sendai virus suggesting that F protein is in fact quite important for proper paramyxovirus budding. To investigate further the relative roles of the HN and F glycoproteins for paramyxovirus budding, a VLP system was developed for SV5 (Schmitt et al. 2002). Here efficient budding of VLPs from transfected cells was observed only on coexpression of multiple SV5 proteins. Thus expression of M protein alone did not lead to substantial particle budding, and neither of the SV5 glycoproteins was found to have an autonomous exocytosis activity. However, coexpression of M protein with NP protein and either the HN or F glycoprotein led to budding of VLPs with an efficiency comparable to that found in virus-infected cells. Budding decreased more than 25-fold when neither of the SV5 glycoproteins were included, and the HN and F proteins were found to be completely interchangeable for VLP budding, with similar and highly efficient budding occurring regardless of whether the glycoprotein being expressed was HN, F, or both HN and F together. This result suggested that the two SV5 glycoproteins might have redundant functions for budding, analogous to the situation in influenza A virus, in which HA and NA glycoproteins have redundant assembly activities. Consistent with this idea, recombinant SV5 lacking both HN and F protein cytoplasmic tails was found to have a greater defect in particle production and release than SV5 lacking only the HN protein cytoplasmic tail (Waning et al. 2002). In VLP experiments, the importance of the glycoprotein cytoplasmic tails for efficient budding was confirmed: VLPs containing wt HN or wt F as the only glycoprotein bud efficiently, whereas those containing only cytoplasmic tail-deleted HN or cytoplasmic tail-deleted F protein bud poorly. VLPs containing both wt HN protein and cytoplasmic tail-deleted F protein bud normally, and this is presumably due to the activity of the wt HN protein. However, VLPs containing both wt F protein and cytoplasmic tail-deleted HN protein bud poorly despite the ability of wt F protein to support efficient budding in the complete absence of HN protein expression (Schmitt et al. 2002). These data suggest that cytoplasmic tail-deleted HN protein is inhibitory for budding, and the data explain the debilitated budding phenotype observed for recombinant virus having a truncated HN cytoplasmic tail. This finding is remarkably similar to that reported for Sendai virus, in which severe truncation of the HN protein cytoplasmic tail was also proposed to have an inhibitory effect on virus budding (Fouillot-Coriou and Roux 2000). These data for SV5 indicate that both HN and F glycoproteins have important and redundant assembly functions and that in both cases this assembly function requires the presence of the glycoprotein cytoplasmic tail.

Additional insight into the roles played by paramyxovirus glycoproteins in virus assembly has been gained through analysis of SSPE measles virus strains. These budding-defective viruses have been found to contain drastic sequence alterations not only in their M genes, as discussed above, but also in their F protein cytoplasmic tail sequences (Cattaneo et al. 1988; Schmid et al. 1992). The F protein ectodomains, however, are not extensively altered in these viruses (Schmid et al. 1992), and F proteins derived from these viruses were found to have normal membrane fusion function (Cattaneo and Rose 1993). These observations provided support for the hypothesis that establishment of lethal measles virus persistence is associated specifically with defects in virus assembly. Experiments using recombinant measles viruses with altered glycoprotein cytoplasmic tails have also been used to assess the importance of these cytoplasmic domains in virus assembly. Truncations or other alterations to the F protein cytoplasmic tail led to more rapid and extensive cell-to-cell fusion of virus-infected cells, consistent with a shift in the mode of virus spread to one that is independent of budding (Cathomen et al. 1998b; Moll et al. 2002). Truncation of the H protein cytoplasmic tail led to varied results. In one case truncation of the 34-residue H protein cytoplasmic tail to 20 residues led to increased cell-to-cell fusion of virus-infected cells (Cathomen et al. 1998b), whereas in another case truncation of the H protein cytoplasmic tail to 14 amino acids led to reduced cell-to-cell fusion (Moll et al. 2002). Double-mutant viruses having alterations in both glycoprotein cytoplasmic tails were found to cause faster and more extensive cell-to-cell fusion than single mutant viruses, and this level of fusion was comparable to that induced by recombinant measles virus lacking the M gene (Cathomen et al. 1998a,b). In cells infected with either of the single-mutant viruses, colocalization of the cytoplasmic tail-truncated glycoprotein with matrix protein failed to occur as it does in wt measles virus-infected cells, although neither single-mutant virus was found to be defective in the production and release of infectious particles (Moll et al. 2002). The relative importance of the H and F glycoproteins for measles virus assembly is presently unclear. SSPE measles virus isolates typically contain truncations or other significant alterations in the F protein cytoplasmic tail and not the H protein cytoplasmic tail. However, natural truncations to the N-terminal H protein cytoplasmic tail would be expected to typically disrupt the rest of the H protein as well, whereas the C-terminal F protein cytoplasmic tail could easily be disrupted by naturally occurring mutations without affecting the rest of the F protein. Thus it has been proposed that the reason for the frequency of F cytoplasmic tail alteration in SSPE viruses is simply the structure of the F and H genes (Cathomen et al. 1998b).

3.4 Contributions Made by the Filovirus Glycoproteins

Filoviruses encode only a single glycoprotein, GP. Ebola virus GP has been found to possess an autonomous exocytosis activity similar to that of rhabdovirus G proteins and the Sendai virus F protein. Particles released on expression of Ebola virus GP alone are variable in size and shape and contain GP spikes (Bavari et al. 2002; Noda et al. 2002; Volchkov et al. 1998). The Ebola virus matrix protein, VP40, when expressed alone can also induce particle release, and these particles were observed to have a filamentous morphology resembling that of spikeless filovirus particles (Noda et al. 2002). Coexpression of VP40 and GP increases budding efficiency compared with expression of VP40 alone, and the released particles are filamentous and contain GP spikes (Bavari et al. 2002; Noda et al. 2002). These findings most closely approximate those described for rhabdoviruses, in which the viral glycoprotein is not required for budding of particles with the correct morphology but may contribute to the efficiency of budding.

4 Interactions Between Viral Components During Assembly

Assembly of negative-strand RNA viruses requires coordinated localization of multiple but distinct virus components, including viral glycoproteins, which are transported to the plasma membrane by the exocytic pathway, and soluble viral components, such as the RNP. This coordination appears to be accomplished through a series of protein-protein and protein-lipid interactions, many of which involve the viral matrix protein that could potentially interact with both glycoproteins via their cytoplasmic tails and with the RNPs in the cytoplasm of the infected cell. Thus many investigations of virus assembly have focused on characterization of the various interactions involving the viral matrix proteins.

4.1
Interactions Between Matrix Proteins and Membranes

Matrix proteins of negative-strand RNA viruses have intrinsic membrane-binding properties, and when expressed alone in living cells, a fraction of M protein is found tightly bound to membranes. For example, about 10% of VSV M protein expressed from cDNA in living cells is associated with the plasma membrane (Chong and Rose 1993; Chong and Rose 1994). Interestingly, although the highly basic M proteins lack hydrophobic stretches that would indicate membrane spanning domains and, in contrast to some retroviral matrix proteins generally do not appear to be modified by fatty acids [an exception being the rabies virus matrix protein, in which a fraction of the molecules are palmitoylated (Gaudin et al. 1991)], the membrane binding of VSV M did not display characteristics typical of a peripheral membrane protein but rather was found to be very stable to salt or high-pH treatments (Chong and Rose 1993). These unusual membrane binding properties have been confirmed for matrix proteins of a variety of other negative-strand RNA viruses in addition to VSV. For example, membrane binding by the Ebola virus matrix protein VP40 expressed in mammalian cells was not disrupted by salt treatment, and detergent partitioning experiments suggested a hydrophobic component to the membrane interaction (Jasenosky et al. 2001). Membrane binding of Marburg virus VP40 expressed in cells was also shown to be resistant to high salt concentrations, as well as high pH and EDTA treatment (Kolesnikova et al. 2002). Similar observations have been made for the influenza virus matrix protein (Kretzschmar et al. 1996; Zhang and Lamb 1996), the Sendai virus matrix protein (Stricker et al. 1994), and the measles virus matrix protein (Riedl et al. 2002).

Membrane binding properties of matrix proteins have also been studied by reconstitution of interactions between purified matrix proteins and synthetic liposomes in vitro. In some cases, such as that of the Newcastle disease virus (NDV) matrix protein, liposome binding was found to occur regardless of the charge of the liposomes and was not prevented by high salt concentrations, suggesting a nonelectrostatic interaction (Faaberg and Peeples 1988). Early work with the influenza A virus M1 protein was similarly interpreted to suggest a nonelectrostatic interaction with liposomes (Bucher et al. 1980; Gregoriades 1980), but it was based on approaches limited by the technology of the time, and more recent work has indicated that the M1 protein associates preferentially with negatively charged liposomes through electrostatic interactions, which could be prevented (although not completely reversed) by high salt concentrations (Ruigrok et al. 2000a). Based on the atomic structure of the N-terminal two-thirds of M1 protein it has been proposed that membrane binding occurs after a conformational change that exposes a hydrophobic surface in M1 protein that could be inserted into the inner leaflet of the lipid bilayer (Sha and Luo 1997). Unfortunately, no evidence has been presented to indicate that a conformational change in the M1 protein takes place on membrane binding, and membrane binding may instead be mediated by positively charged residues that are well oriented to interact with phosphate head groups of the membrane (Arzt et al. 2001; Sha and Luo 1997).

The Ebola virus matrix protein may rely on multiple types of interactions for membrane binding. This protein was found to associate via its C-terminal domain only with lipid bilayers containing negatively charged phospholipids, and the interaction could be reversed by high salt concentrations, indicating an important electrostatic contribution to membrane binding (Ruigrok et al. 2000b). However, the crystal structure of the Ebola virus matrix protein revealed the presence of a large solvent-exposed hydrophobic surface likely involved in membrane binding, and so it was proposed that this protein binds membranes through both hydrophobic and electrostatic interactions (Dessen et al. 2000).

Studies on liposome binding by VSV matrix protein have similarly revealed a complex interaction with lipid that is mediated by multiple distinct membrane binding sites (Gaudier et al. 2002; Ye et al. 1994). One membrane binding site was localized to the highly charged N terminus of the protein, consistent with previous results based on photochemical cross-linking of virions (Lenard and Vanderoef 1990), and membrane binding directed by this site could be prevented or reversed by high salt concentrations (Ye et al. 1994). Membrane binding directed by a second site located in the central portion of the M protein amino acid sequence was insensitive to salt (Ye et al. 1994). Based on the crystal structure of VSV M protein it is thought that lipid binding regions are likely to include multiple positively charged surfaces, as well as a hydrophobic loop

(Gaudier et al. 2002). Thus matrix proteins from a variety of different viruses appear to achieve membrane binding by using a combination of electrostatic and hydrophobic interactions.

4.2 Interactions Between Matrix Proteins and Glycoproteins

Specific interactions between viral glycoproteins and matrix proteins have been inferred from experiments showing enhanced binding of matrix protein to membranes in transfected cells on coexpression of viral glycoproteins. In the case of Sendai virus, either viral glycoprotein (F or HN) was found to be individually sufficient to promote membrane binding of the matrix protein (Sanderson et al. 1993, 1994). However, another group reported membrane binding by the Sendai virus matrix protein in transfected cells to be efficient even in the absence of glycoproteins, with no increase in membrane binding on F protein coexpression (Stricker et al. 1994). Similarly, for influenza A virus one group reported a substantial increase in the membrane binding of matrix protein on coexpression of either the HA or NA glycoproteins (Enami and Enami 1996; Gomez-Puertas et al. 2000), but conflicting results were obtained by others who observed efficient membrane binding of matrix protein to be independent of glycoprotein coexpression (Kretzschmar et al. 1996; Zhang and Lamb 1996). No increase in the membrane binding of VSV M protein was observed in transfected cells on coexpression of the VSV glycoprotein (Chong and Rose 1993).

In recent investigations, this experimental approach has been refined by measuring the association of matrix protein specifically with glycolipid-enriched "raft" membrane microdomains operationally defined as lipids resistant to solubilization by the nonionic detergent Triton X-100 (TX-100) at low temperatures. Influenza A virus assembles and buds from raft membranes (Scheiffele et al. 1999), and the influenza viral glycoproteins are sorted intrinsically to raft membranes when expressed in cells (Kundu et al. 1996; Skibbens et al. 1989). The influenza A virus M1 protein when expressed alone in cells does not bind TX-100-resistant membranes (Ali et al. 2000). Thus it has been proposed that interactions between viral glycoproteins and M1 protein could induce M1 protein to assemble together with the glycoproteins at raft membranes during viral infection. Consistent with this view, coexpression of either the HA or NA glycoproteins together with M1 protein in transfected cells resulted in a substantial increase in the amount of M1 protein bound to detergent-resistant membranes (Ali et al. 2000). Similar results were obtained with the Sendai virus matrix protein, which bound to TX-100-resistant membranes only on coexpression of the Sendai virus HN or F glycoproteins (Ali and Nayak 2000). Additional studies have been carried out with recombinant influenza A viruses having deleted glycoprotein cytoplasmic tails (Zhang et al. 2000b). Both glycoproteins were found to have decreased association with raft membranes on removal of the cytoplasmic tail. Most M1 protein was found associated with raft membranes in wt virus-infected cells, but in virus lacking both glycoprotein cytoplasmic tails M1 protein was no longer raft associated (Zhang et al. 2000b), in agreement with a model in which M1 protein is recruited to raft membranes for virus budding as a result of interactions between M1 protein and the glycoprotein cytoplasmic tails.

Another approach to characterizing M protein-glycoprotein interactions has been to measure M protein incorporation into virus particles. In a recombinant measles virus in which both measles virus glycoproteins were replaced with VSV G, M protein was not incorporated into particles, although M protein was expressed efficiently in the infected cells. When VSV G protein was replaced with a G/F hybrid protein containing the measles virus F protein cytoplasmic tail, incorporation of M protein into particles was restored (Spielhofer et al. 1998). Thus the measles virus cytoplasmic tail sequence was able to recruit measles virus matrix protein to sites of budding, but the VSV G protein was not. Similarly, a recombinant human parainfluenza type 3 virus (hPIV3) with F and HN glycoproteins derived from human parainfluenza type 2 virus (hPIV2) could only be recovered from an infectious clone when either the cytoplasmic domain or both the cytoplasmic and transmembrane domains of the hPIV2 glycoproteins were replaced with those of hPIV3 (Tao et al. 2000).

4.3
Self-Association of Matrix Proteins

Matrix proteins have been found to self-associate into ordered assemblies both as purified proteins in vitro and in virus-infected cells. For example, in Sendai virus-infected cells, patches likely composed of M protein have been observed by electron microscopy that are ordered into crystalline arrays at the inner surface of the plasma membrane (Bachi

1980; Buechi and Bachi 1982) and M protein purified from Sendai virions has been found to assemble in vitro into tubes and sheets consistent morphologically with the structures observed in infected cells (Heggeness et al. 1982). In vitro aggregation of VSV M protein under conditions of low salt concentration has also been observed (Gaudin et al. 1995, 1997; McCreedy et al. 1990), and electron microscopy has shown these aggregates to be made up of fine filaments (Gaudin et al. 1995). In some cases, VSV M protein aggregation could be reversed on increasing the salt concentration (Gaudin et al. 1995). In addition to low salt, M protein polymerization could also be induced by the presence of $ZnCl_2$ even at physiological salt concentrations (Gaudin et al. 1997), suggesting that aggregation is not simply the result of protein misfolding.

Ebola virus VP40 can be induced to multimerize into defined ringlike structures in vitro under certain conditions. VP40 is a two-domain protein, and on trypsin cleavage that separates the two domains the N-terminal domain forms oligomers (Dessen et al. 2000; Ruigrok et al. 2000b). From analysis of the VP40 crystal structure, it has been suggested that the two domains of the wt protein interact weakly and that the membrane-binding C-terminal domain may sterically prevent the formation of multimers through its interaction with the N-terminal domain (Dessen et al. 2000). Oligomerization of VP40 could also be induced by mild truncation of its C-terminal end by seven amino acids (a modification that, based on the VP40 crystal structure, was predicted to destabilize interactions between the two domains) or by urea treatment of VP40 (Scianimanico et al. 2000). Oligomerization was also induced on binding of the protein with synthetic liposomes (Scianimanico et al. 2000). In all of these cases, VP40 oligomerized into ringlike structures that could be visualized by electron microscopy. In light of these observations, it has been suggested that a conformational switch in VP40 may be induced by lipid binding that leads to domain separation, oligomerization, and formation of ringlike structures that could be intermediate structures in virus assembly (Scianimanico et al. 2000).

4.4 Interactions Between Matrix Proteins and Nucleocapsids

Interactions between matrix proteins and viral nucleocapsids are well established and are presumed to be critical for efficient incorporation of genomes into budding virions. Interactions between paramyxovirus nucleocapsids and matrix proteins were demonstrated with a variety of biochemical methods, such as chemical cross-linking (Markwell and Fox 1980; Nagai et al. 1978; Yoshida et al. 1976). Nucleocapsids isolated from influenza A virions are found associated with M1 protein (Murti et al. 1992). Virion-derived M1 protein can be removed from influenza virus nucleocapsids by treatment with high salt concentrations, and the resulting M1-depleted nucleocapsids can then be reconstituted with purified recombinant M1 protein at physiological salt concentrations (Baudin et al. 2001; Watanabe et al. 1996; Ye et al. 1999). With this reassociation assay, nucleocapsid binding activity was mapped to the C-terminal domain of the M1 protein (Baudin et al. 2001), suggesting that the M1 protein, in the form of 6-nm-long rods in virions as visualized by electron microscopy (Ruigrok et al. 2000a), is oriented with its N terminus touching the membrane and its C terminus interacting with nucleocapsids. One caveat should be added with respect to the interpretation of M-RNP reassociation assays, however, because it is difficult to be certain that M1 depletion of nucleocapsids was complete and that a portion of M1 reassociation with RNPs was not due to self-association with residually bound M1 protein. Interestingly, M-RNP interactions were found to be pH dependent, because M1 protein could be removed from virion-derived nucleocapsids not only by high salt concentrations but also by low-pH treatment (Zhirnov 1992) and reassociation of M1 protein with nucleocapsids was efficient at neutral pH but binding was essentially abolished at low pH (Ye et al. 1999). During viral entry, influenza virions that have been internalized by endocytosis are acidified as protons are transported from the endosomal lumen to the interior of the virion via the M2 ion channel, and if ion channel activity is blocked by the antiviral drug amantadine, viral replication is inhibited because of a block in viral uncoating (Hay 1992; Lamb et al. 1994; Martin and Helenius 1991; Takeda et al. 2002). These observations underscore the importance of having control mechanisms to shift the equilibrium of interactions that drive virus assembly versus disassembly. Viruses must be assembled in such a way that they can be disassembled later on during subsequent infections, and in the case of influenza virus this reversibility is at least in part accomplished through the pH dependence of RNP-M interactions.

Interactions between the VSV matrix protein and nucleocapsids in infected cells have been documented by electron microscopy (Odenwald et al. 1986) and by observing colocalization by using immunofluorescence microscopy (McCreedy and Lyles 1989). VSV nucleocapsids are also found associated with M protein when isolated from virions. Virions treated with detergent at physiological salt concentrations yield nucleocapsids with M protein still bound, and these nucleocapsids have a condensed and tightly coiled morphology similar to that of nucleocapsids in intact virions (Barge et al. 1993; Newcomb and Brown 1981). Incubation of these complexes with high salt concentrations leads to irreversible dissociation of M protein and a change in the morphology of the nucleocapsids to an extended and flexible structure (Barge et al. 1993; Lyles and McKenzie 1998; Newcomb and Brown 1981). Thus VSV M protein has a role in maintaining the VSV nucleocapsid in the compact form found in virions. Colocalization between nucleocapsids and M protein can only be detected at the plasma membranes of infected cells, implying that assembly of M protein with nucleocapsids does not occur in the cytoplasm (McCreedy and Lyles 1989). When cytosolic and membrane-derived M proteins were each tested for their abilities to assemble with virion-derived nucleocapsid-M complexes in vitro by an M protein exchange assay, both forms of M protein were found to assemble efficiently with nucleocapsids, indicating that lack of association between these components in the cytoplasm was not due to differences between cytoplasmic and membrane-bound forms of M protein (Flood and Lyles 1999). Instead, lack of association appears to be due to differences in the nucleocapsids, because neither form of M protein was able to bind to isolated intracellular nucleocapsids. Furthermore, the intracellular nucleocapsids exhibited an extended morphology similar to that observed for virion-derived nucleocapsids after salt treatment. These results suggest that M-RNP assembly occurs only after an initiating event that converts intracellular nucleocapsids into a form that is capable of assembling with M protein at the plasma membrane (Flood and Lyles 1999).

5
Selection of a Budding Site

5.1
Signals for Sorting Viral Glycoproteins into Virions

During assembly of negative-strand RNA viruses, viral proteins must be specifically selected and concentrated at budding sites at the plasma membrane of the infected cell, allowing release of virions that are densely packed with viral components. Efforts to study the mechanism by which viral proteins are selected to coalesce at virus assembly sites have been facilitated greatly by the recently available ability to recover negative-strand RNA viruses from cloned DNA by using reverse genetics. For example, a recombinant rabies virus lacking the gene for the glycoprotein G was generated (Mebatsion et al. 1996), which, in addition to allowing direct study of the role of G protein, also facilitated the generation of pseudotyped rabies viruses that could be made on expression of various foreign glycoproteins *in trans* (Mebatsion and Conzelmann 1996). By this approach, requirements for efficient packaging of the foreign glycoproteins into budding rabies virus particles were defined. When the HIV-1 Env protein was expressed *in trans*, pseudotyped particles failed to be released. However, when a chimeric HIV-1 Env protein was expressed that contained the rabies virus G protein cytoplasmic tail in place of the normal HIV-1 Env protein cytoplasmic tail, efficient release of pseudotyped particles containing the chimeric Env protein was restored (Mebatsion and Conzelmann 1996). Efficient incorporation of the chimeric protein was a result of the specific G protein cytoplasmic tail sequence and not the cytoplasmic tail length, because truncation of the 150-amino acid Env protein cytoplasmic tail to the same length as the G protein cytoplasmic tail (44 amino acids) failed to result in incorporation of the truncated protein into particles (Mebatsion and Conzelmann 1996). In similar experiments, the G protein cytoplasmic tail was found to be necessary to direct efficient incorporation of the cellular proteins CD4 and CXCR4 into pseudotyped particles (Mebatsion et al. 1997), indicating that the G protein cytoplasmic tail contains a positive signal sufficient to direct foreign glycoproteins into budding particles. Further evidence for the existence of an incorporation signal in the G protein cytoplasmic tail was obtained by generating a recombinant rabies virus in which the gene encoding the wt G protein was replaced with a gene encoding a mutant G protein lacking its cytoplasmic tail (Mebatsion et al. 1996). This resulted in poor budding of rabies virus particles, and the particles that were released contained significantly less G protein relative to the other viral proteins, indicating that the cytoplasmic tail of G protein is necessary to ensure efficient incorporation of G protein into the virion envelope (Mebatsion et al. 1996). These results are consistent with a model for rabies virus budding in which interactions between the G protein and the soluble matrix protein are important for efficient budding of particles as well as for specific incorporation of G protein into particles.

Requirements for specific packaging of glycoproteins into virus particles have also been studied extensively for VSV, in which a wide variety of recombinant viruses have been generated with reverse genetics that contain one or more extra genes encoding foreign glycoproteins. In contrast to results obtained with rabies virus, these studies indicated that no specific signals are required for incorporation of glycoproteins into VSV particles. For example, even unmodified cellular membrane protein CD4 expressed from a VSV recombinant virus was incorporated efficiently into VSV particles, and replacing the transmembrane and cytoplasmic domains of CD4 with the corresponding regions of VSV G did not enhance the incorporation efficiency (Schnell et al. 1996). These results are in agreement with earlier studies on VSV assembly showing efficient incorporation of CD4 and chimeric CD4 proteins into VSV particles when expressed *in trans* (Schubert et al. 1992). Other unmodified foreign glycoproteins that have been found to be incorporated efficiently into VSV particles when expressed from a VSV recombinant virus include the measles virus H glycoprotein (Schnell et al. 1996), the influenza virus glycoproteins HA and NA (Kretzschmar et al. 1997), the Ebola virus glycoprotein GP (Takada et al. 1997), the cellular membrane protein CXCR4 (Schnell et al. 1997), the RS virus glycoprotein G (Kahn et al. 1999, 2001), and the bovine viral diarrhea virus glycoprotein E2 (Grigera et al. 2000). Many of these recombinant viruses that express foreign antigens were generated so that they could be tested for use as live vaccines, and many have been shown to elicit protective immunity in animal model systems. The fact that such a wide variety of different glycoproteins can be incorporated efficiently into VSV particles argues that no specific incorporation signal is necessary to direct this process. It should be noted that one counterexample has also been described that is in apparent contradiction to these findings. The HIV-1 Env protein was not detectably incorporated into VSV virions when expressed from a VSV recombinant or when expressed *in trans*, but replacing the cytoplasmic tail of Env with the cytoplasmic tail of VSV G led to efficient incorporation into virions (Johnson et al. 1997; Owens and Rose 1993). However, subsequent investigation revealed that high-level incorporation of the chimeric Env protein into VSV particles was not due to a positive signal supplied by the G protein cytoplasmic tail, but rather was due to removal of a negative signal from the Env protein cytoplasmic tail. The negative signal was localized to the membrane-proximal ten amino acids of the Env protein cytoplasmic tail and caused sequestration of Env protein at sites distinct from VSV budding sites (Johnson et al. 1998). Truncation of the Env protein cytoplasmic tail to remove this sequence was sufficient to allow high-level incorporation of Env protein into virions. Poor incorporation of the wt Env protein into virions was not simply a consequence of the large size of the Env protein cytoplasmic tail, because Env proteins with cytoplasmic tails truncated so that they were smaller than the normal G protein cytoplasmic tail yet containing the negative signal still failed to be incorporated into virions (Johnson et al. 1998). In addition, an unrelated chimeric VSV G/GFP protein having a cytoplasmic tail even larger than the Env protein cytoplasmic tail has been constructed, and this protein was found to be incorporated efficiently into VSV virions (Dalton and Rose 2001). Thus, even in the case of HIV-1 Env protein, incorporation into the VSV envelope does not appear to depend on a specific positive incorporation signal.

Additional studies with VSV have been performed in which the viral G protein itself was altered and effects on incorporation of the altered G protein into virions were observed. Mutant G proteins were studied in which the transmembrane domain, the cytoplasmic domain, or both of these domains were replaced with the corresponding parts of human CD4, and all of these chimeric G proteins were incorporated efficiently into virions (Schnell et al. 1998). Furthermore, truncation of the G protein cytoplasmic tail did not substantially affect the efficiency of this protein's incorporation into budding virions except in the case of a very severe truncation that left a cytoplasmic tail of only one amino acid (Schnell et al. 1998; Whitt et al. 1989). A revertant to this cytoplasmic tail-deleted virus was selected and found to encode a new 8-amino acid cytoplasmic tail unrelated in sequence to the original cytoplasmic tail but that allowed efficient budding and efficient incorporation of the mutant G protein into virions (Schnell et al. 1998). Together, these experiments provide strong evidence that incorporation of glycoproteins into VSV virions does not require specific positive signals.

Similar studies to define requirements for glycoprotein incorporation into virus particles have also been carried out with influenza A virus and paramyxoviruses. A foreign polypeptide derived from HIV-1 gp41 has been expressed from a transfectant influenza A virus via a bicistronic NA segment, and this polypeptide was incorporated into influenza virions only when its transmembrane domain and cytoplasmic tail were replaced with those of the influenza virus HA protein (Garcia-Sastre et al.

1994), suggesting that specific signals in these domains are required to direct incorporation. *Trans* expression of HA proteins containing foreign cytoplasmic or transmembrane domains in influenza virus-infected cells results in poor incorporation of the mutant proteins into influenza virions compared with wt HA (Naim and Roth 1993). However, deletion of the HA protein cytoplasmic tail did not prevent efficient incorporation of this protein into virions either when expressed *in trans* or when expressed from a recombinant influenza virus, indicating that signals in the HA protein cytoplasmic tail are not absolutely required for efficient incorporation of this protein into virions (Jin et al. 1994; Naim and Roth 1993). Although deletion of the NA protein cytoplasmic tail in a recombinant virus resulted in poor incorporation of the mutant NA protein into virions, this reduction may have been due to an overall reduced expression level of the mutant protein in infected cells (Garcia-Sastre and Palese 1995; Mitnaul et al. 1996).

For paramyxoviruses, a role for glycoprotein cytoplasmic tails in specific targeting of glycoproteins into virions has been established. In recombinant measles viruses, truncations or other alterations to the cytoplasmic tails of the F or H glycoproteins results in poor incorporation of the mutant proteins into virions (Cathomen et al. 1998b). These virions also contain increased amounts of nonspecifically incorporated cellular proteins, indicating that the F and H protein cytoplasmic tails are important for both specific targeting of these proteins into measles virions as well as exclusion of host proteins from budding sites (Cathomen et al. 1998b). A similar result has been obtained with the paramyxovirus SV5, in which truncation of the HN protein cytoplasmic tail in a recombinant virus led to an increase in nonspecific incorporation of cellular proteins into virions (Schmitt et al. 1999). Further evidence that incorporation of paramyxovirus glycoproteins into virions requires specific signals has been obtained with Sendai virus. A foreign glycoprotein (NDV HN protein) expressed *in trans* was not incorporated into Sendai virions, whereas the Sendai virus HN protein expressed *in trans* was incorporated efficiently (Takimoto et al. 1998). Analysis of chimeric HN proteins led to identification of a specific amino acid sequence, SYWST, in the Sendai virus HN protein cytoplasmic tail that is critical for the incorporation of transfected HN protein into Sendai virions (Takimoto et al. 1998). These results have been confirmed with recombinant Sendai viruses, in which it was shown that absence of the sequence SYWST from the HN protein cytoplasmic tail results in exclusion of the mutant HN protein from budding Sendai virus particles (Fouillot-Coriou and Roux 2000). Thus, at least for Sendai virus, specific sorting of a glycoprotein into virions can be mapped to a signal in the glycoprotein cytoplasmic tail.

5.2
Polarized Budding from Epithelial Cells

For infection of host organisms, many viruses initially infect epithelial body surfaces such as the epithelial cells of the respiratory tract. Epithelial cells are polarized, having distinct apical (externally facing) and basolateral (internally facing) surfaces, and this polarized organization can have important effects on viral replication. For example, many viruses select preferentially either the apical or the basolateral plasma membrane as the site for assembly and budding. Orthomyxoviruses and paramyxoviruses, including influenza A virus, Sendai virus, SV5, and measles virus have been found to bud preferentially from the apical membranes of polarized cells (Blau and Compans 1995; Rodriguez-Boulan and Sabatini 1978). VSV and Marburg virus are examples of viruses that bud almost exclusively from the basolateral cell surface (Rodriguez-Boulan and Sabatini 1978; Sanger et al. 2001). Polarized budding may have important consequences for viral pathogenesis, because budding from the apical surface could favor restriction of the infection to the epithelial cell layer, whereas budding from the basolateral surface allows viral access to underlying tissue and could favor development of a systemic infection. Consistent with this view, Sendai virus and SV5 both produce localized infections of the respiratory tract in vivo, whereas VSV and Marburg virus both produce systemic infections in vivo. Furthermore, a mutant Sendai virus has been characterized in which polarized budding from the apical membrane is lost and virus is instead released in a nonpolar fashion from both the apical and basolateral cell surfaces, and this virus causes a systemic infection and is more virulent than wt Sendai virus (Tashiro et al. 1990). The correlation between pathogenicity and virus budding from the basolateral cell surface is not absolute, however, because measles virus is released apically yet produces a systemic infection in vivo.

For many viruses that exhibit polarized budding in epithelial cells, it has been found that the viral glycoproteins are targeted intrinsically to the same membrane from which virus buds. For example, both of the influenza virus glycoproteins HA (Roth et al. 1983) and NA (Jones et al.

1985), as well as the nonglycosylated ion channel membrane protein M2 (Hughey et al. 1992), are targeted specifically to the apical cell surface when expressed in the absence of other viral proteins. Similarly, the VSV glycoprotein is targeted specifically to the basolateral cell surface when it is expressed in the absence of other viral proteins (Stephens et al. 1986). Both of the influenza virus glycoproteins contain signals for apical sorting in their transmembrane domains (Barman et al. 2001; Barman and Nayak 2000; Kundu et al. 1996; Lin et al. 1998), and deletion of the glycoprotein cytoplasmic tails does not affect apical transport in the context of a recombinant virus (Zhang et al. 2000b). The basolateral sorting signal in the VSV G protein has been identified as a short tyrosine-containing sequence in the G protein cytoplasmic tail (Thomas and Roth 1994).

The observation that many viral glycoproteins when expressed alone are targeted to the same membrane in polarized cells from which virus budding occurs suggests a model in which the site of budding is selected on the basis of glycoprotein localization in the cell. However, a number of recent observations have been made that do not fit this model. For instance, some viral glycoproteins have been found to sort intrinsically to the membrane opposite from the one that is used for virus budding. Marburg virus glycoprotein expressed in a stably transfected MDCKII cell line was shown to be transported almost exclusively to the apical cell surface, despite the fact that release of infectious Marburg virus could only be detected from the basolateral cell surface (Sanger et al. 2001). When the distribution of glycoprotein was examined in Marburg virus-infected cells, a substantial fraction was found to be shifted to the basolateral cell surface where virions assemble, although the majority of glycoprotein was still found at the apical surface even in this case (Sanger et al. 2001). When interpreting these results for Marburg virus, it should be kept in mind that the natural biological reservoir for this virus has not yet been identified, and so it is not clear whether glycoprotein sorting in MDCKII cells is reflective of the situation in a biologically optimal host. Measles virus provides another example in which selection of budding sites likely is not specified by intrinsic sorting signals in the glycoproteins, because this virus buds from the apical cell surface despite the fact that both the H and F viral glycoproteins are localized predominantly at the basolateral surface of polarized cells when expressed alone (Maisner et al. 1998). These results suggest that other factors in addition to the viral glycoproteins can be important for selection of viral assembly sites in polarized cells. For measles virus, evidence suggests it is the matrix protein that drives apical release of virus. In cells infected with standard measles virus, both H and F proteins were found to be predominantly apical, whereas in cells infected with a recombinant measles virus lacking the M gene, both H and F proteins were sorted to the basolateral membrane (Naim et al. 2000). Thus M protein was able to revert transport of the F and H proteins to the apical cell surface where virus assembly takes place. However, not all of the F and H protein is redirected by M protein, and that which remains targeted to the basolateral cell surface may have important consequences for cell-to-cell fusion of polarized cells and systemic spread of the infection beyond the epithelial cell layer (Moll et al. 2001).

Even for viruses such as influenza A virus and VSV in which glycoproteins sort intrinsically to the same surface from which virus buds, budding site selection may not be solely a function of glycoprotein localization. Recent experiments with influenza A virus demonstrate that even when the HA glycoprotein is mutated such that its expression is no longer apical but instead is distributed in a nonpolarized fashion, budding of recombinant virus still occurs almost exclusively from the apical cell membrane (Mora et al. 2002). In VSV-infected cells, although the VSV G protein is transported mainly to the basolateral cell surface, some G protein can also be detected at the apical cell surface, but this does not result in a corresponding amount of budding from that surface (Rindler et al. 1984). When the basolateral targeting signal in the cytoplasmic tail of G protein is mutated so that the protein no longer exhibits polarized expression, release of recombinant virus is still almost completely from the basolateral surface (Zimmer et al. 2002). Thus factors other than VSV G must contribute toward determining the site of virus release, and the VSV matrix protein is a good candidate for performing this role because this protein has been found to associate specifically with the basolateral cell membrane independently of G (Bergmann and Fusco 1988).

5.3 Budding of Viruses from Membrane Rafts

Negative-strand RNA virus assembly occurs as a result of highly organized localizations of multiple separate viral components to discrete sites on the plasma membrane of an infected cell. In some cases, interactions between viral components and microdomains within the plasma membrane could play an important role in defining nucleation points for assembly. It has recently become clear that lipid molecules within the plasma membrane are not distributed homogenously in each leaflet of the bilayer but rather participate in lateral associations to form subcompartments within the membrane. One type of lipid microdomain is the membrane raft, which preferentially contains sphingolipids and cholesterol as well as certain integral membrane proteins (Brown and London 2000; Simons and Ikonen 1997). Membrane rafts can be separated biochemically from other membrane components based on their resistance to solubilization by certain nonionic detergents such as TX-100 at low temperatures. Some viral proteins have been found to be enriched within membrane rafts of infected cells, suggesting that virus assembly can occur on rafts. For example, in influenza A virus-infected cells the glycoproteins HA and NA as well as the matrix protein are found associated predominantly with TX-100-insoluble lipids (Skibbens et al. 1989; Zhang et al. 2000b). Assembly of viral proteins on raft membranes does not appear to be a universal strategy for negative-strand RNA virus assembly, however, because in the cases of VSV and rabies virus infections the viral proteins are found excluded from raft membranes in the infected cells. Paramyxovirus proteins in many cases have been found to be associated with raft membranes in infected cells, including the HN and F glycoproteins of Sendai virus (Sanderson et al. 1995), the measles virus proteins H, F, M, and N (Manie et al. 2000; Vincent et al. 2000), and the RS virus glycoproteins F and G (Brown et al. 2002). The Ebola virus glycoprotein GP and matrix protein VP40 are both found associated mainly with raft membranes in Ebola virus-infected cells (Bavari et al. 2002). Additional evidence in support of the notion that membrane rafts can be used as platforms for virus assembly and budding in some cases has been obtained through biochemical analysis of purified virus particles. HA and NA proteins from purified influenza virions were found associated with membrane that had solubility properties characteristic of membrane rafts, and the lipid composition of influenza virions is consistent with that of membrane rafts (Scheiffele et al. 1999; Zhang et al. 2000b), suggesting that the virion envelope is composed of raft membrane and that the virus therefore buds through rafts. Purified Ebola virions were found to contain the raft marker GM1 ganglioside, suggesting that this virus might also bud through rafts (Bavari et al. 2002).

Specific targeting of viral proteins to raft membranes has been observed not only in virus-infected cells but also in transfected cells expressing these proteins individually. Both of the influenza virus glycoproteins are found associated with raft membranes in transfected cells (Kundu et al. 1996; Skibbens et al. 1989). Signals directing the association of these proteins with lipid rafts have been mapped to the transmembrane domains, particularly the amino acid residues predicted to span the outer leaflets of the lipid bilayers (Barman et al. 2001; Barman and Nayak 2000; Kundu et al. 1996; Lin et al. 1998; Scheiffele et al. 1997). Although signals for the apical transport of these proteins in polarized epithelial cells also reside in the transmembrane domains, the two types of signals were found not to be identical, and it was possible to construct influenza virus glycoproteins with altered transmembrane domains in which raft association was disrupted but apical targeting was retained (Barman et al. 2001; Barman and Nayak 2000; Lin et al. 1998). Glycoprotein cytoplasmic tails have also been found to play a role in the targeting of influenza virus glycoproteins to membrane rafts, because raft association (but not apical targeting) was disrupted on removal of the glycoprotein cytoplasmic tails both in transfection experiments and in the context of a recombinant influenza virus (Zhang et al. 2000b). Intrinsic raft-targeting signals have also been characterized for proteins from other viruses. Both of the Sendai virus glycoproteins are found associated with raft membranes when expressed alone (Sanderson et al. 1995), and the Ebola virus glycoprotein expressed by itself is associated with rafts (Bavari et al. 2002).

Not all viral proteins that assemble at rafts in infected cells possess autonomous raft-targeting signals. Some viral proteins fail to associate with raft membranes when expressed alone and are only recruited to raft membranes when coexpressed with additional viral proteins that are themselves targeted to rafts, presumably as a result of protein-protein interactions. The influenza virus matrix protein, for example, is associated with TX-100-sensitive non-raft membrane when expressed alone but when coexpressed with either the HA or NA glycoprotein is found bound to raft membrane that resists TX-100 solubilization (Ali et al. 2000). Similarly, the Sendai virus matrix protein associates with non-raft membrane when expressed alone but associates with raft membrane when coexpressed with either the HN or F viral glycoprotein (Ali and Nayak 2000). The Ebola virus matrix protein VP40 provides an additional example of a viral protein that appears to be recruited to rafts indirectly by interacting with other viral components, because VP40 is not detected in raft membranes when expressed alone but is mainly detected in raft membranes in Ebola virus-infected cells (Bavari et al. 2002). Furthermore, coexpression of VP40 with glycoprotein GP results in efficient budding of VLPs into the culture media that contain the raft marker GM1, suggesting that the VLPs may be released through rafts (Bavari et al. 2002). A somewhat different situation exists for measles virus, in which a fraction of the matrix protein is found associated with raft membranes even when expressed alone (Vincent et al. 2000). The measles virus F glycoprotein is also associated with rafts when expressed alone, but not the H glycoprotein or the nucleocapsid protein N (Vincent et al. 2000). All four viral proteins are associated with raft membranes in measles virus-infected cells (Manie et al. 2000; Vincent et al. 2000). When coexpressed with F protein, the H protein is recruited to rafts. However, neither the M protein nor the N protein exhibited increased raft association on F protein coexpression (Vincent et al. 2000). In a chimeric measles virus in which the glycoproteins F and H were replaced with the non-raft-associated VSV G protein, the M and N proteins could still attach to rafts, but not VSV G (Manie et al. 2000). These results are consistent with a model for measles virus assembly in which the glycoproteins are targeted to rafts via a signal in the F protein, whereas the soluble M-RNP complexes are targeted independently to rafts via a signal in the M protein (Vincent et al. 2000).

Budding from membrane rafts may provide a mechanism for controlling which viral proteins are incorporated efficiently into virions. Vast differences sometimes exist in the incorporation levels of different viral proteins into virions. For example, the influenza virus M2 membrane protein is significantly underrepresented in virions (5–15 tetramers per virion) compared with the influenza virus HA protein (approximately 500 trimers per virion) despite abundant expression of M2 protein in infected cells (Zebedee and Lamb 1988). M2 protein fails to associate with raft membranes both in transfected cells expressing M2 protein alone and in influenza virus-infected cells (Zhang and Lamb 1996; Zhang et al. 2000b). In a recombinant influenza virus in which raft targeting of the HA and NA proteins was disrupted, preferential selection of these proteins over M2 protein for incorporation into virions was reduced substantially, suggesting that raft targeting may provide a mechanism for the selective inclusion or exclusion of viral proteins into budding virions (Zhang et al. 2000b).

6
Contributions of Host Factors in Virus Budding

Recent evidence suggests that a wide variety of enveloped viruses depend on recruitment of cellular machinery to assist in the late stages of budding. Domains within viral proteins that mediate this recruitment have been termed L domains because in the absence of these domains virus budding is blocked at a late step such that fully assembled virus particles fail to be released by membrane fission and instead accumulate as tethered particles at the cell surface. L domains were first characterized in retroviral Gag proteins, and functionally equivalent domains have now been identified in matrix proteins of some negative-strand RNA viruses as well. Multiple types of L domains having distinct amino acid sequences have been characterized. One type contains the consensus amino acid sequence P(T/S)AP and was first identified in the p6 region of the human immunodeficiency virus type 1 (HIV-1) Gag polyprotein (Gottlinger et al. 1991; Huang et al. 1995), and another type containing the consensus sequence PPxY was identified in the p2b region of the Rous sarcoma virus (RSV) Gag polyprotein (Wills et al. 1994; Xiang et al. 1996). These two types of L domains were shown to be functionally interchangeable. Poor budding caused by disruption of the RSV PPxY-type L domain could be restored by appending the HIV-1 p6-derived PTAP-containing sequence to the defective RSV Gag protein (Parent et al. 1995; Xiang et al. 1996), and poor budding caused by removal of the p6 region from a modified HIV-1 Gag protein could be restored by replacing p6 with the RSV p2b PPxY-containing sequence (Accola et al. 2000).

The matrix proteins of VSV and Ebola virus both contain L domain consensus sequences in their N-terminal regions. For VSV, distinct PPPY and PSAP sequences are present, and this N-terminal domain could functionally replace the L domain of the RSV Gag protein for Gag particle budding (Craven et al. 1999). Mutation of the PPPY motif in the VSV M protein inhibited the budding of VSV VLPs induced by expression of M protein alone (Harty et al. 1999) and caused a late budding defect of a recombinant VSV virus, as shown in Fig. 3 (Jayakar et al. 2000). These results suggest that at least some events in the budding process are conserved among a variety of enveloped viruses that include retroviruses and rhabdoviruses. The Ebola virus matrix protein contains a sequence PTAPPEY having overlapping L domain motifs near its N terminus. The

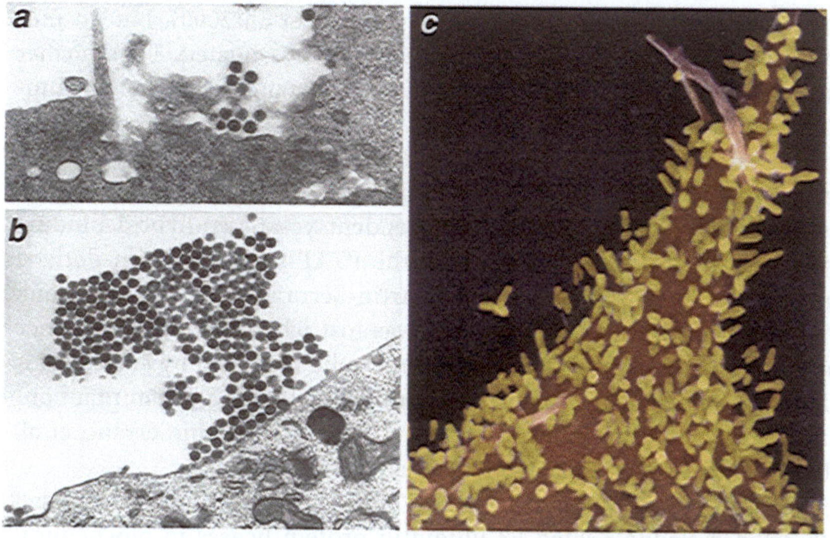

Fig. 3a–c Late budding defects in retroviruses and rhabdoviruses visualized by electron microscopy. **a** Thin section of an RSV-infected cell. **b** Thin section of an RSV-infected cell after treatment with the proteasome inhibitor MG-132. A late budding defect is observed in which virus particles fail to release from the infected cell. **c** Colorized scanning electron micrograph of a BHK-21 cell infected with recombinant VSV in which the PPPY late budding domain has been changed to PPPA, resulting in failure of the virus particles to release from the infected cell. (Electron micrographs for panels **a** and **b** courtesy of John Wills, The Pennsylvania State University. Electron micrograph for panel **c** courtesy of Gopal Murti, St. Jude Children's Research Hospital, Himangi Jayakar, University of Tennessee-Memphis, and Michael Whitt, University of Tennessee-Memphis)

relative contributions of the PTAP and PPxY components of this sequence in virus budding are presently unclear. In one study, the overlapping L domain sequences from Ebola virus were found to be capable of replacing the L domain from HIV-1 Gag and mutating the Ebola virus-derived sequence so that it had an intact PPxY sequence but a disrupted PTAP sequence resulted in poor budding, suggesting that, in this context, the PTAP component is very important (Martin-Serrano et al. 2001). However, it was not determined whether disruption of the PPxY sequence would also affect budding. In another study, VLP budding directed by expression of Ebola virus matrix protein by itself in cells was found to be reduced for a mutant protein in which the PPxY motif (but not the PTAP motif) was altered, indicating that the PPxY component is in fact important for efficient budding (Harty et al. 2000), but no mutants having disrupted PTAP sequences were investigated. Thus further study is needed to determine the relative importance of these two components for Ebola virus budding.

L domains appear to function by interacting with cellular proteins that are recruited to assist in virus budding. The cellular protein Tsg101 has been identified in multiple independent yeast two-hybrid binding assays as a factor that interacts with the PTAP-type L domain derived from HIV-1 Gag (Garrus et al. 2001; Martin-Serrano et al. 2001; VerPlank et al. 2001). Efficient Tsg101 binding was lost when the PTAP sequence was altered. L domain binding was found to be mediated by an N-terminal UEV (ubiquitin E2 variant) domain in the Tsg101 protein that contains a PTAP binding groove (Garrus et al. 2001; Martin-Serrano et al. 2001; Pornillos et al. 2002; VerPlank et al. 2001).

The PPxY-type L domain from RSV Gag has been shown to interact with cellular Nedd4-related E3 ubiquitin protein ligases in biochemical assays (Kikonyogo et al. 2001), and the matrix proteins of VSV and Ebola virus have been shown to interact with modular WW protein-protein interaction domains from the cellular proteins Nedd4 and YAP, as well as with the yeast Nedd4 homolog Rsp5 in in vitro binding assays (Harty et al. 1999, 2000, 2001). Interactions between viral matrix proteins and cellular WW domains were detected only when the matrix protein contained an intact PPxY motif.

The observations that PPxY-type L domains interact with E3 ubiquitin ligases and PTAP-type L domains interact with an E2-like domain of Tsg101 have suggested a possible role for ubiquitin in L domain function and virus budding. Additional evidence in support of this idea has also been obtained. For example, treatment of cells with proteasome inhibitors that deplete cellular levels of free ubiquitin blocks budding of HIV-1 (Schubert et al. 2000), RSV (Patnaik et al. 2000), and rhabdoviruses (Harty et al. 2001). Budding was shown to be blocked at a late stage on proteasome inhibitor treatment such that fully assembled particles accumulated at the cell surface (as shown in Fig. 3), consistent with the phenotype observed on L domain disruption (Patnaik et al. 2000; Schubert et al. 2000; Steinberg 2001). Moreover, genetic fusion of ubiquitin to the C-terminus of RSV Gag rendered the modified Gag partially resistant to the effect of proteasome inhibitor treatment, suggesting that direct attachment of ubiquitin to Gag is important for budding (Patnaik et al. 2000). In support of this notion, evidence has been obtained that viral proteins can be direct targets for ubiquitination. A minimal RSV Gag protein was shown to induce budding of VLPs that contain not only (unmodified) Gag but also Gag-ubiquitin conjugates (Strack et al. 2000). The same mutations to the L domain that caused poor VLP budding also prevented Gag ubiquitination. Ebola virus and rhabdovirus matrix proteins were shown to be substrates for ubiquitination by yeast Rsp5 in in vitro ubiquitination assays (Harty et al. 2000, 2001), and ubiquitination failed to occur in matrix proteins that lack the PPxY L domain sequence. Thus a connection has been made between ubiquitination and L domain function, and viral proteins that participate in budding may themselves be ubiquitination targets. However, it is not yet clear whether ubiquitin is important for virus budding because the ubiquitination of viral proteins is in fact critical to the budding process or whether the ubiquitination of viral proteins is incidental for budding and it is the ubiquitination of other cellular factors that is essential for normal virus budding.

One model for the late steps of enveloped virus budding that could potentially explain ubiquitin involvement proposes that the machinery that normally is used in formation of multivesicular bodies (MVBs) in the cell, and that is known to rely on ubiquitination reactions for normal function, is recruited via L domains for viral budding (Garrus et al. 2001; Patnaik et al. 2000). Budding of cellular vesicles into MVBs could be viewed as a process analogous to budding of virus particles from cells because in both cases cytoplasmic cargo is packed into a vesicle that buds outward from the cytoplasm (the opposite of what occurs in endocytotic budding). Cellular cargo proteins have been found to be subject to ubiquitination during MVB formation, and when these proteins are modified to remove the ubiquitination sites, proper sorting of the proteins into MVBs fails to occur (Hicke 2001). Moreover, in yeast, cargo protein recognition and sorting into MVBs is carried out in part by the ESCRT-1 complex that contains the Tsg101 homolog Vps23p as a critical component (Katzmann et al. 2001). This cellular pathway appears to be involved not just in the budding of viruses that bind directly to Tsg101 via PTAP-type L domains but more generally in the budding of enveloped viruses containing different types of L domains. Disruption of the MVB formation pathway in cells by expression of a dominant-negative Vps4 protein not only inhibits HIV-1 budding but also inhibits budding of murine leukemia virus, which relies on a PPxY-type L domain for budding (Garrus et al. 2001). In both cases, virus budding was blocked at a late step in which fully assembled particles failed to be released from the cell membrane. Thus different viruses that use distinct L domain sequences to interact with different cellular proteins could achieve the common goal of budding from the cell by using the cellular MVB formation pathway.

Conserved L domain sequences and functions have been characterized in widely divergent viruses including retroviruses and negative-strand RNA viruses such as rhabdoviruses and filoviruses. Not all negative-strand RNA virus matrix proteins contain recognizable L domains, however. Some matrix proteins, such as that of the paramyxovirus SV5, do not contain any of the previously characterized L domain sequences. Interestingly, rhabdovirus and filovirus M proteins, like retroviral Gag proteins, induce efficient budding of vesicles from transfected cells when expressed alone, whereas the SV5 M protein directs efficient budding only when coexpressed with additional SV5 proteins. One possibility that needs to be tested is whether the specific mechanism used for recruitment of host factors to facilitate virus budding affects the minimum set of viral proteins that must be expressed for efficient budding to occur. Even in the case of SV5, both virus budding and VLP budding were inhibited by expression of a dominant-negative Vps4 protein (A.P. Schmitt and R.A. Lamb, unpublished observation), suggesting that host machinery does play an important role in SV5 budding. Thus, for many viruses that bud efficiently yet lack well-defined L domain sequences, participation of host machinery may be critically important for budding, although the mechanisms used to recruit this machinery remain undefined.

7
Concluding Remarks

Much has been learned about assembly and budding of negative-strand RNA viruses in the 10 years since reverse genetics technology has become available for the study of these viruses. Roles for matrix proteins and glycoproteins in virus budding have been confirmed. Signals that allow incorporation of foreign glycoproteins into budding virions have been identified. Other aspects of virus assembly and budding are not as well understood. For example, although membrane rafts have been shown to serve as sites for the budding of some viruses, the precise function(s) of these lipid microdomains in virus assembly remain unclear. Although L domain sequences that serve to recruit host factors to assist in the late steps of virus budding have been identified in some cases, such sequences remain undefined for most negative-strand RNA viruses. A clearer understanding of these events can be expected in the future as these topics are further investigated, and as advances in cell biology add to our understanding of the underlying principles upon which these virological processes are based.

Acknowledgements. The authors thank George Leser, Doug Lyles, and Rob Ruigrok for critical reading of the manuscript and George Leser, Michael Whitt, and John Wills for providing electron micrographs. Research in the authors' laboratory was supported in part by research grants R37 AI-20201 and R01 AI-23173 from the National Institute of Allergy and Infectious Diseases. A.P.S. was an Associate and R.A.L. is an Investigator of the Howard Hughes Medical Institute.

References

Accola MA, Strack B, Gottlinger HG (2000) Efficient particle production by minimal Gag constructs which retain the carboxy-terminal domain of human immunodeficiency virus type 1 capsid-p2 and a late assembly domain. J Virol 74:5395–5402

Ali A, Avalos RT, Ponimaskin E, Nayak DP (2000) Influenza virus assembly: effect of influenza virus glycoproteins on the membrane association of M1 protein. J Virol 74:8709–8719

Ali A, Nayak DP (2000) Assembly of Sendai virus: M protein interacts with F and HN proteins and with the cytoplasmic tail and transmembrane domain of F protein. Virology 276:289–303

Arzt S, Baudin F, Barge A, Timmins P, Burmeister WP, Ruigrok RW (2001) Combined results from solution studies on intact influenza virus M1 protein and from a new crystal form of its N-terminal domain show that M1 is an elongated monomer. Virology 279:439–446

Bachi T (1980) Intramembrane structural differentiation in Sendai virus maturation. Virology 106:41–49

Barge A, Gaudin Y, Coulon P, Ruigrok RW (1993) Vesicular stomatitis virus M protein may be inside the ribonucleocapsid coil. J Virol 67:7246–7253

Barman S, Ali A, Hui EK, Adhikary L, Nayak DP (2001) Transport of viral proteins to the apical membranes and interaction of matrix protein with glycoproteins in the assembly of influenza viruses. Virus Res 77:61–69

Barman S, Nayak DP (2000) Analysis of the transmembrane domain of influenza virus neuraminidase, a type II transmembrane glycoprotein, for apical sorting and raft association. J Virol 74:6538–6545

Baudin F, Petit I, Weissenhorn W, Ruigrok RW (2001) In vitro dissection of the membrane and RNP binding activities of influenza virus M1 protein. Virology 281:102–108

Bavari S, Bosio CM, Wiegand E, Ruthel G, Will AB, Geisbert TW, Hevey M, Schmaljohn C, Schmaljohn A, Aman MJ (2002) Lipid raft microdomains: a gate-

way for compartmentalized trafficking of Ebola and Marburg viruses. J Exp Med 195:593–602
Bergmann JE, Fusco PJ (1988) The M protein of vesicular stomatitis virus associates specifically with the basolateral membranes of polarized epithelial cells independently of the G protein. J Cell Biol 107:1707–1715
Blau DM, Compans RW (1995) Entry and release of measles virus are polarized in epithelial cells. Virology 210:91–99
Brown DA, London E (2000) Structure and function of sphingolipid- and cholesterol-rich membrane rafts. J Biol Chem 275:17221–17224
Brown G, Rixon HW, Sugrue RJ (2002) Respiratory syncytial virus assembly occurs in GM1-rich regions of the host-cell membrane and alters the cellular distribution of tyrosine phosphorylated caveolin-1. J Gen Virol 83:1841–1850
Bucher DJ, Kharitonenkov IG, Zakomirdin JA, Grigoriev VB, Klimenko SM, Davis JF (1980) Incorporation of influenza virus M-protein into liposomes. J Virol 36:586–590
Buechi M, Bachi T (1982) Microscopy of internal structures of Sendai virus associated with the cytoplasmic surface of host membranes. Virology 120:349–359
Cathomen T, Mrkic B, Spehner D, Drillien R, Naef R, Pavlovic J, Aguzzi A, Billeter MA, Cattaneo R (1998a) A matrix-less measles virus is infectious and elicits extensive cell fusion: consequences for propagation in the brain. EMBO J 17:3899–3908
Cathomen T, Naim HY, Cattaneo R (1998b) Measles viruses with altered envelope protein cytoplasmic tails gain cell fusion competence. J Virol 72:1224–1234
Cattaneo R, Rose JK (1993) Cell fusion by the envelope glycoproteins of persistent measles viruses which caused lethal human brain disease. J Virol 67:1493–1502
Cattaneo R, Schmid A, Eschle D, Baczko K, ter Meulen V, Billeter MA (1988) Biased hypermutation and other genetic changes in defective measles viruses in human brain infections. Cell 55:255–265
Chong LD, Rose JK (1993) Membrane association of functional vesicular stomatitis virus matrix protein in vivo. J Virol 67:407–414
Chong LD, Rose JK (1994) Interactions of normal and mutant vesicular stomatitis virus matrix proteins with the plasma membrane and nucleocapsids. J Virol 68:441–447
Coronel EC, Murti KG, Takimoto T, Portner A (1999) Human parainfluenza virus type 1 matrix and nucleoprotein genes transiently expressed in mammalian cells induce the release of virus-like particles containing nucleocapsid-like structures. J Virol 73:7035–7038
Craven RC, Harty RN, Paragas J, Palese P, Wills JW (1999) Late domain function identified in the vesicular stomatitis virus M protein by use of rhabdovirus-retrovirus chimeras. J Virol 73:3359–3365
Dalton KP, Rose JK (2001) Vesicular stomatitis virus glycoprotein containing the entire green fluorescent protein on its cytoplasmic domain is incorporated efficiently into virus particles. Virology 279:414–421
Dessen A, Volchkov V, Dolnik O, Klenk HD, Weissenhorn W (2000) Crystal structure of the matrix protein VP40 from Ebola virus. EMBO J 19:4228–4236

Enami M, Enami K (1996) Influenza virus hemagglutinin and neuraminidase glycoproteins stimulate the membrane association of the matrix protein. J Virol 70:6653–6657

Faaberg KS, Peeples ME (1988) Association of soluble matrix protein of Newcastle disease virus with liposomes is independent of ionic conditions. Virology 166:123–132

Flood EA, Lyles DS (1999) Assembly of nucleocapsids with cytosolic and membrane-derived matrix proteins of vesicular stomatitis virus. Virology 261:295–308

Flood EA, McKenzie MO, Lyles DS (2000) Role of M protein aggregation in defective assembly of temperature-sensitive M protein mutants of vesicular stomatitis virus. Virology 278:520–533

Fouillot-Coriou N, Roux L (2000) Structure-function analysis of the Sendai virus F and HN cytoplasmic domain: different role for the two proteins in the production of virus particle. Virology 270:464–475

Garcia-Sastre A, Muster T, Barclay WS, Percy N, Palese P (1994) Use of a mammalian internal ribosomal entry site element for expression of a foreign protein by a transfectant influenza virus. J Virol 68:6254–6261

Garcia-Sastre A, Palese P (1995) The cytoplasmic tail of the neuraminidase protein of influenza A virus does not play an important role in the packaging of this protein into viral envelopes. Virus Res 37:37–47

Garrus JE, von Schwedler UK, Pornillos OW, Morham SG, Zavitz KH, Wang HE, Wettstein DA, Stray KM, Cote M, Rich RL, Myszka DG, Sundquist WI (2001) Tsg101 and the vacuolar protein sorting pathway are essential for HIV-1 budding. Cell 107:55–65

Gaudier M, Gaudin Y, Knossow M (2002) Crystal structure of vesicular stomatitis virus matrix protein. EMBO J 21:2886–2892

Gaudin Y, Barge A, Ebel C, Ruigrok RW (1995) Aggregation of VSV M protein is reversible and mediated by nucleation sites: implications for viral assembly. Virology 206:28–37

Gaudin Y, Sturgis J, Doumith M, Barge A, Robert B, Ruigrok RW (1997) Conformational flexibility and polymerization of vesicular stomatitis virus matrix protein. J Mol Biol 274:816–825

Gaudin Y, Tuffereau C, Benmansour A, Flamand A (1991) Fatty acylation of rabies virus proteins. Virology 184:441–444

Gomez-Puertas P, Albo C, Perez-Pastrana E, Vivo A, Portela A (2000) Influenza virus matrix protein is the major driving force in virus budding. J Virol 74:11538–11547

Gomez-Puertas P, Mena I, Castillo M, Vivo A, Perez-Pastrana E, Portela A (1999) Efficient formation of influenza virus-like particles: dependence on the expression levels of viral proteins. J Gen Virol 80:1635–1645

Gottlinger HG, Dorfman T, Sodroski JG, Haseltine WA (1991) Effect of mutations affecting the p6 gag protein on human immunodeficiency virus particle release. Proc Natl Acad Sci USA 88:3195–3199

Gregoriades A (1980) Interaction of influenza M protein with viral lipid and phosphatidylcholine vesicles. J Virol 36:470–479

Grigera PR, Marzocca MP, Capozzo AV, Buonocore L, Donis RO, Rose JK (2000) Presence of bovine viral diarrhea virus (BVDV) E2 glycoprotein in VSV recombinant particles and induction of neutralizing BVDV antibodies in mice. Virus Res 69:3-15

Harty RN, Brown ME, McGettigan JP, Wang G, Jayakar HR, Huibregtse JM, Whitt MA, Schnell MJ (2001) Rhabdoviruses and the cellular ubiquitin-proteasome system: a budding interaction. J Virol 75:10623–10629

Harty RN, Brown ME, Wang G, Huibregtse J, Hayes FP (2000) A PPxY motif within the VP40 protein of ebola virus interacts physically and functionally with a ubiquitin ligase: implications for filovirus budding. Proc Natl Acad Sci USA 97:13871–13876

Harty RN, Paragas J, Sudol M, Palese P (1999) A proline-rich motif within the matrix protein of vesicular stomatitis virus and rabies virus interacts with WW domains of cellular proteins: implications for viral budding. J Virol 73:2921–2929

Hay AJ (1992) The action of adamantanamines against influenza A viruses: inhibition of the M_2 ion channel protein. Semin Virol 3:21–30

Heggeness MH, Smith PR, Choppin PW (1982) In vitro assembly of the nonglycosylated membrane protein (M) of Sendai virus. Proc Natl Acad Sci USA 79:6232–6236

Hicke L (2001) A new ticket for entry into budding vesicles—ubiquitin. Cell 106:527–530

Huang M, Orenstein JM, Martin MA, Freed EO (1995) p6Gag is required for particle production from full-length human immunodeficiency virus type 1 molecular clones expressing protease. J Virol 69:6810–6818

Hughey PG, Compans RW, Zebedee SL, Lamb RA (1992) Expression of the influenza A virus M2 protein is restricted to apical surfaces of polarized epithelial cells. J Virol 66:5542–5552

Jasenosky LD, Neumann G, Lukashevich I, Kawaoka Y (2001) Ebola virus VP40-induced particle formation and association with the lipid bilayer. J Virol 75:5205–5214

Jayakar HR, Murti KG, Whitt MA (2000) Mutations in the PPPY motif of vesicular stomatitis virus matrix protein reduce virus budding by inhibiting a late step in virion release. J Virol 74:9818–9827

Jin H, Leser G, Lamb RA (1994) The influenza virus hemagglutinin cytoplasmic tail is not essential for virus assembly or infectivity. EMBO J 13:5504–5515

Jin H, Leser GP, Zhang J, Lamb RA (1997) Influenza virus hemagglutinin and neuraminidase cytoplasmic tails control particle shape. EMBO J 16:1236–1247

Johnson JE, Rodgers W, Rose JK (1998) A plasma membrane localization signal in the HIV-1 envelope cytoplasmic domain prevents localization at sites of vesicular stomatitis virus budding and incorporation into VSV virions. Virology 251:244–252

Johnson JE, Schnell MJ, Buonocore L, Rose JK (1997) Specific targeting to CD4+ cells of recombinant vesicular stomatitis viruses encoding human immunodeficiency virus envelope proteins. J Virol 71:5060–5068

Jones LV, Compans RW, Davis AR, Bos TJ, Nayak DP (1985) Surface expression of influenza virus neuraminidase, an amino-terminally anchored viral membrane glycoprotein, in polarized epithelial cells. Mol Cell Biol 5:2181–2189

Justice PA, Sun W, Li Y, Ye Z, Grigera PR, Wagner RR (1995) Membrane vesiculation function and exocytosis of wild-type and mutant matrix proteins of vesicular stomatitis virus. J Virol 69:3156–3160

Kahn JS, Roberts A, Weibel C, Buonocore L, Rose JK (2001) Replication-competent or attenuated, nonpropagating vesicular stomatitis viruses expressing respiratory syncytial virus (RSV) antigens protect mice against RSV challenge. J Virol 75:11079–11087

Kahn JS, Schnell MJ, Buonocore L, Rose JK (1999) Recombinant vesicular stomatitis virus expressing respiratory syncytial virus (RSV) glycoproteins: RSV fusion protein can mediate infection and cell fusion. Virology 254:81–91

Katzmann DJ, Babst M, Emr SD (2001) Ubiquitin-dependent sorting into the multivesicular body pathway requires the function of a conserved endosomal protein sorting complex, ESCRT-I. Cell 106:145–155

Kikonyogo A, Bouamr F, Vana ML, Xiang Y, Aiyar A, Carter C, Leis J (2001) Proteins related to the Nedd4 family of ubiquitin protein ligases interact with the L domain of Rous sarcoma virus and are required for gag budding from cells. Proc Natl Acad Sci USA 98:11199–11204

Knipe DM, Baltimore D, Lodish HF (1977) Maturation of viral proteins in cells infected with temperature-sensitive mutants of vesicular stomatitis virus. J Virol 21:1149–1158

Kolesnikova L, Bugany H, Klenk HD, Becker S (2002) VP40, the matrix protein of Marburg virus, is associated with membranes of the late endosomal compartment. J Virol 76:1825–1838

Kondo T, Yoshida T, Miura N, Nakanishi M (1993) Temperature-sensitive phenotype of a mutant Sendai virus strain is caused by its insufficient accumulation of the M protein. J Biol Chem 268:21924–21930

Kretzschmar E, Bui M, Rose JK (1996) Membrane association of influenza virus matrix protein does not require specific hydrophobic domains or the viral glycoproteins. Virology 220:37–45

Kretzschmar E, Buonocore L, Schnell MJ, Rose JK (1997) High-efficiency incorporation of functional influenza virus glycoproteins into recombinant vesicular stomatitis viruses. J Virol 71:5982–5989

Kundu A, Avalos RT, Sanderson CM, Nayak DP (1996) Transmembrane domain of influenza virus neuraminidase, a type II protein, possesses an apical sorting signal in polarized MDCK cells. J Virol 70:6508–6515

Lamb RA, Holsinger LJ, Pinto LH (1994) The influenza A virus M_2 ion channel protein and its role in the influenza virus life cycle. In Wimmer E (ed) "Receptor-mediated virus entry into cells", pp. 303–321. Cold Spring Harbor Laboratory Press, Cold Spring Harbor, N.Y.

Latham T, Galarza JM (2001) Formation of wild-type and chimeric influenza virus-like particles following simultaneous expression of only four structural proteins. J Virol 75:6154–6165

Lenard J, Vanderoef R (1990) Localization of the membrane-associated region of vesicular stomatitis virus M protein at the N terminus, using the hydrophobic, photoreactive probe 125I-TID. J Virol 64:3486–3491

Leyrer S, Bitzer M, Lauer U, Kramer J, Neubert WJ, Sedlmeier R (1998) Sendai virus-like particles devoid of haemagglutinin-neuraminidase protein infect cells via the human asialoglycoprotein receptor. J Gen Virol 79:683–687

Li Y, Luo L, Schubert M, Wagner RR, Kang CY (1993) Viral liposomes released from insect cells infected with recombinant baculovirus expressing the matrix protein of vesicular stomatitis virus. J Virol 67:4415–4420

Lin S, Naim HY, Rodriguez AC, Roth MG (1998) Mutations in the middle of the transmembrane domain reverse the polarity of transport of the influenza virus hemagglutinin in MDCK epithelial cells. J Cell Biol 142:51–57

Liu C, Air GM (1993) Selection and characterization of a neuraminidase-minus mutant of influenza virus and its rescue by cloned neuraminidase genes. Virology 194:403–407

Liu C, Eichelberger MC, Compans RW, Air GM (1995) Influenza type A virus neuraminidase does not play a role in viral entry, replication, assembly, or budding. J Virol 69:1099–1106

Lyles DS, McKenzie M, Parce JW (1992) Subunit interactions of vesicular stomatitis virus envelope glycoprotein stabilized by binding to viral matrix protein. J Virol 66:349–358

Lyles DS, McKenzie MO (1998) Reversible and irreversible steps in assembly and disassembly of vesicular stomatitis virus: equilibria and kinetics of dissociation of nucleocapsid-M protein complexes assembled in vivo. Biochemistry 37:439–450

Lyles DS, McKenzie MO, Kaptur PE, Grant KW, Jerome WG (1996) Complementation of M gene mutants of vesicular stomatitis virus by plasmid-derived M protein converts spherical extracellular particles into native bullet shapes. Virology 217:76–87

Maisner A, Klenk H, Herrler G (1998) Polarized budding of measles virus is not determined by viral surface glycoproteins. J Virol 72:5276–5278

Manie SN, Debreyne S, Vincent S, Gerlier D (2000) Measles virus structural components are enriched into lipid raft microdomains: a potential cellular location for virus assembly. J Virol 74:305–311

Markwell MA, Fox CF (1980) Protein-protein interactions within paramyxoviruses identified by native disulfide bonding or reversible chemical cross-linking. J Virol 33:152–166

Markwell MA, Portner A, Schwartz AL (1985) An alternative route of infection for viruses: entry by means of the asialoglycoprotein receptor of a Sendai virus mutant lacking its attachment protein. Proc Natl Acad Sci USA 82:978–982

Martin K, Helenius A (1991) Nuclear transport of influenza virus ribonucleoproteins: the viral matrix protein (M1) promotes export and inhibits import. Cell 67:117–130

Martin-Serrano J, Zang T, Bieniasz PD (2001) HIV-1 and Ebola virus encode small peptide motifs that recruit Tsg101 to sites of particle assembly to facilitate egress. Nat Med 7:1313–1319

McCreedy BJ, Jr., Lyles DS (1989) Distribution of M protein and nucleocapsid protein of vesicular stomatitis virus in infected cell plasma membranes. Virus Res 14:189–205

McCreedy BJ, Jr., McKinnon KP, Lyles DS (1990) Solubility of vesicular stomatitis virus M protein in the cytosol of infected cells or isolated from virions. J Virol 64:902–916

Mebatsion T, Conzelmann KK (1996) Specific infection of CD4+ target cells by recombinant rabies virus pseudotypes carrying the HIV-1 envelope spike protein. Proc Natl Acad Sci USA 93:11366–11370

Mebatsion T, Finke S, Weiland F, Conzelmann KK (1997) A CXCR4/CD4 pseudotype rhabdovirus that selectively infects HIV-1 envelope protein-expressing cells. Cell 90:841–847

Mebatsion T, Konig M, Conzelmann K-K (1996) Budding of rabies virus particles in the absence of the spike glycoprotein. Cell 84:941–951

Mebatsion T, Weiland F, Conzelmann KK (1999) Matrix protein of rabies virus is responsible for the assembly and budding of bullet-shaped particles and interacts with the transmembrane spike glycoprotein G. J Virol 73:242–250

Mena I, Vivo A, Perez E, Portela A (1996) Rescue of a synthetic chloramphenicol acetyltransferase RNA into influenza virus-like particles obtained from recombinant plasmids. J Virol 70:5016–5024

Metsikko K, Simons K (1986) The budding mechanism of spikeless vesicular stomatitis virus particles. EMBO J 5:1913–1920

Mitnaul LJ, Castrucci MR, Murti KG, Kawaoka Y (1996) The cytoplasmic tail of influenza A virus neuraminidase (NA) affects NA incorporation into virions, virion morphology, and virulence in mice but is not essential for virus replication. J Virol 70:873–879

Moll M, Klenk HD, Herrler G, Maisner A (2001) A single amino acid change in the cytoplasmic domains of measles virus glycoproteins H and F alters targeting, endocytosis, and cell fusion in polarized Madin-Darby canine kidney cells. J Biol Chem 276:17887–17894

Moll M, Klenk HD, Maisner A (2002) Importance of the cytoplasmic tails of the measles virus glycoproteins for fusogenic activity and the generation of recombinant measles viruses. J Virol 76:7174–7186

Mora R, Rodriguez-Boulan E, Palese P, Garcia-Sastre A (2002) Apical budding of a recombinant influenza A virus expressing a hemagglutinin protein with a basolateral localization signal. J Virol 76:3544–3553

Murti KG, Brown PS, Bean WJ, Jr., Webster RG (1992) Composition of the helical internal components of influenza virus as revealed by immunogold labeling/electron microscopy. Virology 186:294–299

Nagai Y, Yoshida T, Hamaguchi M, Iinuma M, Maeno K, Matsumoto T (1978) Cross-linking of Newcastle disease virus (NDV) proteins. Arch Virol 58:15–28

Naim HY, Ehler E, Billeter MA (2000) Measles virus matrix protein specifies apical virus release and glycoprotein sorting in epithelial cells. EMBO J 19:3576–3585

Naim HY, Roth MG (1993) Basis for selective incorporation of glycoproteins into the influenza virus envelope. J Virol 67:4831–4841

Newcomb WW, Brown JC (1981) Role of the vesicular stomatitis virus matrix protein in maintaining the viral nucleocapsid in the condensed form found in native virions. J Virol 39:295–299

Noda T, Sagara H, Suzuki E, Takada A, Kida H, Kawaoka Y (2002) Ebola virus VP40 drives the formation of virus-like filamentous particles along with GP. J Virol 76:4855–4865

Odenwald WF, Arnheiter H, Dubois-Dalcq M, Lazzarini RA (1986) Stereo images of vesicular stomatitis virus assembly. J Virol 57:922–932

Ono K, Dubois-Dalcq ME, Schubert M, Lazzarini RA (1987) A mutated membrane protein of vesicular stomatitis virus has an abnormal distribution within the infected cell and causes defective budding. J Virol 61:1332–1341

Owens RJ, Rose JK (1993) Cytoplasmic domain requirement for incorporation of a foreign envelope protein into vesicular stomatitis virus. J Virol 67:360–365

Palese P, Tobita K, Ueda M, Compans RW (1974) Characterization of temperature sensitive influenza virus mutants defective in neuraminidase. Virology 61:397–410

Parent LJ, Bennett RP, Craven RC, Nelle TD, Krishna NK, Bowzard JB, Wilson CB, Puffer BA, Montelaro RC, Wills JW (1995) Positionally independent and exchangeable late budding functions of the Rous sarcoma virus and human immunodeficiency virus Gag proteins. J Virol 69:5455–5460

Patnaik A, Chau V, Wills JW (2000) Ubiquitin is part of the retrovirus budding machinery. Proc Natl Acad Sci USA 97:13069–13074

Patterson JB, Cornu TI, Redwine J, Dales S, Lewicki H, Holz A, Thomas D, Billeter MA, Oldstone MB (2001) Evidence that the hypermutated M protein of a subacute sclerosing panencephalitis measles virus actively contributes to the chronic progressive CNS disease. Virology 291:215–225

Pattnaik AK, Brown DJ, Nayak DP (1986) Formation of influenza virus particles lacking hemagglutinin on the viral envelope. J Virol 60:994–1001

Pornillos O, Alam SL, Rich RL, Myszka DG, Davis DR, Sundquist WI (2002) Structure and functional interactions of the Tsg101 UEV domain. EMBO J 21:2397–2406

Portner A, Marx PA, Kingsbury DW (1974) Isolation and characterization of Sendai virus temperature-sensitive mutants. J Virol 13:298–304

Portner A, Scroggs RA, Marx PS, Kingsbury DW (1975) A temperature-sensitive mutant of Sendai virus with an altered hemagglutinin-neuraminidase polypeptide: consequences for virus assembly and cytopathology. Virology 67:179–187

Riedl P, Moll M, Klenk HD, Maisner A (2002) Measles virus matrix protein is not cotransported with the viral glycoproteins but requires virus infection for efficient surface targeting. Virus Res 83:1-12

Rindler MJ, Ivanov IE, Plesken H, Rodriguez-Boulan E, Sabatini DD (1984) Viral glycoproteins destined for apical or basolateral plasma membrane domains traverse the same Golgi apparatus during their intracellular transport in doubly infected Madin-Darby canine kidney cells. J Cell Biol 98:1304–1319

Robison CS, Whitt MA (2000) The membrane-proximal stem region of vesicular stomatitis virus G protein confers efficient virus assembly. J Virol 74:2239–2246

Rodriguez-Boulan E, Sabatini DD (1978) Asymmetric budding of viruses in epithelial monolayers: a model system for study of epithelial polarity. Proc Natl Acad Sci USA 75:5071–5075

Rolls MM, Webster P, Balba NH, Rose JK (1994) Novel infectious particles generated by expression of the vesicular stomatitis virus glycoprotein from a self-replicating RNA. Cell 79:497–506

Roth MG, Compans RW, Giusti L, Davis AR, Nayak DP, Gething MJ, Sambrook J (1983) Influenza virus hemagglutinin expression is polarized in cells infected with recombinant SV40 viruses carrying cloned hemagglutinin DNA. Cell 33:435–443

Roux L, Waldvogel FA (1982) Instability of the viral M protein in BHK-21 cells persistently infected with Sendai virus. Cell 28:293–302

Ruigrok RW, Barge A, Durrer P, Brunner J, Ma K, Whittaker GR (2000a) Membrane interaction of influenza virus M1 protein. Virology 267:289–298

Ruigrok RW, Schoehn G, Dessen A, Forest E, Volchkov V, Dolnik O, Klenk HD, Weissenhorn W (2000b) Structural characterization and membrane binding properties of the matrix protein VP40 of Ebola virus. J Mol Biol 300:103–112

Sakaguchi T, Uchiyama T, Fujii Y, Kiyotani K, Kato A, Nagai Y, Kawai A, Yoshida T (1999) Double-layered membrane vesicles released from mammalian cells infected with Sendai virus expressing the matrix protein of vesicular stomatitis virus. Virology 263:230–243

Sanderson CM, Avalos R, Kundu A, Nayak DP (1995) Interaction of Sendai viral F, HN, and M proteins with host cytoskeletal and lipid components in Sendai virus-infected BHK cells. Virology 209:701–707

Sanderson CM, McQueen NL, Nayak DP (1993) Sendai virus assembly: M protein binds to viral glycoproteins in transit through the secretory pathway. J Virol 67:651–663

Sanderson CM, Wu HH, Nayak DP (1994) Sendai virus M protein binds independently to either the F or the HN glycoprotein in vivo. J Virol 68:69–76

Sanger C, Muhlberger E, Ryabchikova E, Kolesnikova L, Klenk HD, Becker S (2001) Sorting of Marburg virus surface protein and virus release take place at opposite surfaces of infected polarized epithelial cells. J Virol 75:1274–1283

Scheiffele P, Rietveld A, Wilk T, Simons K (1999) Influenza viruses select ordered lipid domains during budding from the plasma membrane. J Biol Chem 274:2038–2044

Scheiffele P, Roth MG, Simons K (1997) Interaction of influenza virus haemagglutinin with sphingolipid-cholesterol membrane domains via its transmembrane domain. EMBO J 16:5501–5508

Schmid A, Spielhofer P, Cattaneo R, Baczko K, ter Meulen V, Billeter MA (1992) Subacute sclerosing panencephalitis is typically characterized by alterations in the fusion protein cytoplasmic domain of the persisting measles virus. Virology 188:910–915

Schmitt AP, He B, Lamb RA (1999) Involvement of the cytoplasmic domain of the hemagglutinin-neuraminidase protein in assembly of the paramyxovirus simian virus 5. J Virol 73:8703–8712

Schmitt AP, Leser GP, Waning DL, Lamb RA (2002) Requirements for budding of paramyxovirus simian virus 5 virus-like particles. J Virol 76:3952–3964

Schnell MJ, Buonocore L, Boritz E, Ghosh HP, Chernish R, Rose JK (1998) Requirement for a non-specific glycoprotein cytoplasmic domain sequence to drive efficient budding of vesicular stomatitis virus. EMBO J 17:1289–1296

Schnell MJ, Buonocore L, Kretzschmar E, Johnson E, Rose JK (1996) Foreign glycoproteins expressed from recombinant vesicular stomatitis viruses are incorporated efficiently into virus particles. Proc Natl Acad Sci USA 93:11359–11365

Schnell MJ, Johnson JE, Buonocore L, Rose JK (1997) Construction of a novel virus that targets HIV-1-infected cells and controls HIV-1 infection. Cell 90:849–857

Schnitzer TJ, Dickson C, Weiss RA (1979) Morphological and biochemical characterization of viral particles produced by the tsO45 mutant of vesicular stomatitis virus at restrictive temperature. J Virol 29:185–195

Schubert M, Joshi B, Blondel D, Harmison GG (1992) Insertion of the human immunodeficiency virus CD4 receptor into the envelope of vesicular stomatitis virus particles. J Virol 66:1579–1589

Schubert U, Ott DE, Chertova EN, Welker R, Tessmer U, Princiotta MF, Bennink JR, Krausslich HG, Yewdell JW (2000) Proteasome inhibition interferes with gag polyprotein processing, release, and maturation of HIV-1 and HIV-2. Proc Natl Acad Sci USA 97:13057–13062

Scianimanico S, Schoehn G, Timmins J, Ruigrok RH, Klenk HD, Weissenhorn W (2000) Membrane association induces a conformational change in the Ebola virus matrix protein. EMBO J 19:6732–6741

Sha B, Luo M (1997) Structure of a bifunctional membrane-RNA binding protein, influenza virus matrix protein M1. Nat Struct Biol 4:239–244

Simons K, Ikonen E (1997) Functional rafts in cell membranes. Nature 387:569–572

Skibbens JE, Roth MG, Matlin KS (1989) Differential extractability of influenza virus hemagglutinin during intracellular transport in polarized epithelial cells and nonpolar fibroblasts. J Cell Biol 108:821–832

Spielhofer P, Bachi T, Fehr T, Christiansen G, Cattaneo R, Kaelin K, Billeter MA, Naim HY (1998) Chimeric measles viruses with a foreign envelope. J Virol 72:2150–2159

Steinberg D (2001) Virus budding linked to ubiquitin system. The Scientist 15:16

Stephens EB, Compans RW, Earl P, Moss B (1986) Surface expression of viral glycoproteins is polarized in epithelial cells infected with recombinant vaccinia viral vectors. EMBO J 5:237–245

Strack B, Calistri A, Accola MA, Palu G, Gottlinger HG (2000) A role for ubiquitin ligase recruitment in retrovirus release. Proc Natl Acad Sci USA 97:13063–13068

Stricker R, Mottet G, Roux L (1994) The Sendai virus matrix protein appears to be recruited in the cytoplasm by the viral nucleocapsid to function in viral assembly and budding. J Gen Virol 75:1031–1042

Stricker R, Roux L (1991) The major glycoprotein of Sendai virus is dispensable for efficient virus particle budding. J Gen Virol 72:1703–1707

Takada A, Robison C, Goto H, Sanchez A, Murti KG, Whitt MA, Kawaoka Y (1997) A system for functional analysis of Ebola virus glycoprotein. Proc Natl Acad Sci USA 94:14764–14769

Takeda M, Pekosz A, Shuck K, Pinto LH, Lamb RA (2002) Influenza a virus M_2 ion channel activity is essential for efficient replication in tissue culture. J Virol 76:1391–1399

Takimoto T, Bousse T, Coronel EC, Scroggs RA, Portner A (1998) Cytoplasmic domain of Sendai virus HN protein contains a specific sequence required for its incorporation into virions. J Virol 72:9747–9754

Takimoto T, Murti KG, Bousse T, Scroggs RA, Portner A (2001) Role of matrix and fusion proteins in budding of sendai virus. J Virol 75:11384–11391

Tao T, Skiadopoulos MH, Davoodi F, Riggs JM, Collins PL, Murphy BR (2000) Replacement of the ectodomains of the hemagglutinin-neuraminidase and fusion glycoproteins of recombinant parainfluenza virus type 3 (PIV3) with their counterparts from PIV2 yields attenuated PIV2 vaccine candidates. J Virol 74:6448–6458

Tashiro M, Yamakawa M, Tobita K, Seto JT, Klenk HD, Rott R (1990) Altered budding site of a pantropic mutant of Sendai virus, F1-R, in polarized epithelial cells. J Virol 64:4672–4677

Thomas DC, Roth MG (1994) The basolateral targeting signal in the cytoplasmic domain of glycoprotein G from vesicular stomatitis virus resembles a variety of intracellular targeting motifs related by primary sequence but having diverse targeting activities. J Biol Chem 269:15732–15739

Timmins J, Scianimanico S, Schoehn G, Weissenhorn W (2001) Vesicular release of ebola virus matrix protein VP40. Virology 283:1-6

Tuffereau C, Portner A, Roux L (1985) The role of haemagglutinin-neuraminidase glycoprotein cell surface expression in the survival of Sendai virus-infected BHK-21 cells. J Gen Virol 66:2313–2318

VerPlank L, Bouamr F, LaGrassa TJ, Agresta B, Kikonyogo A, Leis J, Carter CA (2001) Tsg101, a homologue of ubiquitin-conjugating (E2) enzymes, binds the L domain in HIV type 1 Pr55(Gag). Proc Natl Acad Sci USA 98:7724–7729

Vincent S, Gerlier D, Manie SN (2000) Measles virus assembly within membrane rafts. J Virol 74:9911–9915

Volchkov VE, Volchkova VA, Slenczka W, Klenk HD, Feldmann H (1998) Release of viral glycoproteins during Ebola virus infection. Virology 245:110–119

Waning DL, Schmitt AP, Leser GP, Lamb RA (2002) Roles for the cytoplasmic tails of the fusion and hemagglutinin-neuraminidase proteins in budding of the paramyxovirus simian virus 5. J Virol 76:9284–9297

Watanabe K, Handa H, Mizumoto K, Nagata K (1996) Mechanism for inhibition of influenza virus RNA polymerase activity by matrix protein. J Virol 70:241–247

Whitt MA, Chong L, Rose JK (1989) Glycoprotein cytoplasmic domain sequences required for rescue of a vesicular stomatitis virus glycoprotein mutant. J Virol 63:3569–3578

Wills JW, Cameron CE, Wilson CB, Xiang Y, Bennett RP, Leis J (1994) An assembly domain of the Rous sarcoma virus Gag protein required late in budding. J Virol 68:6605–6618

Wong TC, Ayata M, Ueda S, Hirano A (1991) Role of biased hypermutation in evolution of subacute sclerosing panencephalitis virus from progenitor acute measles virus. J Virol 65:2191–2199

Xiang Y, Cameron CE, Wills JW, Leis J (1996) Fine mapping and characterization of the Rous sarcoma virus Pr76gag late assembly domain. J Virol 70:5695–5700

Ye Z, Liu T, Offringa DP, McInnis J, Levandowski RA (1999) Association of influenza virus matrix protein with ribonucleoproteins. J Virol 73:7467–7473

Ye Z, Sun W, Suryanarayana K, Justice P, Robinson D, Wagner RR (1994) Membrane-binding domains and cytopathogenesis of the matrix protein of vesicular stomatitis virus. J Virol 68:7386–7396

Yoshida T, Nagai Y, Maeno K, Iinuma M, Hamaguchi M, Matsumoto T, Nagayoshi S, Hoshino M (1979) Studies on the role of M protein in virus assembly using a ts mutant of HVJ (Sendai virus). Virology 92:139–154

Yoshida T, Nagai Y, Yoshii S, Maeno K, Matsumoto T (1976) Membrane (M) protein of HVJ (Sendai virus): its role in virus assembly. Virology 71:143–161

Zebedee SL, Lamb RA (1988) Influenza A virus M2 protein: monoclonal antibody restriction of virus growth and detection of M2 in virions. J Virol 62:2762–2772

Zhang J, Lamb RA (1996) Characterization of the membrane association of the influenza virus matrix protein in living cells. Virology 225:255–266

Zhang J, Leser GP, Pekosz A, Lamb RA (2000a) The cytoplasmic tails of the influenza virus spike glycoproteins are required for normal genome packaging. Virology 269:325–334

Zhang J, Pekosz A, Lamb RA (2000b) Influenza virus assembly and lipid raft microdomains: a role for the cytoplasmic tails of the spike glycoproteins. J Virol 74:4634–4644

Zhirnov OP (1992) Isolation of matrix protein M1 from influenza viruses by acid-dependent extraction with nonionic detergent. Virology 186:324–330

Zimmer G, Zimmer KP, Trotz I, Herrler G (2002) Vesicular stomatitis virus glycoprotein does not determine the site of virus release in polarized epithelial cells. J Virol 76:4103–4107

Accessory Genes of the *Paramyxoviridae*, a Large Family of Nonsegmented Negative-Strand RNA Viruses, as a Focus of Active Investigation by Reverse Genetics

Y. Nagai[1] · A. Kato[2]

[1] Toyama Institute of Health, 17-1 Nakataikouyama, Kosugi-machi, 939-0363, Toyama, Japan
E-mail: yoshiyuki.nagai@pref.toyama.lg.jp

[2] National Institute of Infectious Diseases, Gakuen 4-7-1, Musashi-Murayama, 208-0011, Tokyo, Japan

Abstract The Paramyxoviridae, a large family of nonsegmented negative-strand RNA viruses, comprises several genera each containing important human and animal pathogens. They possess in common six basal genes essential for viral replication and, in addition, a subset of accessory genes that are largely unique to each genus. These accessory genes are either encoded in one or more alternative overlapping frames of a basal gene, which are accessed transcriptionally or translationally, or inserted before or between the basal genes as one or more extra genes. However, the question of how the individual accessory genes contribute to actual viral replication and pathogenesis remained unanswered. It was not even established whether they are dispensable or indispensable for the viral life cycle. The plasmid-based reverse genetics of the full-length viral genome has now come into wide use to demonstrate that most, if not all, of these putative accessory genes can be disrupted without destroying viral infectivity, conclusively defining them as indeed dispensable accessory genes. Studies on the phenotypes of the resulting gene knockout viruses have revealed that the individual accessory genes greatly contribute specifically and additively to the overall viral fitness both in vitro and in vivo.

Abbreviations *aMPV* Avian metapneumovirus · *bPIV3* Bovine parainfluenza virus 3 · *bRSV* Bovine respiratory syncytial virus · *CDV* Canine distemper virus · *CPE* Cytopathic effect · *DDB1* The 127-kDa subunit of the damage-specific DNA binding protein · *GAF* Gamma-activated factor · *GAS* Gamma-activated sequence · *HeV* Hendra virus · *HIV-1* Human immunodeficiency virus 1 · *hMPV* Human metapneumovirus · *hPIV1* Human parainfluenza virus 1 · *hPIV2* Human parainfluenza virus 2 · *hPIV3* Human parainfluenza virus 3 · *hRSV* Human respiratory syncytial virus ·

IFN Interferon · *IRF* Interferon regulatory factor · *ISG* Interferon-stimulated gene · *ISGF* Interferon-stimulated gene factor · *ISRE* Interferon-stimulated response element · *JAK* Janus kinase · LD_{50} 50% Lethal dose · *MeV* Measles virus · *MuV* Mumps virus · *NDV* Newcastle disease virus · *NiV* Nipah virus · *ORF* Open reading frame · *PDV* Phocine distemper virus · *RACK* Receptor for activated C kinase · *PKR* RNA-dependent protein kinase · *RNP* Ribonucleoprotein · *RPV* Rinderpest virus · *RV* Rabies virus · *SeV* Sendai virus · *STAT* Signal transducer and activator of transcription · *SV5* Simian virus 5 · *SV41* Simian virus 41 · *TNF* Tumor necrosis factor · *TPMV* Tupaia paramyxovirus · *VSV* Vesicular stomatitis virus · *VV* Vaccinia virus

1 Introduction

The family *Paramyxoviridae* contains a variety of important human and animal pathogens, including hPIVs, MeV, MuV, hRSV, CDV, and NDV, which have been well recognized since early studies in virology. The family also includes newly emergent agents such as HeV, NiV, PDV, and hMPV (Tables 1, 2). The family is classified into two subfamilies, the *Paramyxovirinae* and the *Pneumovirinae*. *Paramyxovirinae* comprises five genera, the *Respirovirus*, *Morbillivirus*, *Rubulavirus*, *Avulavirus*, and *Henipavirus* (Table 1). The latter two were approved by the latest International Committee on Taxonomy of Viruses (ICTV) in 2002. Yet unassigned to a genus is TPMV, which is tentatively categorized in "TPMV-like viruses." *Pneumovirinae* has two genera, *Pneumovirus* and *Metapneumovirus* (Table 2). The *Paramyxoviridae* are enveloped viruses with a linear, nonsegmented negative-sense (−) RNA genome of approximately 15 kb. As exceptions, the genomes of Henipaviruses and TPMV are relatively long, ranging from 17.9 to 18.2 kb, and those of the *Metapneumovirus*, hMPV and aMPV, are relatively short (13.4 kb). A genome consisting of a single molecule of negative-sense RNA also is characteristic of three other families, *Rhabdoviridae*, *Filoviridae*, and *Bornaviridae*, together constituting the order *Mononegavirales*.

The individual members of the *Paramyxoviridae* have their own unique histories of coevolution with the respective hosts but share common features regarding genome structure and function. They have six basal genes encoding the genomic RNA-binding N (nucleocapsid) protein, the phospho (P) protein (the smaller subunit of RNA polymerase),

Table 1 Products of the P and SH genes and their length (number of amino acids) in the subfamily *Paramyxovirinae*[a]

Genus, species [GenBank accession no.]	P	V	D	W*/ R†/I#	C	SB	X	SH
Respirovirus								
SeV [X00087]	568	316/68	–	316/2*	215(C'), 204(C), 181(Y1), 175(Y2)	–	95	–
hPIV1 [M74080]	568	–*	–	–	218(C'), 204(C), 181(Y1)	–	–	–
bPIV3 [Y00115]	596	240/171	240/127	–	201	–	–	–
hPIV3 [X04721]	602	240/8, –*	240/133	–	199	–	–	–
Morbillivirus								
MeV [M10456]	507	230/69	–	230/6* 294/5†	186	–	–	–
RPV [Z30697]	507	230/69	–	230/4*	177	–	–	–
CDV [M32418]	507	230/69	–	230/13*	174	–	–	–
PDV [X75960]	507	230/69	–	230/6*	174	–	–	–
Avulavirus								
NDV [Z26249]	395	133/106	–	133/94*	–	96	–	–
"TPMV-like viruses"								
TPMV [AF079780]	527	228/54	–	228/6*	153	77	–	–
Henipavirus								
HeV [AF017149]	707	404/53	–	404/4*	166	65	–	–
NiV [AF212302]	709	406/50	–	406/44*	166	–	–	–
Rubulavirus								
SV5 [J03142]	162/229	222	–	162/9#	–	–	–	44
hPIV2 [M37751]	161/233	225	–	161/4#	–	–	–	–
SV41 [X64275]	161/233	225	–	161/5#	–	–	–	–
MuV [D00352]	153/238	224	–	153/17#	–	–	–	57

[a] The numbers of amino acids were derived from GenBank with the individual accession numbers given in brackets.
When frame shifting is involved, numbers upstream before shifting and those downstream after shifting are separately shown with a slash.
–, Not encoded; –*, V relic is present. For details, see text.

Table 2 Numbers of amino acids of the accessory proteins in the subfamily *Pneumovirinae*[a]

Genus, species [GenBank accession no.]	NS1	NS2	SH	M2-1	M2-2
Pneumovirus					
hRSV [AF013254]	139	124	65	195	90
bRSV [AF092942]	136	124	81	186	90
Metapneumovirus					
hMPV [AF371337]	–	–	183	187	71
aMPV [AJ492378]	–	–	175	186	73

[a] Data were derived as described in the footnote of Table 1; –, not encoded.

the matrix (M) protein, the fusion glycoprotein F, an attachment glycoprotein HN, H, or G, and the L protein (the larger and catalytic subunit of polymerase), arranged in that order from the 3′ terminus (Fig. 1). The *Pneumovirus* is exceptional in that the order of the two glycoprotein genes is reversed (Fig. 1, in parentheses). The attachment glycoprotein is termed hemagglutinin-neuraminidase or HN for the *Respirovirus*, *Avulavirus* and *Rubulavirus*, hemagglutinin or H for *Morbillivirus* and TPMV, and simply G for the *Henipavirus* genus and the *Pneumovirinae* subfamily.

The extracistronic 3′ leader and 5′ trailer regions are about 50 nucleotides in length and contain a *cis*-acting element essential for replication. The genomic (−)RNA is tightly associated with N proteins and is further complexed with the RNA polymerase comprising the L and P proteins (Hamaguchi et al. 1983), forming a helical (−)RNP. The 3′-terminal 12 nucleotides of the genome and antigenome of each genus are nearly identical, suggesting their roles as the replication promoter. An additional internal element, which is aligned with the 3′-terminal region to the same surface of helical RNP, is also thought to be important for promoter function for the *Paramyxovirinae*, suggesting a bipartite model of promoter (reviewed in Lamb and Kolakofsky 2001). Transcription initiates at the 3′ end of the genome, and genes are copied sequentially in their 3′ to 5′ order. By recognizing the stop (termination/polyadenylation) and restart signals present at each gene boundary, the polymerase generates each mRNA. Transcription has a polarity gradient such that the efficiency of transcription decreases with the increasing distance of the various genes from the promoter. These transcription-regulatory

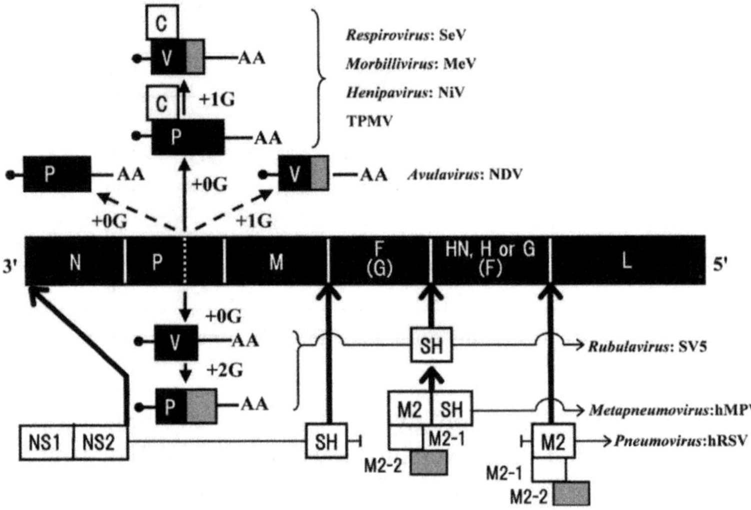

Fig. 1 Gene maps of the representatives of different genera of the *Paramyxoviridae* with emphasis on the acquisition and expression of the putative accessory genes. In common, paramyxovirus genomes possess six basal genes encoding the N, P, M, F, HN (or H or G), and L proteins in this order from the 3′ terminus, which are essential for the viral life cycle. In the *Pneumovirus*, the order of F and G is reversed as shown in *parentheses*. The viruses appear to have evolved the putative accessory genes via two distinct pathways. One pathway involves encoding the genetic information into one or more alternative overlapping reading frames that are accessed transcriptionally or translationally, and the other involves the addition (*bold arrows*) of information as one or more extra genes with a new set of transcription start and stop signals before or between the basal genes. *+1G* and *+2G*, cotranscriptional insertion of 1G and 2Gs, respectively, at the specific genome locus (editing site, *dashed line* in the P gene) into the transcripts; *+0G*, no G addition. ●, 5′ cap and AA, 3′ poly A of mRNAs. The gene size is not drawn to scale. For details see the text, Table 1 and Table 2

units are highly conserved in each virus species. Nevertheless, subtle sequence differences that occur naturally in these transcription signals can affect the efficiency of transcription or reinitiation at the individual gene junctions and thus can further moderate the efficiency of downstream gene expression (Kato et al. 1999). Between the stop and restart signals are the intergenic regions that are not copied into mRNAs. They are conserved as trinucleotides in some genera and in others are highly divergent in length and sequence. As an exception, the M2 and L genes of hRSV overlap by 68 nucleotides, and the start signal of the L gene is located inside of, but not after, the M2 gene (reviewed in Collins et al. 2001). After translation of these mRNAs and accumulation of the products, genome replication begins. Here, the same polymerase copies the same RNP template but now ignores the successive stop signals and reads through gene boundaries to generate a full-length antigenomic, positive-sense (+)RNP. A precondition for this read-through is thought to be the ongoing association of newly synthesized N subunits with the nascent RNA chain, which initiates at or very close to the 5′ end of the nascent chain. The (+)RNP, in turn, serves as the template for the synthesis of (−)RNP. The (−)RNP is incorporated into the progeny or used as the template for the next round of replication in the same cell. The viral components meet together at the plasma membrane, and the mature virions are formed and released by budding from the plasma membrane. The paramyxovirus life cycle is entirely cytoplasmic and has no nuclear phase.

The paramyxovirus genomes characterized by the presence of the above-described six basal genes and *cis*-acting elements appear to have already evolved considerable divergence between the two subfamilies and among the genera. Further diversification has been made by acquiring a variety of additional genes that are largely unique to each genus (Fig. 1). Historically, these additional genes were detected as a virus-specific product expressed in infected cells and/or as an ORF in the viral genome nucleotide sequences. They have been regarded as accessory proteins, which may not always be essential for the viral life cycle. However, the question of how the accessory genes contribute to actual viral replication remained unanswered. It was not even established whether they are dispensable or indispensable for the viral life cycle. These questions central to the *Paramyxoviridae* could be addressed only by reverse genetics of the full-length viral genome because the definition of a viral accessory gene is that its ORF can be disrupted without destroying viral infectivity and because the phenotypes of the resulting gene knockout viruses in host cells and organisms provide crucial clues to establishing the roles of the gene in question. The technology to recover infectious virus entirely from cDNA and to allow plasmid-based reverse genetics has come into wide use for the *Paramyxoviridae* (reviewed in Nagai 1999; Nagai and Kato 1999), highlighting the putative accessory genes as the focus of active investigation.

The (−) and (+)RNP complexes, not the naked RNA, are functional as the template for transcription and antigenome synthesis and the template for genome synthesis, respectively. Such a functional RNP must be generated to initiate the viral life cycle from cDNA. Thus the plasmid encoding the full-length viral genome or antigenome is transfected to a cell and the N, P and L proteins must be supplied *in trans* in the cell. These supporting proteins are usually expressed from the respective cotransfected plasmids. All these plasmids are under the control of the bacteriophage T7 RNA polymerase promoter and are driven in cells infected with a recombinant VV expressing the T7 polymerase. Some modifications have been made, which include the use of cell transformants stably expressing the T7 polymerase, or one or more of the viral proteins (for details see the chapter by Conzelmann, this volume). At the beginning, rescue efficiency of paramyxoviruses (MeV, hRSV, SeV) as well as rhabdoviruses (RV, VSV) was generally low, being one infectious unit (IU) from approximately 10^7 transfected cells, but was remarkably high in our case of SeV (at least 1 IU per 10^5 cells), possibly because of better tuning of the various factors in the system (Kato et al. 1996 and references therein). Generally, low rescue efficiency might be largely due to CPE caused by the helper VV. Thus SeV rescue efficiency was increased further as much as 100-fold (1 IU per 10^3 cells) by using VV that had been irradiated with long-wave UV in the presence of psoralen and thus exhibited impaired replication and reduced CPE without losing much of the capacity to express the T7 polymerase (Li et al. 2000).

As demonstrated for hPIV3, rescue does not always require the N plasmid and was achieved after transfection of only two supporting plasmids encoding the P and L proteins (Hoffman and Banerjee 1997). Here, the N gene could be expressed in sufficient amounts from the T7-directed, full-length antigenomic RNA transcript, probably because its start codon represented the first AUG in the transcript. The M2-1 protein of *Pneumovirus* is a potent transcription elongator and antiterminator (see Sect. 2.5). Initially, inclusion of M2-1 plasmid as an additional support plasmid was reported to be required for hRSV rescue (Collins et al. 1995). Subsequently, the N, P, and L plasmids were found to be sufficient and M2-1 expression did not significantly increase rescue efficiency (Jin et al. 1998; see the chapter by Conzelmann, this volume). As described above, the synthesis of RNP complexes in the natural life cycle is thought to progress strictly coupled with the chain elongation under the continuous supply of de novo synthesized N proteins. Therefore, preformed genome or antigenome RNA was believed to be incapable of assembling into a functional RNP template to initiate the viral life cycle.

However, this was disproved by the success of SeV recovery after the transfection of preformed genome or antigenome RNA into cells expressing the N, P, and L proteins (Kato et al. 1996).

2
Strategies to Express the Accessory Genes

2.1
Two Principal Pathways for Expression

There are two principally distinct pathways for expressing the accessory genes (Fig. 1). One of the pathways involves encoding the genetic information in an alternative, overlapping reading frame. The alternative frame can be expressed by several distinct mechanisms, including: (a) independent ribosomal access of the frame ("ribosomal choice"), exemplified by C of *Respirovirus* and *Morbillivirus*; (b) expression of a downstream ORF by ribosomal reinitiation after translation of an upstream frame, exemplified by M2-2 of hRSV and R of MeV; and (c) transcriptional frameshift ("RNA editing") to access an internal ORF, exemplified by V of *Respirovirus*, *Avulavirus*, and *Morbillivirus*. The other principal pathway involves the insertion of one or more additional genes, each with its own set of transcription start and stop signals, before or between the six basal genes, as exemplified by the SH, M2, NS1, and NS2 genes of the *Pneumovirus* genus (Fig. 1). The two principally different pathways appear to clearly characterize the two subfamilies, *Paramyxovirinae* and *Pneumovirinae*. Of the *Paramyxovirinae*, *Rubulavirus* maps closer to the *Pneumovirinae* than others because it not only uses an alternative frame but also has inserted an additional gene (Fig. 1).

2.2
V, W, D, R, X, and I Proteins of the Subfamily *Paramyxovirinae*

Of the paramyxovirus accessory proteins, the V protein is the most widely distributed, being encoded by all five genera and TPMV in the subfamily *Paramyxovirinae*, and is expressed as one of the multiple products from the P gene (Fig. 1, Table 1). As detailed below, a fraction of mRNA transcribed from P is a faithful copy of the gene, whereas the remainder can be modified by a process called RNA editing, in which the polymerase stutters at an editing motif located midway down the gene and directs the insertion of one, two, or more G residues, which shift the reading frame and access one or more internal ORFs. This co-transcriptional RNA editing creates chimeric proteins, where the N-terminal domain encoded by the P gene is fused to that encoded by the accessed ORF.

In SeV, the type species of the *Respirovirus*, the V protein of 384 amino acids is encoded by an edited mRNA, which has an insertion of a single nontemplated G residue (+1G), whereas the P mRNA, the faithful copy (+0G) of the P gene, encodes the P protein of 568 residues (Vidal et al. 1990a,b) (Fig. 1, Table 1). The +1G editing to generate the V mRNA occurs for about 25% (Pelet et al. 1991; Kato et al. 1997b) of the transcripts at a specific P gene locus containing the editing motif 3′-UUUUUUCCC-5′ (in genome sense). The RNA polymerase likely pauses at the run of 3 Cs in this stretch, allowing for separation of the nascent transcript from the template. A shift (slippage) upstream by one nucleotide during the subsequent realignment of the transcript with the template can occur, allowing the polymerase to copy one C template again and insert a single G residue (Vidal et al. 1990a). This +1G editing shifts the reading frame to fuse the upstream P/V common ORF encoding 316 residues to the downstream −1 ORF of V-unique 68 residues (Table 1). Essentially the same strategy of one G insertion to express the V protein comprising the P/V common and V-unique regions is used by all genera except the *Rubulavirus* genus, including the *Morbillivirus* MeV (Cattaneo et al. 1989) and the *Avulavirus* NDV (Fig. 1) as well as the other members of *Respirovirus* (Table 1). Exceptions are hPIV1, which undergoes no editing and possesses a V ORF interrupted with several stop codons, and hPIV3, which possesses a closed *trans* V frame by two stop codons present between the editing site and V ORF of 89 residues (Durbin et al. 1999). Thus the V frame soon terminates after editing (Table 1). The presence of such V relics for hPIV1 and hPIV3 (Table 1) may suggest that these viruses once expressed a V-like protein.

In bPIV3 and hPIV3, +2G mRNAs are also generated and the products are called D with an extension of 127 and 133 amino acids from the editing site, respectively (Table 1). In addition, one to six Gs are added at roughly equal frequencies for bPIV3 so that mRNAs encoding all three overlapping frames are generated (Pelet et al. 1991). Insertion of two Gs also occurs for SeV, albeit much less frequently and actually for less than 5% of the SeV transcripts (Kato et al. 1997b), generating the W protein. This shifts the frame to terminate two codons later (Table 1). According to the sequence database, a similar W-like protein is potentially synthesized by many other viruses (Table 1). NDV and NiV appear to be exceptional in that their Ws may have an unusually long C-terminal extension (Table 1). As an additional complexity, two products of the P or V ORF are expressed by mechanisms that do not involve RNA editing. The R protein of MeV, which has not been identified for any other virus, has an extended P ORF and a deletion of large part of V. It is produced by ribosomal frame shifting that occurs during translation of P ORF to access the V ORF 5 codons upstream of the V termination codon (Liston and Briedis 1995) (Table 1). The X protein of SeV (Table 1), which has not been identified for any other virus, is expressed from the P ORF using the initiation codon positioned more than 1,500 nucleotides downstream of the P mRNA and presumably involves internal ribosome entry (Curran and Kolakofsky 1987).

The paramyxovirus cotranscriptional editing was first disclosed for SV5 in the *Rubulavirus* (Thomas et al. 1988). In this genus, however, it is the P mRNA that is edited, whereas the V mRNA is the faithful copy (Fig. 1, Table 1). The editing site is 3′-AAAUUCUCCC-5′, inserting two Gs. One G insertion also occurs for SV5, generating the I protein that is soon terminated like the W protein of SeV and hence essentially represents the upstream module of P (P/V common segment) (Table 1). However, because of the high stringency of the editing mechanism, a single G insertion appears to occur rarely. The I protein may thus be expressed from +4G mRNA rather than +1G mRNA (reviewed in Lamb and Kolakofsky 2001). Other rubulaviruses share the same pattern of editing (Table 1).

Whereas the N-terminal domain that is common to P and V is the least conserved in amino acid sequence among the paramyxovirus proteins, the C-terminal V-unique regions are the best conserved (Nagai 1999). Of the 15 perfectly conserved residues, 7 are cysteines. They are clustered in the extreme C terminus and bind two atoms of zinc (Liston and Briedis 1994; Paterson et al. 1995). However, the domain appears to be distinct in amino acid sequence and in the spacing of cysteine residues from any other known zinc-binding motifs (Ulane and Horvath 2003). The V protein of SV5 and other rubulaviruses as well as the *Avulavirus* NDV is a structural protein associated with the nucleocapsid and is incorporated into virions, whereas it is a nonstructural protein absent in the virions of respiro- and morbilliviruses (reviewed in Lamb and Kolakofsky 2001).

2.3 C Proteins of the *Paramyxovirinae*

The C protein is expressed by three genera (*Respiro-*, *Morbilli-*, and *Henipavirus*) and TPMV of the *Paramyxovirinae* from a separate ORF that overlaps the upstream half of the P/V common ORF in the +1 frame and is upstream of the editing site and hence undisturbed and present in both edited and unedited mRNAs (Fig. 1). C is expressed as a single or multiple C-coterminal species, depending on the virus, because of the use of alternative translational start sites. SeV C ORF produces a nested set of four C-coterminal proteins, C′, C, Y1, and Y2, initiating at a non-AUG codon, ACG/81, and AUGs/114, 183, 201, respectively (Curren and Kolakofsky 1988, 1989) (Table 1, also see Fig. 2A). These are collectively called C proteins. Among them, the 204-residue-long C is the major spe-

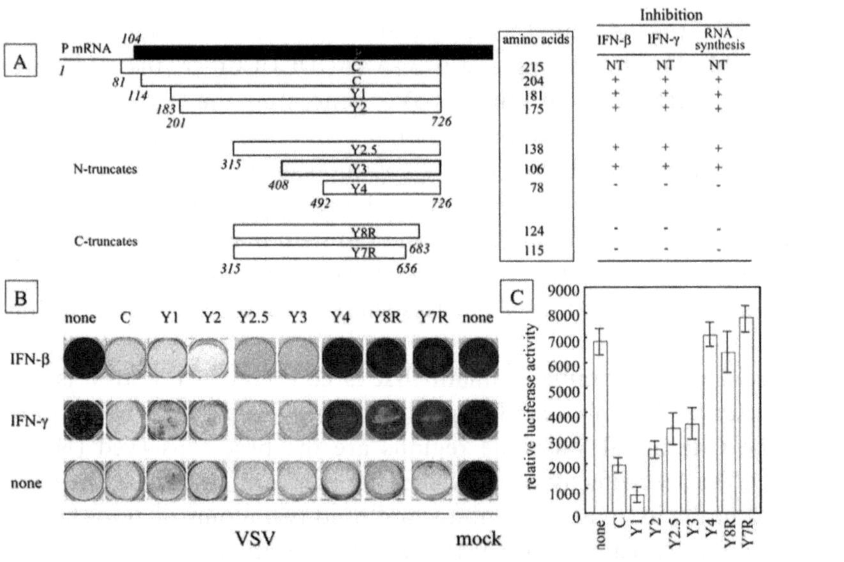

Fig. 2 Naturally occurring C proteins and artificial truncates constitutively expressed in HeLa cells (**A**) and their capacity to antagonize the anti-VSV action induced by IFN-β and IFN-γ (**B**) and to inhibit viral RNA synthesis from an SeV minigenome (**C**). Cells that survived challenge infection with VSV or were mock infected and were attached to the plates were fixed and stained (**B**). The expression of the reporter luciferase gene from the mini-genome was measured (**C**). *NT* in **A**, not tested; *none* in **B** and **C**, no IFN addition or no C protein expression. (Adapted from Kato et al. 2001, 2002 with permission)

cies that is expressed in infected cells at a molar ratio severalfold higher than the other three. For hPIV1 three C proteins (C′, C, and Y1) have been identified (Table 1) (Power et al. 1992). All the other viruses have been thought to express a single C protein. However, there are numerous methionine codons in their C ORFs. These predict, for instance, a possible second C protein of 157 residues, 156 residues, and 136 residues for hPIV3, bPIV3, and CDV, respectively. In view of the importance of the C-terminal half (106 residues) for the multifunctional SeV C protein of 204 residues (see Sects. 3.4.1 and 3.4.2), it should be determined experimentally whether or not those predicted additional C proteins are expressed in cells. HeV and TPMV contain another ORF positioned midway along the C ORF, which putatively encodes a short basic (SB) protein (Wang et al. 1998; Tidona et al. 1999), but this small ORF is not conserved in NiV (Harcount et al. 2000) (Table 1). An analogous SB protein ORF was identified in the P/V ORF of NDV (Lamb and Kolakofsky 2001) and is listed in Table 1, although NDV does not encode the C protein.

The SeV C proteins are expressed abundantly in infected cells but are present only in trace amounts in the virions and thus are essentially nonstructural proteins. Paramyxovirus C protein homology is considerably high within each genus and quite low between different genera. In common, the C proteins are relatively small (containing some 150–200 amino acids) (Table 1) and highly basic, with an isoelectric point of around 10 (reviewed in Nagai 1999).

2.4
SH Proteins of the Genus *Rubulavirus* and the Subfamily *Pneumovirinae*

Two rubulaviruses, SV5 and MuV, have a small gene located between the F and HN genes, which is expressed as a separate mRNA encoding a small hydrophobic (SH) protein (Hiebert et al. 1985; Takeuchi et al. 1991) (Fig. 1, Table 1). The SV5 SH protein is a 44-residue-long type II integral membrane protein. It is expressed at the cell surface and incorporated into the virions (reviewed in Lamb and Kolakofsky 2001). A gene encoding a comparable SH protein is found in members of the *Pneumovirinae* (Table 2): In the *Pneumovirus* genus SH is inserted between the M and G genes, and in the *Metapneumovirus* genus SH is inserted, together with the M2 gene, between the F and G genes (Fig. 1). The hRSV SH protein is also a type II membrane protein expressed at the cell surface and packaged into virions. It is present in infected cells in four biochemically different forms: SHo contains 65 amino acids and is an abundant, full-length, nonglycosylated species; SHg additionally contains a single N-glycan ; SHp has a high-mannose type N-glycan further modified by adding polylactosaminoglycan ; and SHt is a species derived from a second initiation codon of the ORF and nonglycosylated (reviewed in Collins et al. 2001). The metapneumovirus SH proteins are considerably longer than those of pneumoviruses as well as SV5 and MuV (Tables 1, 2).

2.5
NS1, NS2, M2-1, and M2-2 Proteins of the *Pneumovirus*

The nonstructural (NS) proteins 1 and 2 are unique to the members of the *Pneumovirus* genus, and no counterpart is found in other paramyxoviruses including the *Metapneumovirus* (Fig. 1, Table 2). They are each expressed as a separate mRNA from genes that are mapped to the very 3′ terminus of the genome upstream of the N gene, which is always localized in the closest proximity to the 3′ end in other paramyxoviruses (Fig. 1). However, the assignment "nonstructural" is provisional because the virions cannot be sufficiently purified to conclusively define whether they are structural or nonstructural (Collins et al. 2001).

The M2 gene of the *Pneumovirus* hRSV is located between the F and L genes (Fig. 1) and is expressed as a separate mRNA with two overlapping ORFs. The M2-1 ORF is 5′ proximal and encodes a protein that is critically required for the RSV life cycle. The M2-1 protein is required for full processivity of transcription and in addition acts as a transcription antitermination factor that enhances read-through of intergenic junctions (Fearns and Collins 1999; Hardy and Wertz 1998; Hardy et al. 1999). Intracellularly, M2-1 appears to be associated with RNP (Garcia-Barreno et al. 1996). The second, downstream ORF encodes the M2-2 protein, which appears to be a nonessential regulatory factor for RNA synthesis (Bermingham and Collins 1999; Jin et al. 2000). The M2-1 and M2-2 ORFs slightly overlap each other (Fig. 1). After completion of the translation of M2-1, the ribosomes likely reinitiate at any of the three closely spaced AUG codons that are localized just before the M2-1 termination codon (reviewed in Collins et al. 2001). A comparable M2 gene with comparable sizes of M2-1 and M2-2 ORFs is mapped, together with the downstream SH gene, between the F and G genes of the *Metapneu-*

movirus hMPV and aMPV, according to the sequence database (Fig. 1, Table 2).

3
Strong Requirement of Accessory Proteins for the Overall Fitness of SeV

3.1
SeV-Mouse System as an Experimental Paradigm

Viral accessory genes are generally dispensable for viral replication under standard cell culture conditions. Their functions thus are often brought into relief in the context of in vivo growth and pathogenesis in the susceptible host organisms. SeV, also called hemagglutinating virus of Japan (HVJ), causes fatal pneumonia in mice, although its natural ecology is full of enigma (see Sect. 8). Thus the outcomes of SeV reverse genetics can be assessed with relative ease in laboratory mice. In contrast, in vivo studies of most other paramyxoviruses are less easy because their natural hosts are often humans and large animals and because mouse or small animal models do not satisfactorily mimic natural infection and disease. Thus the SeV-mouse system represents a good experimental paradigm for carrying out reverse genetics to understand the roles of putative accessory genes in viral replication and pathogenesis in vivo.

3.2
Disruption of the V and C ORFs Without Destroying Viral Infectivity

Two nucleotide changes were introduced into the 6Us-3Cs editing motif without affecting the P ORF in the plasmid pSeV(+) that generates full-length SeV antigenome, and an infectious virus, named V(−), was recovered, which undergoes no 1G or 2Gs insertion and expresses no V or W protein (Kato et al. 1997a). Remarkably, the V(−) mutant was as high as the parental wild-type SeV in efficiency of recovery from cDNA and as active as the wild type in replication in various cell lines and embryonated chicken eggs. These results unequivocally demonstrated for the first time that the predicted editing motif indeed directs correct editing and that the V protein (as well as the W protein) is an accessory protein, which is almost completely dispensable for replication in vitro and in ovo. Alternatively, the V-unique C-terminal region was deleted by introducing stop codons just downstream of the editing site, generating VΔC virus, which displayed a normal efficiency of editing and expressed, instead of the V protein, a W-like protein in great excess. VΔC was as active as the wild type or V(−) in vitro and in ovo (Kato et al. 1997b; Delenda et al. 1997).

In many but not all of the cell types tested, V(−) replicated even faster and to higher titers than the wild type, with intracellular accumulation of higher levels of genome, antigenome, transcripts, and the final gene products (Kato et al. 1997a,b). These unique phenotypes of V(−) are compatible with the notion that SeV V protein downmodulates RNA replication (Curran et al. 1991). This downmodulation appeared to be caused by both the V and W proteins, probably through binding to unassembled free N protein, preventing its interaction with P protein and making the N protein unavailable for encapsidation (Horikami et al. 1996). Consistent with this, VΔC expressing the W-like molecules in excess did not display the augmented in vitro phenotypes seen for V(−) (Kato 1997b). Thus the augmented phenotypes of V(−) can simply be attributable to the resulting lack of V or W protein to compete for binding with unassembled N protein. All transcripts from the P gene were the P mRNA for editing minus V(−), whereas the P mRNA accounted for ~75% for editing plus VΔC or the wild type, resulting in an increased level of P protein in cells infected with V(−) (Kato et al. 1997a,b). This increased level of P protein might also be responsible for the augmented phenotype characteristic of V(−), because the predicted coiled-coil for tetramerization, the L binding site and the N:RNA binding site, which are required for RNA polymerase function, were all mapped to the P-unique, C-terminal 40% of the P protein but not to the P/V common region (reviewed in Lamb and Kolakofsky 2001).

It was extremely difficult for us to silence the expression of all four C proteins, C′, C, Y1, and Y2, by disrupting the individual initiation codons and/or introducing stop codons just downstream of the initiation codons without affecting the P ORF. We therefore initially thought that at least one of the four C proteins might be essential for the SeV life cycle. However, after repeated attempts in a system finely tuned to maximize rescue efficiency, we succeeded in generating a strongly attenuated but still viable all four C knockout clone, 4C(−), unequivocally establishing that SeV C proteins also fall in the category of nonessential accessory gene products (Kurotani et al. 1998). The difficulty was obviously due to

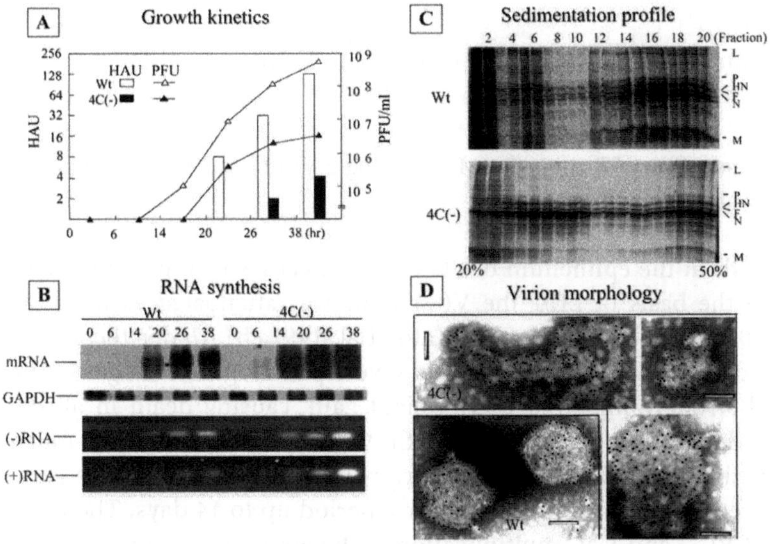

Fig. 3 Growth kinetics (**A**) and RNA synthesis (**B**) of 4C(−) and the wild type (wt) of SeV virus in CV 1 cells and the sedimentation profiles through a 20%–50% sucrose gradient (**C**) and morphology (immunoelectron microscopy; *bar*, 100 nm) (**D**) of the progeny virions. *GAPDH*, glyceraldehyde-3-phosphate dehydrogenase. Viral particles in each fraction of the sucrose gradient were detected as viral proteins resolved on an SDS-PAGE. (Adapted from Hasan et al. 2000 with permission)

a strong reduction of infectivity by nearly four logs in ovo, probably closest to the limit of virus rescue even in our fine-tuned system. The replication of 4C(−) in tissue culture cells was also strongly impaired (Hasan et al. 2000) (see Fig. 3A). It was relatively easy to generate C′ and C double knockout, C/C′(−) viruses and other partial C knockout viruses because they replicated better and reached higher titers than 4C(−) (Kurotani et al. 1998). The C/C′(−) viruses overexpress the Y1 and Y2 proteins, probably because of translation upregulation by the stop codons placed close to the initiation codons for the Y1 and Y2 ORFs to silence C′ and C expression (Kurotani et al. 1998 and references therein). Despite such overproduction of Y1 and Y2, the infectivities of C/C′(−) viruses were still reduced by about two logs in ovo compared with the wild type, indicating that the shorter Y1 and Y2 cannot fully compensate for the loss of the longer C′ and C (Kurotani et al. 1998).

3.3 The V Protein Plays a Role in Pathogenesis In Vivo

3.3.1 Requirement of the V Protein for Maintaining a Viral Load High Enough to Cause Fatal Pneumonia in Mice

SeV spreads at the epithelium of the lung and causes fatal pneumonia in mice. On the basis of LD_{50} the V(−) virus was attenuated as much as about 160-fold, and VΔC 25-fold (Kato et al. 1997a,b). Under the conditions used, the wild-type SeV induced severe pathological changes of the lung and strongly impaired body weight gain, causing death in all the mice by day 10 p.i. In contrast, both V(−) and VΔC induced much milder pathological changes, only marginally affected body weight gain, and allowed all the mice to survive for a period up to 14 days, The main feature of V(−) and VΔC multiplication in the lung was that they grew as efficiently as the wild type in the initial 1 or 2 days of infection but then were rapidly cleared, in contrast to the wild type, which maintained a high viral load throughout. Thus, although almost completely dispensable in vitro and in ovo, the SeV V protein appeared to encode a luxury function specialized for maintaining a viral load high enough to induce fatal pneumonia in mice. Natural killer (NK) cells and IFNs are generally important for host defense early in viral infections. However, rapid eradication of V(−) and VΔC virus after vigorous growth was still clearly seen in beige mice defective in NK response and IFN-α/β receptor knockout mice (unpublished data). Thus the reason why V mutants are rapidly eradicated from the body remains to be elucidated.

Delenda et al. (1998) also generated an editing minus (V edit-minus) mutant analogous to our V(−) and observed its significantly reduced pathogenicity for mice, being compatible with our V(−) phenotype. However, because this V edit-minus virus was created by deleting six Us at the editing site and hence caused two lysines to be deleted from the P protein, it is unfortunately difficult to attribute its phenotypes solely to abolishing editing. The VΔC viruses generated by Delenda et al. (1997) and by ourselves (Kato et al. 1997b) were directly compared, reaching the conclusion that they were as pathogenic as the wild type (Delenda et al. 1998). In this study, however, the loss of initial body weight (~20 g) of 4- to 5 week-old mice was the only indicator of pathogenicity. To discriminate VΔC from the wild type, it probably would be necessary to use younger mice with a body weight of ~10 g and to compare not only body weight but also the viral loads in the lung, degree of lung lesion, and LD_{50} (Kato et al. 1997a,b).

3.3.2
Importance of the Conserved Residues in the V Protein for Zinc Binding and Pathogenicity

The attenuation of not only V(−) but also VΔC suggested the importance of the V-unique C-terminal region for SeV pathogenesis. This 68-residue-long C-terminal region is characterized by the presence of 15 residues that are perfectly conserved in all the other paramyxovirus V-unique segments so far sequenced. They are clustered in subdomains I, II, and III (Tidona et al. 1999; Fukuhara et al. 2002). Domain I comprises four consecutive charged residues just downstream of the editing site. Domain II comprises six residues and contains two of the conserved cysteine residues. Only the fifth residue is variable in this domain. The 13-residue-long domain III possesses the remaining 5 cysteines and a proline. Recombinant SeVs with single, double, or triple point mutations in each of the three domains were all capable of full replication in tissue culture cells but were attenuated in mice to a degree that was less than V(−) but similar to VΔC (Huang et al. 2000; Fukuhara et al. 2002). When the mutated C-terminal peptides were fused to glutathione-*S*-transferase and expressed in *Escherichia coli*, each displayed more or less impaired zinc-binding capacity. These data indicate that all three domains play critical roles in zinc binding and in causing fatal pneumonia in mice. It is noteworthy that mutations of noncysteine residues in domain I and those cysteines in domains II and III were equally effective in impairing zinc binding, suggesting that full zinc binding is highly dependent on the conformation of the entire V-unique segment. In addition, the nucleotides encoding the domain I just downstream of the editing site appeared to affect the efficiency of editing itself (Fukuhara et al. 2002).

3.4
Versatility of the C Proteins

Not only 4C(−) but also C/C′(−) viruses were totally incapable of growing productively in the mouse lung and thus were virtually nonpathogenic (Kurotani et al. 1998). Other phenotypes revealed by our C/

C′(−) and other partial C knockout viruses so far have been intermediate between those of the two extremes, 4C(−) and the wild type, being not very informative. Thus the versatility of SeV C proteins has been revealed mainly by comparisons between the wild-type and 4C(−) viruses.

3.4.1
Capacity to Antagonize the Antiviral Action of IFNs

Studies made in the 1960s showed that preinfection or persistent infection of cells with SeV or hPIV3 enhanced the growth of other IFN-sensitive viruses such as NDV and VSV, suggesting that SeV and hPIV3 somehow render cells unresponsive to autocrine IFN (Hermodsson 1963; Maeno et al. 1966; Valle and Cantell 1965). It was further noted that SV5 infection rendered the cells unresponsive to exogenously added IFN (unpublished experiments cited in Choppin and Compans, 1975). Didcock et al. (1999a) were the first to demonstrate that the paramyxoviruses (SeV and SV5) encode the capacity to block IFN signaling, opening up a new field of IFN antagonism by paramyxoviruses (reviewed in Goodbourn et al. 2000; Gotoh et al. 2001, 2002).

SeV infection rendered both human and murine cells unresponsive to IFN-α/β (Didcock et al; 1999a; Garcin et al. 1999; Gotoh et al. 1999). Infection of HeLa cells with the wild-type or V(−) virus was able to circumvent the IFN-α-mediated anti-VSV state, allowing full replication of VSV in the presence of IFN-α. In contrast, cells infected with 4C(−) virus remained fully responsive to IFN-α like mock-infected cells, exhibiting a strong anti-VSV state (Gotoh et al. 1999). Even in the absence of exogenously added IFN-α, VSV replication was restricted when cells were preinfected with 4C(−). This indicates that 4C(−)-infected cells are highly sensitive to the feedback of autocrine IFN induced by 4C(−) itself. In good agreement with these differences in biological phenotypes between the viruses, induction of ISG products such as STAT1, STAT2, IRF-9 (p48), and PKR was suppressed in either wild-type or V(−) infected cells, but not in 4C(−)-infected cells. These results strongly suggested that the SeV C proteins inhibit IFN-α signaling and abrogate the anti-VSV state induced by IFN-α. This IFN antagonism was seen when SeV infection preceded IFN treatment by at least 1 h and more preferably by several hours but was not appreciable when IFN was added before or simultaneously with SeV infection (Komatsu et al. 2000; Gotoh et al. 2002).

To decide whether all four C proteins are necessary or whether one or some of them are sufficient for IFN antagonism, and whether the IFN antagonism requires an additional viral gene product(s), we established HeLa cell lines constitutively expressing the various C versions and examined them for anti-IFN capacity (Kato et al. 2001). It was found that the 175-residue-long Y2, the smallest of the four C proteins, is fully capable of circumventing activation of ISGs and induction of anti-VSV state by both type I (β) and type II (γ) IFNs (Fig. 2B). Also with stable transformants, various truncates from the N and C termini of Y2 were tested for IFN antagonism, revealing that a relatively large N-terminal truncation close to the middle of the Y2 was tolerated, whereas a small deletion from the carboxyl end resulted in the loss of IFN antagonism (Fig. 2A, B) (Kato et al. 2002). The region responsible was thus narrowed down to the 106-residue-long C-terminal half named Y3 (Fig. 2A).

IFN signaling is mediated through the JAK-STAT pathways (Aaronson and Horvath 2002, also see the chapter by García-Sastre, this volume). Binding of IFN-α/β to the cell surface type I IFN receptor activates the Janus kinases, JAK1 and Tyk2, which in turn phosphorylate STAT1 and STAT2 at tyrosine 701 and 690, respectively. The tyrosine-phosphorylated (pY) STAT1 and STAT2 then form a heterodimer, migrate into the nucleus, and combine with 1RF-9 to form the 1SGF3 complex; this complex targets ISRE to activate the transcription of ISGs. IFN-γ signaling involves the activation of type II IFN receptor-associated JAK1 and JAK2. The pY-STAT1s then formed become dimerized to form the GAF complex, which migrates into the nucleus and eventually targets the GAS elements.

When the extracts were prepared under low-salt conditions from HeLa cells treated with IFN-γ at 16–20 h after infection with SeV and fractionated, both unphosphorylated STAT1 and pY-STAT1 were found in aberrant high-molecular-weight complexes (HMWCs) of over 200 MDa (Takeuchi et al. 2001). Such STAT1-HMWCs were not induced in 4C(−) infected cells but were always found when the C protein, but none of the other viral proteins, was expressed from recombinant vaccinia viruses. The C protein itself was also present in the HMWCs. In all situations in which HMWCs were not formed, pY-STAT1 was present in an activated homodimer form, whereas no such dimer was seen in situations in which HMWCs were present. In the *Rubulavirus* SV5 that does not have the C protein, the V protein encodes the anti-IFN capacity (Didcock et al. 1999b). The SV5 V protein also targets STAT1 and induces proteasome-mediated degradation of STAT1 (see Sect. 6.2.1). In contrast, the SeV C protein did not cause any degradation of STAT1 in a variety of human and mouse cell lines including HeLa, U118, 2fTGH, HEC1B, and BF (Garcin et al. 1999; Gotoh et al. 1999; Komatsu et al. 2000; Komatsu et al. 2002; Takeuchi et al. 2001; Young et al. 2000). Moreover, SeV-infected HeLa cell extracts were capable of converting the pY-STAT1 homodimer to pY-STAT1-HMWCs, and this activity was no longer seen after removal of the C proteins from the extracts. In addition, both STAT1 and pY-STAT1 could be coprecipitated with C proteins by using anti-C antibodies. These molecular interactions captured in test tubes suggest that the SeV C proteins physically associate with STAT1 and pY-STAT1 and somehow interrupt IFN signaling.

This C-STAT1 interaction could result in the suppression of phosphorylation of both STAT1 and STAT2 in response to short-term (~30 min) IFN-α stimulation, which was seen early (~2 h p.i.) in infection (Komatsu et al. 2000). However longer-term (6 h or more) exposure to IFN-α of SeV-infected cells caused a dramatic increase of pY-STAT1 level (Komatsu et al. 2002; Saito et al. 2002). This increase of pY-STAT1 was in part IFN independent and SeV dependent but was largely due to longer-term exposure to IFN-α (Komatsu et al. 2002). In any event, inhibition of STAT1 activation cannot fully account for the strong inhibition of IFN-α signaling by the SeV C protein. With a panel of SeV C mutant proteins, a strong correlation was demonstrated between the STAT1 binding capacity, the inhibition of IFN-α signaling, and the inhibition of activation of STAT2 but not STAT1 (Gotoh et al. 2003). STAT2 activation thus appears to be a crucial target in the inhibition of IFN-α signaling by the SeV C protein. The C protein may participate in this inhibitory process via the interaction with STAT1.

Garcin and coworkers also assigned anti-IFN capacity of SeV to the C proteins based on the results obtained with C mutant viruses and transiently expressed C proteins with various mutations. Remarkably, only a single point mutation of phenylalanine 170 to serine (F170S) (within the 106-residue-long C-terminal half), which was discovered as one of the two point mutations (the other was in the L protein) associated with attenuation of a highly virulent SeV isolate from a mouse through adaptation to growth in LLC-MK2 cells (Itoh et al. 1997), caused complete ablation of IFN-β antagonism (Garcin et al. 1999). At the same time, however, the authors emphasized the specific importance of the 114AUG-initiated C protein relative to the smaller Y1 and Y2, from the observations that the C knockout virus still expressing the Y1 and Y2 failed to abrogate IFN-β-stimulated induction of STAT1 and to counteract the anti-VSV action of the IFN. On the other hand, however, they indicated that the Y1 and Y2 could be as effective as the longer C′ and C proteins at halting IFN-α/β signaling to ISRE, when expressed either independent of or within a recombinant SeV (Garcin et al. 2000, 2001). Therefore, the authors suggested that IFN-α/β signaling to induce an anti-VSV state can occur independently of the well-established JAK/STAT/ISGF3 pathway and that it is this parallel pathway that is targeted by the longer C proteins (Garcin et al. 2001). It was further reported that the larger C′ and C but not shorter the Y1 and Y2 could induce mono- but not polyubiquitination and degradation of STAT1 at least in a mouse fibroblast line, MEF, and human 2fTGH cells (Garcin et al. 2002, 2003). As described above, STAT1 degradation by SeV in the latter cells was not observed by others (Young et al. 2000). Thus, regarding IFN antagonism by SeV, the experimental data actually obtained and the mechanisms proposed by Garcin and coworkers appear to differ considerably from those described above. Under these circumstances it may be misleading to generalize that the AUG114-initiated C protein is specifically required to interfere with IFN signaling (Lamb and Kolakofsky 2001). However, as the AUG114-initiated C is expressed at a molar ratio several fold higher than C′, Y1, and Y2 in the viral life cycle, its actual contribution is greater than that of the other three in natural infection, even though the four different C proteins are equivalent in their capacity to block IFN signaling and induction of antiviral state.

3.4.2
Effects on Viral RNA Synthesis

The SeV C protein was initially recognized as a downmodulator of viral mRNA synthesis and genome amplification using synthetic mini-genomes as templates (Curran et al. 1992; Cadd et al. 1996). Therefore, to rescue SeV from cDNA, the C ORF had to be disrupted in the plasmid supplying the P protein (Garcin et al. 1995). The inhibition of genome amplification was stronger on an internally deleted defective interfering (DI) genome than on a copy-back DI genome, suggesting some but not absolute promoter selectivity (Cadd et al. 1996; Tapparel et al. 1997), and could be attributable, at least in part, to the capacity of the C protein to bind to the L polymerase (Horikami et al. 1997).

RNA synthesis inhibition by the C protein was supported by the finding that the intracellular accumulation of viral messenger and genome RNAs late in infection is greater with 4C(−) than with the wild-type SeV (Fig. 3B) (Hasan et al. 2000). The HeLa cell lines constitutively expressing the various C proteins individually were equally active in inhibiting SeV RNA synthesis from both a mini-genome (Fig. 2C) and a full-length genome, indicating that Y2, the smallest of the 4 C proteins, is sufficient for viral RNA synthesis inhibition (Kato et al. 2001). This is in contrast to the earlier notion that RNA synthesis inhibition is attributable to C′, C, or both but not to Y1 and/or Y2 (Curran et al. 1992) but agrees with the observations that little increased RNA synthesis late in infection was caused by C/C′(−) viruses that overexpressed Y1 and Y2 (Kurotani et al. 1998; Latorre 1998).

The truncation from the N and C termini of Y2 indicated that the Y3 representing the C-terminal 106 residues was required for RNA synthesis inhibition (Fig. 2C). Thus the requirements were similar in IFN antagonism and RNA downregulation (Fig. 2A). The C protein likely targets the viral L protein in RNA downregulation and STATs in IFN antagonism. Whereas a critical requirement of F170 of the C protein for the latter was shown (Garcin et al. 1999), the residues important for interaction with the L protein appear to be distributed throughout, many of which are within the 106-residue-long C-terminal half (Grogan and Moyer 2001). The 106-residue-long Y3 region of SeV C protein shows a striking similarity (89.5%) to the counterpart of hPIV1 but only moderate similarity (48.6%) to that of hPIV3, although the similarity of the entire C protein is 69.1% between SeV and hPIV1 and 44.1% between SeV and hPIV3. SeV Y2 strongly inhibited the growth of hPIV1 but not at all that of hPIV3 (Kato et al. 2002). These results also indicated the importance of the overall sequence of the C-terminal half for the downregulation of viral RNA synthesis.

The fact that the C/C′ double-knockout viruses sometimes displayed delays of viral RNA synthesis early in infection (Kurotani et al. 1998; Latorre et al. 1998) may suggest an accelerating function of RNA synthesis for the C and C′ but can also be explained by overproduction of the Y1 and Y2 proteins from the C/C′ double-knockout viruses (Kurotani et al. 1998) that are fully competent in RNA synthesis inhibition. Whether or not the promoter selectivity of genome replication by the C proteins is exerted on the full-length genome also remains to be defined.

3.4.3
Contribution to Viral Morphogenesis

That the central organizer in paramyxovirus morphogenesis is the M protein has been established by a variety of studies. For instance, earlier studies demonstrated that the M protein of SeV and NDV condenses sparsely distributed envelope proteins on the cell surface into a patch of virus-specific membrane, an immediate precursor of the viral envelope, cross-links the viral envelope and internal components, and drives budding (Nagai et al. 1975, 1976b; Yoshida et al. 1976, 1979; reviewed in Matsumoto 1982). Further support was added by recent reverse genetics that involved deleting the M protein from MeV (Cathomen et al. 1998) and SeV (Inoue et al. 2003) or introducing point mutations into the SeV M protein (Sakaguchi et al. 2002). In addition to these viral protein-protein interactions, the assembly pathway involves viral interactions with cellular proteins, cellular membranes, and membrane microdomains called rafts (for review see Chazal and Gerlier 2003). The SeV C proteins have been believed to have nothing to do with virus assembly, because they are essentially nonstructural components that are expressed abundantly in infected cells but are present in trace amounts in mature virions (see Sect. 2.3). Thus we were surprised to find that in cultured cells 4C(−) virus was produced largely as noninfectious particles with a sedimentation profile distributing throughout the sucrose gradient (Fig. 3C) and with highly anomalous morphology (Fig. 3D) (Hasan et al. 2000). This suggests the contribution of the C proteins to virion assembly. How a protein that is essentially nonstructural plays this role is difficult to conceptualize.

Intracellularly, the SeV C proteins were well colocalized with the M protein as well as the glycoproteins (Hasan et al. 2000). Unexpectedly, colocalization of the basic C proteins and the acidic P protein was poor. The C proteins should associate with the RNP via the interaction with the L protein, but this association could not be visualized, probably because of a low copy number of intracellular and RNP-associated L protein. Because the M proteins are, in large part, phosphorylated intracellularly (Lamb and Choppin. 1977), this negative charge might account for their colocalization with the basic C proteins. However, recombinant SeV engineered to ablate phosphorylation of M protein replicated normally in cells, indicating that M protein phosphorylation itself has no important role but is just an epiphenomenon (Sakaguchi et al. 1997). A comparison of incorporation kinetics of intracellular viral proteins into virions in the culture supernatants between 4C(−) and wild-type infections has shown a delay of incorporation of the M and glycoproteins, but not the internal N and P proteins, in the absence of C proteins (Hasan et al. 2000). Our speculation is that the SeV C protein may interact with and convert the M protein as well as envelope proteins from putative assembly-incompetent forms to assembly-competent forms or, alternatively, may guide these proteins to a presumptive assembly pathway. The possible involvement of such a chaperone-like molecule in the assembly of enveloped viruses is reminiscent of the Vif of HIV-1, which is required during virus assembly to facilitate the formation of particles competent for the early step of infection but is present in trace amounts in the virion itself (Trono 1995). In the morphogenesis of the *Rhabdovirus* VSV, the M protein plays a dual role. It not only assembles viral structural units at the plasma membrane but also converts loosely coiled filamentous RNP, which is transcriptionally active, to a tightly coiled, transcriptionally inactive structure, resulting in shut-off of viral RNA synthesis (reviewed in Rose and Whitt 2001). No RNA synthesis-inhibitory function is known for paramyxovirus M protein. It is tempting to speculate that the SeV C proteins may contribute to morphogenesis via inhibiting or shutting off viral RNA synthesis.

3.4.4
Suppression of Apoptosis

SeV triggers the apoptosis death cascade through activation of caspase-8 and caspase-3 in various cell lines (Bitzer et al. 1999). However, the virus was poorly capable of inducing apoptosis in some cell lines under conditions in which VSV was highly apoptotic (Koyama et al. 2001), although there may be strain-specific differences (Gotoh, B., personal communication). The attenuated SeV with the F170S mutation in the C proteins has shown a loss of IFN antagonism (see Sect. 3.4.1) and at the same time has displayed increased apoptosis (or, more precisely, a reduced antiapoptotic activity) in the airway epithelial cells of mice (Itoh et al. 1998). Now, strong perinuclear condensation and fragmentation of chromatins as well as strong CPE were manifested in Hep2 cells, which are known to be highly susceptible to apoptosis inducers, after infection with 4C(−) SeV, but no such apoptotic phenotype was appreciated for the wild-type SeV (strain Z) in the same cells during the observation period up to 50–60 h (Koyama et al. 2003). That the C protein indeed encodes antiapoptosis function was confirmed by the observation that a HeLa cell line expressing the C protein is more resistant to VSV-induced apoptosis than normal HeLa cells (unpublished data). IFN-stimulated biochemical pathways affect cell growth and metabolism including apoptosis induction (Biron and Sen 2001). IFNs elicit an antiviral state in uninfected cells, whereas they induce apoptosis in virus-infected cells (Tanaka et al. 1998). SeV infection itself induced a certain degree of apoptosis in some cells such as human embryonic kidney 293 and human Jurkat T cells, and this apoptosis was augmented by supplemental coexpression of IRF-3 (Heylbroeck et al. 2000). Thus there remain numerous issues to be resolved. These include to what degree the phenomena of IFN antagonism and apoptosis inhibition by SeV C proteins are related and whether or not SeV C proteins target IRF-3 in apoptosis inhibition. As IRF-3 is a key player not only in apoptosis induction but also IFN-α/β induction, it also must be determined whether SeV C proteins can block IFN production (see Sect. 6.2.2). In this context, it must be noted that persistent infection with SeV rendered HeLa cells not only unresponsive to IFN but also incapable of IFN production (Maeno et al. 1966). So far, however, SeV appears to have little effect on IRF-3 activation by phosphorylation and its nuclear translocation (Lin et al. 1998b; Talon et al. 2000, also see Sect. 6.2.2) in contrast to the influenza A virus NS1 protein capable of inhibiting IRF-3 activation (Talon et al. 2000).

4
Accessory Proteins of Another Respirovirus, hPIV3

The C knockout virus of hPIV3 was significantly attenuated both in tissue culture and in rodents and primates (Durbin et al. 1999). The F164S mutant (corresponding to F170S of SeV C) normally replicated in tissue culture but was attenuated in hamsters and monkeys (Durbin et al. 1999). The attenuation was greater in the upper than the lower respiratory tract. As noted above, the V ORF of hPIV3 is likely inactive because of the presence of stop codons between the editing site and V-unique cysteine-rich region (see Sect. 2.2). Interruption of the D and V ORFs individually by introducing stop codons into the respective ORFs did not affect hPIV3 replication in vitro and in vivo, but interruption of both together attenuated replication in vivo (Durbin et al. 1999). Thus the V or V-like protein may be expressed naturally by an unknown mechanism.

hPIV3 editing at a secondary downstream site or ribosomal frame shifting may occur, the latter being comparable to that used by MeV to express the R protein (see Sect. 2.2). It is important to define whether those phenotypes of C, D, and V knockout hPIV3s are caused via alteration of IFN signaling, because infection with hPIV3 renders cells unresponsive to both IFN-α and IFN-γ (Young et al. 2000).

5
V and C proteins of the *Morbillivirus* and *Henipavirus*

Both the V(−) and C(−) mutants of *Morbillivirus* MeV strain Edmonston that were engineered by reverse genetics (Radecke and Billeter 1996; Schneider et al. 1997) showed significantly fewer and milder clinical symptoms and a lower mortality in a model of MeV-induced disease of the central nervous system in YAC-CD46 transgenic mice (Patterson et al. 2000). The V(−) MeV showed delayed replication, allowing thymocyte survival in human thymus/liver engrafted into SCID mice (SCID-hu thy/liv) (Valsamakis et al. 1998) and grew to lower titers in the lung of cotton rats (Tober et al. 1998). This latter study additionally used a mutant engineered to overexpress the V protein (Schneider et al. 1999) and provided evidence suggesting that V acts as a downregulator of MeV gene expression in a human glioblastoma cell line. The C(−) MeV replicated more slowly and reached lower titers, compared with the parental virus, in human peripheral blood mononuclear cells (PBMC) (Escoffier et al. 1999) and SCID-hu thy/liv (Valsamakis et al. 1998). Either V(−) or C(−) grew well, like the parental virus in Vero cells (Radecke et al. 1996; Schneider et al. 1997). This cell line is IFN nonproducing, whereas human PBMC as well as HeLa cell line are IFN competent. The V(−) and wild-type viruses appear to be equally sensitive to type I IFN in HeLa cells and capable of producing IFN in human PBMC (data not shown in Patterson et al. 2000). It remains to be defined whether MeV C protein has the capacity to block IFN signaling and/or IFN induction and whether the above-described phenotypes of C(−) in vivo, in primary cultures, and in other cell lines are attributable to the loss of this putative anti-IFN capacity of the C protein. More recently, MeV was found to suppress IFN-α signaling, but not IFN-γ signaling, probably via association of the V and/or C proteins with the type I IFN receptor complex in the human epithelioid carcinoma lines SiHa and CaSki (Yokota et al. 2003).

Deletion of the V gene from RPV had little effect on tissue culture replication and rather resulted in an increased genome and antigenome replication and a change in cytopathic effect to a more syncytium-inducing phenotype (Baron and Bannet 2000). In contrast, deleting C expression resulted in defective growth in all the cell lines tested, which correlated with a decreased transcription. The V and C double-knockout virus was further generated and displayed the sum of these positive and negative phenotypes. RPV is the causative agent of an important disease of cattle and wild bovids. The parent for all knockout viruses of RPV described above is actually avirulent, and so it is likely that a major effect of the V(−) mutation would not be fully evident in vivo, because any change in in vivo phenotype potentially caused by the V deletion might be masked by the preexisting attenuation of the virus (Baron and Bannet 2000).

Reverse genetics is not yet available for the *Henipavirus*. However, important information suggesting the roles of accessory proteins has been obtained by conventional plasmid-based expression (Rodriguez et al. 2002). Specifically, NiV V protein was shown to inhibit cellular responses to both type I and type II IFNs by forming HMWCs with both STAT1 and STAT2. Consequently, tyrosine phosphorylation of the STAT proteins and their nuclear translocation were prevented in cells expressing the NiV V protein. Park et al. reported that not only the V protein but also the W protein (P/V commpom region) and C protein exhibited a partial (C) or nearly complete (V and W) blocking of IFN signaling in chicken fibroblasts and the primate Vero cell line (Park et al. 2003).

6
V Protein of the *Rubulavirus* and *Avulavirus* and the SV5 SH Protein

6.1
Replication Constraint Caused by V Deletion

For the *Rubulavirus* the V mRNA is the faithful transcript, whereas the P mRNA is generated by +2G editing (see Sect. 2.2 and Fig. 1). Rescue of a recombinant SV5 that would express only the P mRNA but not the V mRNA has been unsuccessful despite many attempts, including the supply of the V protein *in trans* from a cotransfected expression plasmid (Lin et al. 2000; He et al. 2002). Thus deletion of the C-terminal unique domain was attempted to generate VΔC recombinant viruses from the genus *Rubulavirus*. hPIV2 VΔC was first rescued after introduction of a stop codon immediately downstream of the editing site and was shown to have critically impaired infectivity in cell cultures with highly anomalous virion morphology (Kawano et al. 2001). The reduced infectivity and replication capability could only be partially attributable to the V protein function to inhibit IFN signaling (see Sect. 6.2.1) because neutralization of the endogenously induced IFN by a specific antibody added to the culture medium was only partially able to restore the replication capacity. More recently, SV5 VΔC was generated by changing the sixth and eighth codons downstream of the editing site to stop codons (He et al. 2002). This mutant also grew poorly in IFN-producing cells such as the BHK line but grew well in IFN-nonproducing Vero cells. Interestingly, however, the two stop codons introduced could not be stably maintained during several passages even in Vero cells and were converted to sense codons, although not identical to the original ones, thus giving rise to a pseudo-wild type with a full-length V protein (He et al. 2002). This strongly suggests that deletion of the V-unique region is deleterious to SV5 replication and that there is a strong pressure to repair the truncated V to a full-length V.

The *Avulavirus* NDV V mRNA is generated by +1G editing (Fig. 1). The NDV mutants lacking either six of the eight nucleotides at the editing site or the V-unique C-terminal region were so strongly impaired in replication capacity that they could not productively grow under routine conditions in 9- to 10-day-old chicken embryoneated eggs (Mebatsion et al. 2001). They produced as many as 5,000-fold fewer infectious progeny in cell cultures or 200,000-fold fewer in 6-day-old eggs compared with the parental wild type. However, the former mutant caused a two-amino acid deletion from the P protein, and thus its phenotypes could not be attributable solely to the lack of editing. Thus a second mutant was prepared that contained a single nucleotide change introduced into the editing site. In this virus, editing was dramatically reduced (from ~30% to ~1%) but was not completely abolished, and the virus exhibited a 100-fold reduction of infectivity in 10-day-old eggs (Mebatsion et al. 2001). Thus the results obtained by NDV reverse genetics did not yet answer whether the full-length V protein can be deleted without completely destroying infectivity. After all, the requirement for the V protein appears to be very strong for infectivity of the members of *Rubulavirus* and *Avulavirus* and more marginal in importance for the *Respirovirus* and *Morbillivirus*. This difference may be related to the fact that the former V proteins are structural whereas the latter are nonstructural (see Sect. 2.2).

The interactions of SV5 V protein with various cellular proteins, including DDB1, xeroderma pigmentosum group E binding factor (XPE-BF), and the hepatitis B virus X-associated protein 1 (XAP-1), have been demonstrated (Lin et al. 1998). These interactions appear to be via the C-terminal V-unique region or to require both the P/V common and V-unique modules, because these interactions do not occur with the P protein. Their biological significance, however, had remained largely unclear except that the interaction with DDB1 may slow progression of the cell cycle with a delayed transition from the G_1 to the S phase and prolonged progression through the S phase (Lin et al. 2000) but now has become greater because of the discovery of DDB1 involvement in IFN antagonism (see Sect. 6.2.1).

6.2 Impairment of the IFN system and Suppression of Apoptosis by the *Rubulavirus* V Protein

Nearly four decades ago, Choppin (1964) reported that SV5 continued to multiply in rhesus monkey kidney cells to high titers with minimal CPE for periods up to 30 days. The SV5-infected cells supported the full growth of a variety of superinfecting viruses including VSV that were highly sensitive to IFN. The author's notion that SV5 could inhibit both the production and action of IFN (Choppin and Compans 1975) was highly foresighted, as has recently been confirmed directly.

6.2.1 Inhibition of IFN Signaling Through Proteasome-Mediated Degradation of STAT Proteins

In the *Rubulavirus* genus, which does not encode the C protein, it is the V protein that blocks IFN signaling, and this blockade appears to involve proteasome-mediated degradation of STATs. This was originally demonstrated by plasmid-based V expression systems for SV5, which targets STAT1 for degradation (Didcock et al. 1999a,b), and then for others, including MuV and SV41 targeting STAT1 and hPIV2 targeting STAT2 (Andrejeva et al. 2002a; Kubota et al. 2001; Nishio et al. 2001; Parisien et al. 2001; Young et al. 2000), and were confirmed in the context of viral replication by the lack of such anti-IFN capacity for hPIV2 VΔC and SV5 VΔC (Kawano et al. 2001; He et al, 2002). More recently, the *Avulavirus* NDV V joined the group of IFN antagonists (Park et al. 2003). Evidence that STAT1 degradation by SV5 or MuV is mediated via the proteasome was provided by demonstrating that the proteasome inhibitors MG132 and lactacystin blocked the degradation process (Didcock et al. 1999b; Andrejeva et al. 2000b; Yokosawa et al. 2002). However, the effect of the inhibitors on MuV- and SV41-induced STAT1 reduction did not always appear to be very clear (Kubota et al. 2001; Nishio et al. 2001), suggesting that degradation might be independent of the proteasome. Selective degradation of STAT1 by SV5 requires STAT2 and that of STAT2 by hPIV2 requires STAT1, but these occur independently of IFN-α/β signaling (Pariesen et al. 2002b).

Detailed biochemical analyses have been made to understand the process of IFN signaling block and STAT 1 degradation by SV5. The SV5 V protein likely induces the formation of a large multiprotein complex in which V, STAT1, STAT2, and DDB1 are essential components, although the complex has not yet been isolated and characterized. The presumptive complex formation then leads to selective degradation of STAT1 in the proteasome. Evidence supporting the involvement of DDB1 is as follows (Andrejeva et al. 2002a): (a) the V protein of an SV5 mutant (CPI−), which fails to target STAT1 for degradation because of three point mutations (tyrosine to histidine at position 26 and leucine to proline at 50 and 102) in the P/V common region compared with CPI+ isolate capable of STAT1 degradation (Chatziandreou et al. 2002), does not bind to DDB1; (b) alanine substitutions for the cysteine residues in the V-unique region as well as deletions from both the N and C termini of V, which abolish the binding of V to DDB1, always prevent V from blocking IFN signaling; and (c) treatment of HeLa cells constitutively expressing the SV5 V protein with short interfering (si) RNAs specific for DDB1 reduces DDB1 levels and concomitantly increases STAT1 levels with a restoration of IFN signaling. DDB1 is known to bind to various viral and cellular components including cullin 4A, which, as part of ubiquitin E3 ligases, plays an essential role in targeting proteins for degradation (reviewed in Pickart 2001). It has been hypothesized that the complex formed then binds, via DDB1, to a ubiquitin E3 ligase complex, resulting in STAT1 ubiquitination and degradation (Andrejeva et al. 2002a). Because the hPIV2 V protein also binds to DDB1 (Lin et al. 1998), SV5 and hPIV2 may share the same DDB1-dependent STATs degradation mechanism.

Direct evidence for polyubiquitination of STAT1 and STAT2 by the V proteins of SV5 and hPIV2, respectively, in transfected cells was provided by Ulane and Horvath (2003). Furthermore, the authors demonstrated that both of the V proteins possess intrinsic ubiquitin ligase activity that defines their function as E3 enzymes and that the V-interaction partners include STAT1, STAT2, DDB1, and cullin 4A. It was also shown that interference with not only DDB1 but also cullin 4A by the specific si RNAs reduced the efficiency of STAT1 targeting by SV5 infection (Ulane and Horvath 2003).

The above studies on the SV5 V and DDB1 interaction using various V mutants suggested the importance of both the P/V common and V-unique regions for the interaction. Thus the inability of hPIV2 VΔC to block IFN signaling (Kawano et al. 2001) could be due to the failure of this V mutant to bind to DDB1. The selectivity of degradation (either STAT1 or STAT2 depending on the virus) appears to determine the selectivity of inhibition (either both type I and type II IFN signaling or only the type I IFN signaling depending on the virus). Then, how the selectivity of degradation is determined remains to be elucidated.

The MuV V protein also binds to DDB1 (Lin et al. 1998). The V-unique region alone of MuV was sufficient to abolish the ISG expression and antiviral state induced by IFN-α and IFN-γ (Kubota et al. 2001). If MuV shares the same anti-IFN mechanism with SV5, its V-unique C-terminal half is sufficient for interacting with DDB1. Alternatively, MuV may use an as yet unknown strategy for exhibiting IFN antagonism. Indeed, a more recent study has dissected a novel aspect in IFN signaling block by MuV V protein. The study employed two-hybrid screening using the V-unique region as a bait and other conventional biochemical methods and revealed that the MuV V protein has a strong affinity to RACK1 (Kubota et al. 2002). RACK1 is a 36-kDa protein, acts as an adaptor to mediate the interaction between the human IFN-α/β receptor and STAT1, and thus is essential for recruitment of STAT1 in preparation for its activation on IFN treatment (Crooze et al. 2000; Mochly-Rosen et al. 1991; Ron et al. 1994; Usacheva et al. 2001). The affinity of the MuV V protein to RACK1 was stronger than that of STAT1. Thus the association of RACK1 and STAT1 as well as that of IFN-α receptor β-subunit (IFN-αRβ) and RACK1 was greatly disturbed by the MuV V protein at least in vitro (Kubota et al. 2002). These results suggest that the MuV V protein inhibits IFN-α/β receptor-RACK1-STAT1 interaction and thereby interrupts IFN signaling at the very upstream of the pathway. It is not clear how close or how far this presumable inhibition mechanism and STAT1 degradation are related to each other. In MuV-infected or V plasmid-transfected cells, the V protein level would be sufficiently high to play a dual role, disruption of STAT1-RACK1-IFN receptor interaction and targeting STAT1 for proteasome-mediated degradation. Although it may not be an essential prerequisite for the degradation, polyubiquitination was clearly shown for STAT1 in cells expressing the MuV V protein (Yokosawa et al. 2002).

An additional remarkable issue regarding IFN antagonism by rubulaviruses is that the SV5 V protein causes degradation of STAT1 in human cells but not in murine cells. This species specificity is determined by a single amino acid difference (asparagine or aspartic acid) at position 100 in the P/V common region (Young et al. 2001). STAT2 has been proven to be the host determinant for this species specificity (Parieson et al. 2002a).

6.2.2
Inhibition of IFN Production and Suppression of Apoptosis

SV5 VΔC not only failed to cause STAT1 degradation and block IFN signaling but also induced a large amount of IFN-β in HeLa cells (He et al. 2002). The amount of IFN-β induced by SV5 VΔC reached as high as over 700 pg/ml under conditions in which the wild-type SV5 induced hardly any detectable IFN-β; the actual level of IFN-β induced by the wild type was less than 1% of that induced by VΔC, or in other words, inhibition by the wild type of IFN-β production was greater than 99%. In addition, VΔC infection was highly apoptotic, whereas the wild type induced little apoptosis (He et al. 2002). Thus the V-unique C terminal region or the entire V was suggested to encode strong anti-IFN-inducing capacity and antiapoptosis capacity.

Induction of IFN-β by viral infection as well as by dsRNA such as poly IC involves the activation of certain transcription factors. Of these, NF-κB and members of the IRF family, such as IRF-3, are thought to be particularly important (reviewed in Taniguchi et al. 2001). Studies using reporter assays with deletions of the respective sequences targeted by the transcription factors demonstrated activation of NF-κB and IRF-3 by SV5 VΔC but not by the wild-type SV5 (Poole et al. 2002). That the

SV5V protein has a specific function that directly blocks IFN-β production was confirmed by inhibited induction of IFN-β promoter on poly IC treatment of cells transiently expressing the SV5 V protein (Poole et al. 2002). On viral infection or exposure to poly IC the cytoplasmic IRF-3 is phosphorylated, dimerized, and translocated to the nucleus to act as the transcriptional activator for IRF genes. This nuclear translocation was prevented by wild-type SV5 infection or transient V expression but not by SV5 VΔC infection or by expression of the P protein, which shares the N-terminal 162 residues with the V protein. All these findings could be related to the observed lack of antiapoptosis capacity of SV5 VΔC (He et al 2002).

Deletion of the N-terminal 125 residues from the SV5 V protein had no effect on inhibition of IFN production, whereas the C-terminal 48 residues are required for inhibition. CPI- had no effect on inhibition of IFN production, either (Poole et al. 2002). In addition, the level required for 50% inhibition of IFN production appears to be as much as two orders of magnitude higher, on the basis of transfected amounts of V-expressing plasmid, than that necessary for IFN signaling block to abrogate the antiviral state. Thus the inhibition of IFN signaling and the inhibition of IFN production are mechanistically distinct from each other.

Targeting of STAT1 for degradation by the SV5 V protein is species-specific, as described above, the mouse STAT1 being resistant. Thus mice homologous for a targeted disruption of STAT1 ($STAT^{-/-}$) served as an animal model for SV5 pathogenicity studies (He et al. 2002a). SV5 VΔC was found to be attenuated in this model, consistent with the notion of clearance of apoptotic cells in vivo (He et al. 2002).

Remarkably, the V proteins of not only hPIV2 V but also SeV expressed from the respective plasmids were able to inhibit IFN-β promoter activation by poly IC to essentially the same extent as seen in SV5 V protein expression (Poole et al. 2002). This may be related to the earlier observation that persistent SeV infection rendered the cells incapable of IFN production (Maeno et al. 1966). However, when the wild-type SeV and V(−) or VΔC SeV were compared for IFN-inducing capacity in mouse L cells and mouse lung, the wild type did produce IFN in amounts clearly detectable as anti VSV state induction to the levels only about threefold lower than those reached by either mutant (Kiyotani et al., unpublished data). This is in contrast with the virtually all (by VΔC) or none (less than 1% by the wild type) IFN production in SV5 infection (see above). These results suggest the existence of additional factors encoded by the SV5 V protein but not SeV V protein, which limits IFN induction, when expressed in the context of whole virus infection or that generation of much greater amounts of intracellular IFN inducers occurs during the course of SV5 VΔC infection than in SeV V(−) or VΔC infection. In addition, activation and nuclear translocation of IRF-3 appeared to occur normally in SeV-infected cells (Lin et al. 1998) or were not inhibited under conditions in which activation was blocked by the influenza A virus NS1 protein (Talon et al. 2000) (see Sect. 3.4.4).

6.3 SV5 SH Protein

An SH-deleted SV5 (SV5ΔSH) was generated by reverse genetics and found to be very similar to the parental virus in all aspects tested in tissue culture cells, including growth rate, infectivity, and plaque size as well as virion morphology (He et al. 1998). In MDBK cells, SV5 continues reproduction for long periods up to 40 days without causing strong CPE. This remarkable feature of SV5 was explained by the antiapoptotic function of the SH protein for this particular cell line (He et al. 2001). SV5ΔSH appeared to induce cell death by activating TNF-α expression (Lin et al. 2003). SV5ΔSH caused less mortality than the wild type in *STAT1*$^{-/-}$ mice, again consistent with the notion of clearance of apoptotic cells in vivo. Thus two antiapoptotic proteins, V and SH, have been identified for SV5. It is slightly puzzling that deletion of only one of these caused such remarkable in vitro and in vivo phenotypes (also see Sect. 6.2.2). They may target different cell species (He et al. 2001, 2002) or may not be individually fully competent yet and possibly act in concert both in vitro and in vivo.

7 Specific Roles of the NS1, NS2, SH, M2-1, and M2-2 Proteins of the *Pneumovirus* and Their Cooperation

The attempt to delete the M2-1 gene from pneumoviruses has been unsuccessful, and thus it has not been established whether M2-1 represents an essential or nonessential gene product. The other putative accessory genes encoding NS1, NS2, SH, and M2-2 proteins can be deleted individually from hRSV and/or bRSV without destroying infectivity, generating ΔNSI, ΔNS2, ΔSH, and ΔM2-2 viruses, respectively (Bukreyev et al.

1997; Whitehead et al. 1999; Buchholz et al. 1999; Teng and Collins 1999; Jin 2000a,b; Schlender et al. 2000). Of these deletion mutants, ΔSH appeared to be least attenuated in HEp2 cells, whereas the greatest attenuation was found for ΔM2-2 (Jin et al. 2000b). ΔNS1 and ΔNS2 were in between. M2-2 is thought to act as a regulator during RNA synthesis, possibly by mediating the switch from transcription to genome replication to facilitate virus morphogenesis (Bermingham and Collins, 1999). Indeed, ΔM2-2 exhibited severely reduced levels of viral genomic and antigenomic RNAs (Jin et al. 2000a).

It was possible to delete simultaneously two and sometimes three genes from hRSV, generating ΔNS1/NS2, ΔM2-2/NS2, ΔM2-2/SH, ΔSH/NS1, ΔSH/NS2 and ΔSH/NS1/NS2 (Jin et al. 2000b). ΔM2-2/NS2 was still more attenuated than ΔM2-2, and ΔM2-2/NS1 could not be rescued so far, again suggesting greater contribution of M2-2 to the viral life cycle, compared with SH, NS1, or NS2. The varying degrees of tissue-culture replication were not seen as clearly in an IFN-incompetent Vero cell line as in IFN-competent human HEp-2 line. Indeed, bRSV NS1 and NS2 were found to cooperatively antagonize IFN-α/β-induced antiviral response (Schlender et al. 2000). The hRSV NS1 and NS2 could only partially substitute for the bRSV counterparts, as a chimeric bRSV with NS genes from hRSV was attenuated in IFN-competent bovine cells. Yet it was much less attenuated than the NS1/NS2 double-deletion mutant (Bossert and Conzelmann 2002). Thus the *Pneumovirus* NS1 and NS2 proteins appear to represent major determinants of viral host range. The impaired replication of the various hRSV deletion mutants in HEp-2 cells further suggests an antiapoptotic role for one or more of the RSV accessory proteins because of high sensitivity of this cell line to apoptosis inducers.

hRSV targets the lower respiratory tract to produce pneumonia in humans. ΔSH is not attenuated in the upper respiratory tract but is moderately attenuated in the lower respiratory tract of chimpanzees, a host that is permissive to hRSV infection (Whitehead et al. 1999). All other deletion mutants described above exhibited varying degrees of attenuation in the lower respiratory tract of cotton rats (Jin et al. 2000b). In this latter study, ΔSH/NS1/NS2 was clearly more attenuated than ΔSH or ΔNS1/NS2 and attenuation in vivo did not always correlate with attenuation in HEp-2 cells. The requirement for M2-2 was stronger than the other three both in vivo and in vitro.

The individual accessory proteins of pneumoviruses thus appear to contribute specifically and additively to the viral life cycle and pathogenesis.

8
Conclusions and Perspectives

Nucleotide sequencing of the entire genomes of the *Paramyxoviridae* was completed in the 1980s for many classic members. One of the important issues revealed by sequencing was identification of ORFs encoding putative accessory proteins. These were found to be conserved in newly emergent agents. The establishment of plasmid-based reverse genetics was given the highest priority in the postsequencing era and has been achieved for most but not all prototypic paramyxoviruses. As illustrated here by SeV and several other members, the technology has been applied successfully not only to define unequivocally that most of the putative accessory genes are indeed accessory and can be deleted without destroying infectivity, but also to demonstrate their strong requirement for the overall viral fitness both in vitro and in vivo.

The failure to delete the V protein from *Rubulavirus* and *Avulavirus*, or the M2-1 protein from *Pneumovirus* presumes but does not verify that these proteins are indispensable because the rescue efficiency might not be high enough to recover a critically impaired but still viable progeny. In addition, several important pathogens such as henipaviruses and metapneumoviruses await the development of reverse genetics technology. Under these situations, conventional plasmid-based expression has dissected many novel aspects regarding the roles of paramyxovirus accessory genes, as best exemplified by the SV5 V protein. Reverse genetics and plasmid-based expression thus are highly complementary with each other.

Thus far, individual paramyxoviruses appear to have acquired in common an accessory gene that encodes the capacity to block IFN signaling that induces an antiviral state. At least some of the accessory genes were further demonstrated to encode the capacity to block IFN production. These results suggest that of the various aspects of innate and adaptive immunity generally important for host defense against viral infections, antiviral IFN action is of prime importance as the front line to halt virus spread from the site of virus entry and that paramyxovirus evolution would have given the highest priority to the acquisition of the capacity to evade the IFN system. The notion that the induction of apoptosis by a virus is another limiting factor for the virus spreading in vivo was strongly supported by the demonstration of antiapoptosis function for various paramyxovirus accessory genes. The paramyxovirus accessory genes are now regarded as useful tools to study the cellular metabolism and signaling as well as host defense in vivo.

Apparently the same or similar functions, such as IFN antagonism or antiapoptosis activity, are encoded by the C protein in *Respirovirus* SeV and by the V protein in *Rubulavirus* SV5. On the other hand, *Henipavirus* NiV appears to encode anti-IFN capacity in both the V and C proteins (Park et al. 2003) and a similar possibility remains for *Morbillivirus* MeV (Yokota et al. 2003) (see Sect. 5). These differences in coding strategy of "accessory" functions may be related to the fact that *Henipavirus* and *Morbillivirus* are phylogenetically close to each other and map between the *Respirovirus* and *Rubulavirus* (Lamb and Kolakofsky 2001). That the W protein (P/V common region) of NiV was as active as the full-length V in counteracting antiviral action of IFN is surprising (Park et al. 2003) because the V-unique region is not always sufficient but essential to carry out this function in all rubulaviruses so far tested and NDV (see Sect. 6.2.1). This may be related to the facts that paramyxovirus P/V common (W) regions are extremely variable and, in addition, that those of henipaviruses are unusually long (Table 1), and it raises the possibility that even the P protein can block IFN response in henipaviruses. The *Filovirus* Ebola virus P protein (VP35) is itself sufficient to inhibit IFN production (Basler et al. 2000). The ancestral P gene of *Mononegavirales* would have been given considerable options in the process of acquiring IFN antagonism; encoding it in the P ORF itself, in the alternative V or C ORF, or in both.

To address gene functions by reverse genetics in the context of in vivo pathogenesis, the origin of the viral genome incorporated into the plasmid must be taken into consideration. For instance, the wild-type SeV genome backbone used here was derived from a laboratory strain, which was attenuated significantly for mice because of numerous rounds of passage in an unnatural host, the chick embryo (Kiyotani et al. 2001). The findings generated by its reverse genetics are of considerable relevance because the virus still retains remarkable pathogenicity for mice but may not directly be applicable to natural SeV disease. MeV isolates into marmoset B cell lines such as B95a reproduce human measles in monkeys, whereas the prototypic strain Edmonston or the isolates into

Vero cells and Vero cell-adapted strains fail to do so (Kobune et al. 1990). Thus it is the former but not the latter that represents naturally circulating MeV (reviewed in Nagai 1999 and Yanagi et al. 2002). Furthermore, differences in receptor usage have been demonstrated between the B95 isolates and the Edmonston strain (Tatsuo et al. 2000). Additional genetic and biological data have discriminated between the B95a isolates and their Vero cell-adapted forms or the Edmonston strain itself, part of which appeared to be relevant to accessory genes (Takeda et al. 1998, 1999). Nevertheless, reverse genetics studies have been limited to those based on the Edmonston strain. Many other paramyxoviruses, including RPV and CDV, for whose isolation the use of B95a cells is also recommended (Kobune et al. 1991; Kai et al. 1993), may be in a similar situation in that their cDNAs for virus rescue have originated from something other than field strains. An avian/chicken pathogen, NDV, may be an exception because it has been maintained in the same species (chick embryo). Indeed, the in vitro correlate of in vivo pathogenicity (Nagai et al. 1976; reviewed in Nagai 1993) was verified to be a real pathogenicity determinant by reverse genetics (Peeters et al. 1999). Similarly, hRSV does not appear to have become attenuated by passage in vitro and the wild-type hRSV recovered from cDNA is virulent in chimpanzees. In view of these considerations the establishment of reverse genetics in the genome backbones of an SeV wild-strain highly virulent for mice (Fujii et al. 2002) and an MeV B95 isolate capable of reproducing measles in monkeys (Takeda 2000) will be of considerable value.

However, an additional complexity exists for SeV. We have encountered SeV-induced pneumonia only for mice in animal experimental facilities but not in nature. It is not known whether some natural reservoirs existed that transmitted SeV to laboratory mice or the mice had been chronically infected with SeV before their introduction to laboratories. The latter possibility has been suggested; even such an apparently marginal stress as dropping several tens of microliters of normal allantoic fluid of chick embryo (and perhaps even saline) into the nose could have activated the persistently infecting SeV to spread in the lung to cause fatal pneumonia (Nishikawa 1997). If so, it is not always easy for us to define what "SeV field isolates" actually are nor to assess the feasibility of using an acute infection protocol of young mice with either a laboratory strain or a fresh isolate. Indeed, the great contribution of the V protein to pathogenesis in such young mice (see Sect. 3.3.1) was not seen (or was masked) in an infection model using the recombinant viruses engineered in the backbone that had been derived from a highly virulent field isolate (Sakaguchi et al. 2003). The roles of SeV accessory genes thus should be clarified in relation to a broader scenario including the maintenance of the virus in nature and the mechanisms of persistent infection and its activation.

Reverse genetics of paramyxoviruses and other negative-strand RNA viruses has opened the possibility of using these viruses as a novel class of expression vectors for the delivery and expression of therapeutic genes and vaccine antigen genes (reviewed in the chapter by von Messling and Cattaneo, this volume; also see Yonemitsu et al. 2000 for SeV). Introduction of predetermined attenuating mutations into the wild-type strain is a straightforward approach to designing a live attenuated vaccine. In this context, paramyxovirus accessory genes represent promising targets for mutation because their deletion often resulted in attenuation in vivo without significantly impairing the ability of the viruses to replicate in vitro, which is critical for vaccine production. Moreover, deletion of some accessory genes such as the SeV V protein resulted in augmented expression of a foreign gene inserted into the virus (Yu et al 1997) and thus is advantageous for production of the foreign gene product in greater amounts in vitro and in vivo. Studies on the accessory genes, therefore, are also of importance from a technological point of view.

Acknowledgements. We thank H.-D. Klenk, P. Collins, K.-C. Conzelmann, R. E. Randall, B. Gotoh, and T. Sakaguchi for critical reading of the manuscript and valuable suggestions and discussions and R. A. Lamb for providing preprints. We also thank I. Higashiyama for preparing the manuscript. The work conducted by the authors was supported by research grants from the Ministry of Education, Sports, Culture, Science, and Technology, and from the Bio-oriented Technology Research Advancement Institution (BRAIN), Japan. Finally, we would like to dedicate this paper to Dr. Rudolf Rott (1926–2003) and to state that his demise leaves a large scientific and personal void, but his memory will remain with all of us in the research field of negative-strand RNA viruses.

References

Aaronson DS, and Horvath CM (2002) A road map for those who don't know JAK-STAT. Science 296, 1653–1655

Andrejeva J, Poole E, Young DF, Goodbourn S, and Randall RE (2002a) The p127 subunit (DDB1) of the UV-DNA damage repair binding protein is essential for the targeted degradation of STAT1 by the V protein of the paramyxovirus simian virus 5. J. Virol. 76, 11379–11386

Andrejeva J, Young DF, Goobourn S, and Randall RE (2002b) Degradation of STAT1 and STAT2 by the V proteins of simian virus 5 and human parainfluenza virus type 2, respectively: consequences for virus replication in the presence of alpha/ beta and gamma interferons. J. Virol. 76, 2159–2167

Baron MD, and Barrett T (2000) Rindsepest viruses lacking the C and V proteins show specific defects in growth and transcription of viral RNAs. J. Virol. 74, 2603–2611

Basler CF, Wang X, Mühlberger E, Vulchkov V, Paragas J, Klenk HD, Garcia-Sastre A, and Palese P (2000) The Ebola virus VP35 protein functions as a type 1 IFN antagonist. Proc. Natl. Acad. Sci. USA 97, 12289–12294

Bermingham A, and Collins PL (1999) The M2-2 protein of human respiratory syncytial virus is a regulatory factor involved in the balance between RNA replication and transcription. Proc. Natl. Acad. Sci. USA 96, 11259–11264

Biron CA, and Sen GC (2001) Interferons and other cytokines. In Fields Virology 4th ed. pp. 321–351. Lippincott Williams & Wilkins, Philadelphia

Bitzer M, Prinz F, Bauer M, Spiegel M, Neubert WJ, Gregor M, Schulze-Osthoff K, and Lauer U (1999) Sendai virus infection induces apoptosis through activation of caspase-8 (FLICE) and caspase-3 (CPP32). J. Virol. 73, 702–708

Bossert B, and Conzelmann KK (2002) Respiratory syncytial virus (RSV) nonstructural (NS) proteins as host range determinants: a chimeric bovine RSV with NS genes from human RSV is attenuated in interferon-competent bovine cells. J. Virol. 76, 4287–4293

Buchholz UJ, Finke S, and Conzelmann, KL (1999) Generation of bovine respiratory syncytial virus (BRSV) from cDNA: BRSV NS2 is not essential for virus replication in acts as a functional BRSV genome promoter. J. Virol. 73, 251–259

Burkreyev A, Whitehead SS, Murphy BR, and Collins PL (1997) Recombinant respiratory syncytial virus from which the SH gene has been deleted grows efficiently in cell culture and exhibits site specific attenuation in the respiratory tract of the mouse. J. Vitrol. 71, 8973–8982

Cadd T, Garcin D, Tapparel C, Itoh M, Homma M, Roux L, Curren J, and Kolakofsky D (1996) The Sendai paramyxovirus accessory C proteins inhibit viral genome amplification in a promoter-specific fashion. J. Virol. 70, 5067–5074

Cathomen T, Mrkic B, Spehner D, Drillien R, Naef R, Pavlovic J, Aguzzi A, Billeter MA, and Cattaneo R (1998) A matrix- less measles virus is infectious and elicits extensive cell fusion: consequences for propagation in the brain. EMBO J. 17, 3899–3908

Cattaneo R, Kaelin K, Baczko K, and Billeter MA (1989) Measles virus editing provides an additional cysteine-rich protein. Cell 56, 759–764

Chanock RM, Miuphy BR, and Collins PL (2001) Parainfluenza viruses. In Fields Virology 4th ed. pp. 1341–1379. Lippincott Williams & Wilkins, Philadelphia

Chatziandreu N, Young D, Andrejeva J, Goodbourn S, and Randall RE (2002) Differences in interferon sensitivity and biological properties of two related isolates of simian virus 5: a model for viral persistence. Virology 293, 234–242

Chazal N, and Gerlier D. (2003). Virus entry, assembly, budding, and membrane rafts. Microbiol Mol Biol Rev. 67, 226–237

Choppin PW (1964) Multiplication of a myxovirus (SV5) with minimal cytopathic effects and without interference. Virology 23, 224–233

Choppin PW, and Compans RW (1975) Reproduction of paramyxoviruses. In: H Fraenkel-Conrat, and RR Wagner, eds. Comprehensive Virology vol. 4, pp. 95–179 Plenum Press, New York

Collins PL, Hill MG., Camargo E, Grosfeld H, Chanock RM, and Murphy BR (1995) Production of infectious human respiratory syncytial virus from cloned cDNA confirms an essential role for the transcription elongation factor from the 5' proximal open reading frame of the M2 mRNA in gene expression and provides a capability for vaccine development. Proc. Natl. Acad. Sci. USA 92, 11563–11567

Collins PL, Chanock RM, and Murphy, BR (2001) Respiratory syncytial virus. In Fields Virology 4th ed. pp. 1443-1485. Lippincott Williams & Wilkins, Philadelphia

Crooze E, Usacheva A, Asarnow D, Minshall RD, Perez HD, and Colamonici O (2000) Receptor for activated C-kinase (RACK-1), a WD motif containing protein, specifically associates with the human type I IFN receptor. J. Immunol. 165, 5127–5132

Curran J, and Kolakofsky D (1987) Identification of an additional Sendai virus non-structural protein encoded by the P/C mRNA. J. Gen. Virol. 68, 2515–2519

Curran J, and Kolakofsky D (1988) Ribosomal initiation from an ACG codon in the Sendai virus P/C mRNA. EMBO J. 7, 245–251

Curran J, and Kolakofsky D (1989) Scanning independent ribosomal initiation of the Sendai virus Y proteins in vitro and in vivo. EMBO J. 8, 521–526

Curran J, Boeck R, and Kolakofsky D (1991) The Sendai virus P gene expresses both an essential protein and an inhibitor of RNA synthesis by shuffling modules via mRNA editing. EMBO J. 10, 3079–3085

Curran J, Marq JB, and Kolakofsky D (1992) The Sendai virus nonstructural C proteins specifically inhibit viral mRNA synthesis. Virology 189, 647–656

Delenda C, Hausmann S, Garcin D, and Kolakofsky D (1997) Normal cellular replication of Sendai virus without the trans-frame, nonstructural V protein. Virology 228, 55–62

Delenda C, Taylor G, Hausmann S, Garcin D, and Kolakofsky D (1998) Sendai viruses with altered P, V, and W protein expression. Virology 242, 327–337

Didcock L, Young DF, Goodbourn S, and Randall RE (1999a) Sendai virus and simian virus 5 block activation of interferon-responsive genes: importance for viral pathogenesis. J. Virol. 73, 3125–3133

Didcock L, Young DF, Goodbourn S, and Randall RE (1999b) The V protein of simian virus 5 inhibits interferon signaling by targeting STAT1 for proteasome-mediated degradation. J. Virol. 73, 9928–9933

Durbin AP, McAuliffe JM, Collins PL, and Murphy BR (1999) Mutations in the C, D, and V open reading frames of human parainfluenza virus type 3 attenuate replication in rodents and primates. Virology 261, 319–330

Escoffier C, Manié S, Vincent S, Muller CP, Billeter M, and Gerlier D (1999) Non-structural C protein is required for efficient measles virus replication in human peripheral blood cells. J. Virol. 73, 1695–1698

Fearns R, Collins PL (1999) Role of the M2-1 transcription anti-termination protein of respiratory syncytial virus in sequential transcription. J. Virol. 73, 5852–5864

Fujii Y, Sakaguchi T, Kiyotani K, Huang C, Fukuhara N, Egi Y. and Yoshida T (2002) Involvement of the leader sequence in Sendai virus pathogenesis revealed by recovery of pathogenic field isolate from cDNA. J. Virol. 76, 8540–8547

Fukuhara N, Huang C, Kiyotani K, Yoshida T, and Sakaguchi T (2002) Mutational analysis of the Sendai virus V protein: importance of the conserved residues for Zn binding, virus pathogenesis and efficient RNA editing. Virology 299, 172–178

Garcia-Barreno B, Delgado T, and Melero JA (1996) Identification of protein regions involved in the interaction of human respiratory syncytial virus phosphoprotein and nucleoprotein: Significance for nucleocapsid assembly and formation of cytoplasmic inclusions. J. Virol. 70, 801–808

Garcin D, Pelet T, Calain P, Roux L, Curran J, and Kolakofsky D (1995) A highly recombinogenic system for the recovery of infectious Sendai paramyxovirus from cDNA: generation of a novel copy-back nondefective interfering virus. EMBO J. 14, 6087–6094

Garcin D, Latore P, and Kolakofsky D (1999) Sendai virus C proteins counteract the interferon-mediated induction of an antiviral state. J. Virol. 73, 6559–6565

Garcin D, Curran J, and Korakofsky D (2000) Sendai virus C proteins must interact directly with cellular components to interfere with interferon action. J. Virol. 74, 8823–8830

Garcin D, Curran JM, Itoh D, and Kolakofsky (2001) Longer and shorter forms of Sendai virus C proteins play different roles in modulating the cellular antiviral response. J. Virol. 75, 6800–6807

Garcin D, Marq JB, Stahle L, Le Mercier P. and Kolakofsky D (2002) All four Sendai Virus C proteins bind STAT1, but only the larger forms also induce its monoubiquitination and degradation. Virology 295, 256–265

Garcin D, Marq JB, Goodbourn S, and Kolakofsky D (2003) The amino-terminal extensions of the longer Sendai virus C proteins modulate pY701-Stat1 and bulk Stat1 levels independently of interferon signaling. J.Virol. 77, 2321–2329

Goodbourn S, Didcock L, and Randall RE (2000) Interferons: Cell signaling, immune modulation, antiviral response and virus countermeasures. J. Gen. Virol. 81, 2341–2364

Gotoh B, Takeuchi K, Komatsu T, Yokoo, J, Kimura Y, Kato A, Kurotani, and Nagai Y (1999) Knockout of the Sendai virus C genes eliminates the viral ability to prevent the interferon- α/β mediated responses. FEBS Lett. 459, 205–210

Gotoh B, Komatsu T, Takeuchi K, and Yokoo J (2001) Paramyxovirus accessory proteins as interferon antagonists. Microbiol Immunol 45, 787–800

Gotoh B, Komatsu T, Takeuchi K, and Yokoo J (2002) Paramyxovirus strategies for evading the interferon response. Rev. Med. Virol. 12, 337–357

Gotoh B, Takeuchi K, Komatsu T, and Yokoo J (2003) The STAT2 activation process is a crucial target of Sendai virus C protein for the blockade of alpha interferon signaling. J. Virol. 77, 3360–3370

Grogan CC. and Moyer SA (2001) Sendai virus wild-type and mutant C protein show a direct correlation between L polymerase binding and inhibition of viral RNA synthesis. Virology 288, 96–108

Hamaguchi M, Yoshida T, Nishikawa K, Naruse H, and Nagai Y (1983) Transcriptive complex of Newcastle disease virus. I. Both L and P proteins are required to constitute an active complex. Virology 128, 105–117

Harcourt BH, Tamin A, Ksiazek TG, Rollin PE, Anderson LJ, Bellini WJ, and Rota PA (2000) Molecular characterization of Nipah virus, a newly emergent paramyxovirus. Virology 271, 334–349

Hardy RW, and Wertz GW (1998) The product of the respiratory syncytial virus M2 gene ORF1 enhances readthrough of intergenic junctions during viral transcription. J. Virol. 72, 520–526

Hardy RW, Harman SB, and Wertz GW (1999) Diverse gene junctions of respiratory syncytial virus modulate the efficiency of transcription termination and respond differently to M2-mediated antitermination. J. Virol. 73, 170–176

Hasan Mk, Kato A, Muranaka, M, Yamaguchi R, Sakai Y, Hatano I, Tashiro M. and Nagai Y (2000) Versatility of the accessory C proteins of Sendai virus. Contribution to virus assembly as an additional role. J. Virol. 74, 5619–5628

He B Leser, GP Paterson RG, and Lamb RA (1998) The paramyxovirus SV5 small hydrophobic (SH) protein is not essential for virus growth in tissue culture cells. Virology 250, 30–40

He B, Lin GY, Durbin JE, Durbin RK, and Lamb RA (2001) The SH integral membrane protein of the paramyxovirus simian virus 5 is required to block apoptosis in MDBK cells. J. Virol. 75, 4068–4079

He B, Paterson RG, Stock N, Durbin JE, Durbin RK, Goodbourn S, Randall RE, and Lamb RA (2002) Recovery of paramyxovirus simian virus 5 with a V protein lacking the conserved cysteine-rich domain: the multifunctional V protein blacks both interferon-β induction and interferon signaling. Virology. 303, 15–32

Hermodsson S (1963) Inhibition of interferon by an infection with parainfluenza virus type 3 (PIV-3) Virology 20, 333–343

Heylbroeck C, Balachandran S, Servant MJ, Deluca G, Barber GN, Lin R, and Hiscott J (2000) IRF-3 transcription factor mediates Sendai virus induced apoptosis. J. Virol. 74, 3781–3792

Hiebert SW, Paterson RG, and Lamb RA (1985) Identification and predicted sequence of a previously unrecognized small hydrophobic protein, SH, of the paramyxovirus simian 5. J. Virol. 55, 744–751

Hoffman MA, Banerjee AK (1997) An infectious clone of human parainfluenza virus type 3. J. Virol. 71, 4272–4277

Horikami SM, Smallwood S, and Moyer SA (1996) The Sendai virus V protein interacts with the NP protein to regulate viral genome RNA replication. Virology 222, 383–390

Horikami SM, Hector RE, Smellwood S, and Moyer SA (1997) The Sendai virus C protein binds the L polymerase protein to inhibit vital RNA synthesis. Virology 235, 261–270

Huang C, Kiyotani K, Fujii Y, Fukuhara N, Kato A, Nagai Y, Yoshida T, and Sakaguchi T (2000) Involvement of the zinc-binding capacity of Sendai virus V protein in viral pathogenesis. J. Virol. 74, 7834–7841

Inoue M, Tokusumi Y, Ban H, Kanaya T, Shirakura M, Tokusumi T, Hirata T, Nagai Y, Iida A, and Hasegawa M (2003) A new type of Sendai virus vector deficient in the matrix gene has lost virus particle formation and gained extensive cell-to-cell spreading. J. Viral. 77, 6419–6429

Itoh M, Isegawa Y, Hotta H, and Homma M (1997) Isolation of an avirulent mutant of Sendai virus with two amino acid mutations from a highly virulent field strain through adaptation to LLC-MK2 cells. J. Gen. Virol. 78, 3207–3215

Itoh M, Hotta H, and Homma M (1998) Increased induction of apoptosis by a Sendai virus mutant is associated with attenuation of mouse pathogenicity. J. Virol. 72, 2977–2934

Jin H, Clarke D, Zhou HZ, Cheng X, Coelingh K, Bryant M, Li S (1998) Recombinant human respiratory syncytial virus (RSV) from cDNA and construction of subgroup A and B chimeric RSV. Virology 251, 206–214

Jin H, Cheng X, Zhou HZ, Li S, and Seddiqui A (2000a) Respiratory syncytial virus that lacks open reading frame 2 of the M2 gene (M2-2) has altered growth characteristics, and is attenuated in rodents. J. Virol. 74, 74–82

Jin H, Zhou H, Cheng X, Tang R, Munoz M, and Nguyen N (2000b) Recombinant respiratory syncytial virus with deletions in the NS1, NS2, SH, and M2-2 genes are attenuated in vitro and in vivo. Virology 273, 210–218

Kai C, Ochikubo F, Okita M, Linuma T, Mikami T, Kobune F, and Yamanouchi K (1993) Use of B95a cells for isolation of canine distemper virus form clinical cases. J. Vet. Med. Sci. 55, 1067–1070

Kato A, Sakai Y, Shioda T, Kondo T, Nakanishi M, and Nagai Y (1996) Initiation of Sendai virus multiplication from transfected cDNA or RNA with negative or positive sense. Genes Cells 1, 569–579

Kato A, Kiyotani K, Sakai Y, Yoshida T, and Nagai Y (1997a) The paramyxovirus, Sendai virus, V protein encodes a luxury function required for viral pathogenesis. EMBO J. 16, 578–587

Kato A, Kiyotani K, Sakai Y, Yoshida T, Shioda T, and Nagai Y (1997b) Importance of the cysteine-rich carboxyl-terminal half of V protein for Sendai virus pathogenesis. J. Virol. 71, 7266–7272

Kato A, Kiyotani K, Hasan MK, Shioda T, Sakai Y, Yoshida T, and Nagai Y (1999) Sendai virus gene start signals are not equivalent in reinitiation capacity: Moderation at the F gene. J. Virol. 73, 9237–9246

Kato A, Ohnishi Y, Kohase M, Saito S, Tasgiro M, and Nagai Y (2001) Y2, the smallest of Sendai virus C proteins is fully capable of counteracting the anti-viral action of interferons and inhibiting viral RNA synthesis. J. Virol. 75, 3802–3810

Kato A, Ohnishi Y, Hishiyama M, Saito S, Tashiro M. and Nagai Y (2002) The amino-terminal half of Sendai virus C protein is not responsible for either counteracting the antiviral action of interferons or down-regulating viral RNA synthesis. J. Virol. 76, 7114–7124

Kawano M, Kaito M, Kozuka Y, Komada H, Noda N, Nanba K, Tsurudome M, Ito M, Nishio M, and Ito Y (2001) Recovery of infectious human parainfluenza type 2 virus from cDNA clones and properties of the defective virus without V-specific cysteine-rich domain. Virology 284, 99–112

Kiyotani K, Sakaguchi T, Fujii Y, and Yoshida T (2001) Attenuation of a field Sendai virus isolate through egg-passages is associated with an impediment of viral genome replication in mouse respiratory cells. Arch. Virol 146, 893–908

Kobune F, Sakata H, and Sugiura A (1990) Marmoset lymphoblastoid cells as a sensitive host for isolation of measles virus. J. Virol. 64, 700–705

Kobune F, Sakata H, Sugiyama M, and Sugiura A (1991) B95a, a marmoset lymphoblastoid cell line, as a sensitive host for rinderpest virus. J. Gen. Virol. 72, 687–692

Komatsu T, Takeuchi K, Yokoo J, and Gotoh B (2002) Sendai virus C protein impairs both phosphorylation and dephosphorylation processes of Stat1. FEBS Lett. 511, 139–144

Komatsu T, Takeuchi K, Yokoo J, Tanaka Y, and Gotoh B (2000) Sendai virus blocks alpha interferon signaling to signal transducers and activators of transcription. J. Virol. 74, 2477–2480

Koyama AH, Ogawa M, Kato A, Nagai Y. and Adachi A (2001) Lack of apoptosis in Sendai virus-infected Hep-2 cells without participation of viral antiapoptosis gene. Microbes Infect. 3, 1115–1121

Koyama AH, Irie H, Kato A, Nagai Y, and Adachi A (2003) Virus multiplication and induction of apoptosis by Sendai virus: role of the C proteins. Microbes Infect. 5, 373–378

Kubota T, Yokosawa N, Yokota S. and Fujii N (2001) C terminal Cys-rich region of mumps virus structural V protein correlates with block of interferon alpha and gamma signal transduction through decrease of STAT1-alpha. Biochem Biophys Res Commun 283, 255–259

Kubota T, Yokosawa N, Yokota S. and Fujii N (2002) Association of mumps virus V protein with RACK1 results in dissociation of STAT1 from the alpha interferon receptor complex. J. Virol. 16, 12676–12682

Kurotani A, Kiyotani K, Kato A, Shioda T, Sakai Y, Mizumoto K, Yoshida T, and Nagai Y (1998) Sendai virus C proteins are categorically nonessential gene products but silencing their expression severely impairs viral replication and pathogenesis. Genes Cells 3, 111–124

Lamb RA, and Choppin PW (1977) The synthesis of Sendai virus polypeptides in infected cells. III. Phosphorylation of polypeptides. Virology 81, 382–397

Lamb RA, and Kolakofsky D (2001) Paramyxoviridae: The viruses and their replication, pp. 1305-1340. In Fields Virology 4th ed. Lippincott Williams & Wilkins, Philadelphia

Latorre P, Cadd T, Itoh M, Curran J, and Kolakofsky D (1998) The various Sendai virus C proteins are not functionally equivalent, and exert both positive and negative effects on viral RNA accumulation during the course of infection. J. Virol. 72, 5984–5993

Li H-O, Zhu YF, Asakawa M, Kuma H, Hirata T, Ueda Y, Lee YS, Fukumura M, Iida A, Kato A, Nagai Y, and Hasegawa M (2000) A cytoplasmic RNA vector from nontransmissible Sendai virus with efficient gene transfer and expression. J. Virol 74, 6564–6569

Lin GY, Paterson RG, Richardson CD, and Lamb RA (1998a) The V protein of the paramyxovirus SV5 interacts with damage-specific DNA binding protein. Virology 249, 189–200

Lin R, Heylbroeck C, Pitha PM, and Hiscott J (1998b) Virus-dependent phosphorylation of the IRF-3 transcription factor regulates nuclear transport, transcription potential, and proteasome-mediated degradation. Mol. Cell. Biol. 18, 2986–2996

Lin GY, and Lamb RA (2000) The paramyxovirus simian virus 5 V protein slows progression of the cell cycle. J. Virol. 74, 9155–9166

Lin Y, Bright AC, Rothermel TA, and He B (2003) Induction of apoptosis by paramyxovirus simian virus 5 lacking a small hydrophobic gene. J. Virol. 77, 3371–3383

Liston P, and Briedis DJ (1994) Measles virus V protein binds zinc. Virology 198, 399–404

Liston P, and Briedis DJ (1995) Ribosomal frameshifting during translation of measles virus P protein mRNA is capable of directing synthesis of a unique protein. J. Virol. 69, 6742–6750

Maeno K, Yoshii S, Nagata I, and Mastumoto T (1966) Growth of Newcastle disease virus in a HVJ carrier culture of HeLa cells. Virology 29, 255–263

Matsumoto T (1982) Assembly of paramyxoviruses. Microbiol. Immunol. 26, 285–320

Mebatsion T, Verstegen S, De Vaan LT, Romer-Oberdorfer A, and Schrier CC (2001) A recombinant Newcastle disease virus with low-level V protein expression is immunogenic and lacks pathogenicity for chicken embryos. J. Virol. 75, 420–428

Mochly-Rosen D, Khaner H, and Lopez J (1991) Identification of intracellular receptor proteins for activated protein kinase C. Proc. Natl. Acad. Sci. USA 88, 3997–4000

Nagai Y (1993) Protease-dependent virus tropism and pathogenicity. Trends Microbiol 1, 81–87

Nagai Y (1999) Paramyxovirus replication and pathogenesis. Reverse genetics transforms understanding. Rev. Med. Virol. 9, 83–99

Nagai Y, Yoshida T, Yoshii S, Maeno K, and Matsumoto T (1975) Modification of normal cell surface by smooth membrane preparations from BHK-21 cells infected with Newcastle disease virus. Med. Microbiol. Immunol. 161, 175–188

Nagai Y, Klenk HD, and Rott R (1976a) Proteolytic cleavage of the viral glycoproteins and its significance for the virulence of Newcastle disease virus. Virology 72, 494–508

Nagai Y, Ogura H, and Klenk HD (1976b) Studies on the assembly of the envelope of Newcastle disease virus. Virology 69, 523–538

Nagai Y, and Kato A (1999) Paramyxovirus reverse genetics is coming of age. Microbial Immunol 43, 613–624

Nishikawa F (1997) On HIJ. Virus 47, 261–265 (in Japanese)

Nishio M, Tsurudome M, Ito M, Kawano M, Komada H, and Ito Y (2001) High resistance of human parainfluenza type2 virus protein-expressing cells to the antiviral and anti-cell proliferative activities of α/β interferons: cysteine-rich v- specific domain is required for high resistance to the interferons. J. Virol. 75, 9165–9176

Parisien JP, Lau JF, Rodriguez JJ, Sullivan BM, Moscona A, Parks G.D, Lamb RA, and Horvath CM (2001) The V protein of human parainfluenza virus 2 antagonizes type I interferon responses by destabilizing signal transducer and activator of transcription 2. Virology 283, 230–239

Parisien JP, Lau JF, Horvath CM (2002a) STAT2 acts as a host range determinant for species-specific paramyxovirus interferon antagonism and simian virus 5 replication. J. Virol. 76, 6435–6441

Parisien JP, Lau JF, Rodriguez JJ, Ulane CM, and Horvath CM (2002b) Selective STAT protein degradation induced by paramyxoviruses requires both STAT1 and STAT2 but is independent of α/β interferon signal transduction. J. Virol. 76, 4190–4198

Park MS, Shaw ML, Munoz-Jordan J, Cros JF, Nakaya T, Bouvier N, Palese P, Garcia-Sastre A, and Basler CF (2003) Newcastle disease virus (NDV)-based assay demonstrates interferon-antagonist activity for the NDV V protein and the Nipah virus V, W, and C proteins. J. Virol. 77, 1501–1511

Paterson R, Leser G, Shaughnessy M, and Lamb R (1995) The paramyxovirus SV5 V protein binds two atoms of zinc and is a structural component of virions. Virology 208, 121–131

Patterson JB, Thomas D, Lewicki H, Billeter MA, and Oldstone MB (2000) V and C proteins of measles virus function as virulence factors in vivo. Virology 267, 80–89

Peeters BPH, DE Leeuw OS, Koch G, and Gielkens ALJ (1999) Rescue of Newcastle disease virus from cloned cDNA: evidence that cleavability of the fusion protein is a major determinant for virulence. J. Virol. 73, 5001–5009

Pelet T, Curran J, and Kolakofsky D (1991) The P gene of bovine parainfluenza virus 3 expresses all three reading frames from a single mRNA editing site. EMBO J. 10, 443–448

Pikart CM (2001) Mechanisms underlying ubiquitination. Annu. Rev. Biochem. 70, 503–533

Poole E, He B, Lamb RA, Randal RE, and Goodbourn S (2002) The V proteins of simian virus 5 and other paramyxoviruses inhibit induction of interferon-β. Virology. 303, 33–46

Power UF, Ryan KW, and Portner A (1992) The P genes of human parainfluenza virus type 1 clinical isolates are polycistronic and microheterogeneous. Virology 189, 340–343

Radecke F, and Billeter MA (1996) The nonstructural C protein is not essential for multiplication of Edmonston B strain measles virus in cultured cells. Virology 217, 418–421

Rodriguez JJ, Parisien JP, and Horvath CM (2002) Nipah virus V protein evades alpha and gamma interferons by preventing STAT1 and STAT2 activation and nuclear accumulation. J.Virol. 76, 11476–11483

Ron D, Chen C-H, Caldwell J, Jamieson L, Orr E, and Mochly-Rosen D (1994) Cloning of an intracellular receptor for protein kinase C: a homolog of the beta subunit of G protein. Proc. Natl. Acad. Sci. USA 91, 839–843

Rose KJ, and Whitt MA (2001) Rhabdoviridae: the viruses and their replication. pp. 1221–1244. In Fields Virology 4th ed. Lippincott Williams & Wilkins, Philadelphia

Saito S, Ogino T, Miyajima N, Kato A, and Kohase M (2002) Dephosphorylation failure of tyrosine-phosphorylated STAT1 in IFN-stimulated Sendai virus C protein-expressing cells. Virology 293, 205–209

Sakaguchi T, Kiyotani K, Kato A, Asakawa M, Fujii Y, Nagai Y, and Yoshida T (1997) Phosphorylation of the Sendai virus M protein is not essential for virus replication either in vitro or in vivo. Virology 235, 360–366

Sakaguchi T, Kiyotani K, Watanabe H, Huang C, Fukuhara N, Fujii Y, Shimazu Y, Sugahara F, Nagai Y, and Yoshida T (2003) Masking of the contribution of V protein to Sendai virus pathogenesis in an infection model with a highly virulent field isolate. Virology 313, 581–587

Sakaguchi T, Uchiyama T, Huang C, Fukuhara N, Kiyotani K, Nagai Y, and Yoshida T (2002) Alteration of Sendai virus morphogenesis and nucleocapsid incorporation due to mutation of cysteine residues of the matrix protein. J. Virol. 76, 1682–1690

Schneider H, Kaelin K, and Billeter MA (1997) Recombinant measles viruses defective for RNA editing and V protein synthesis are viable in cultured cells. Virology 227, 314–322

Schlender J, Bossert B, Buchholz U, and Conzelmann KK (2000) Bovine respiratory syncytial virus nonstructural proteins NS1 and NS2 cooperatively antagonize alpha/beta interferon-induced antiviral response. J. Virol. 74 8234–8242

Takeda M, Kato A, Kobune F, Sakata H, Li Y, Shioda T, Sakai Y, Asakawa M, and Nagai Y (1998) Measles virus attenuation associated with transcriptional impediment and a few amino acid changes in the polymerase and accessory proteins. J. Virol. 72, 8690–8696

Takeda M, Kato A, Sakaguchi T, Kobune F, Li Y, and Nagai Y (1999) The genome nucleotide sequence of a contemporary wild-strain of measles virus and its comparison with the classical Edmonston strain genome. Virology 256, 340–350

Takeda M, Takeuchi K, Miyajima N, Kobune F, Ami Y, Nagata N, Suzuki Y, Nagai Y, and Tashiro M (2000) Recovery of pathogenic measles virus from cloned cDNA. J. Virol. 74, 6643–6647

Takeuchi K, Tanabayashi K, Hishiyama M, et al (1991) Variation of nucleotide sequences and transcription of the SH gene among mumps virus strains. Virology 181, 364–366

Takeuchi K, Komatsu T, Yokoo J, Kato A, Shioda T, Nagai Y, and Gotoh B (2001) Sendai virus C protein physically associates with Stat1. Genes Cells 6, 545–557

Talon J, Horvath CM, Polley R, Basler CF, Muster T, Palese P, and Garcia-Sastre A (2000) Activation of interferon regulatory factor 3 is inhibited by the influenza A virus NS1 protein. J. Virol. 74, 7989–7996

Tanaka N, Sato M, Lamphier MS, Nozawa H, Oda E, Noguchi S, Schreiber RD, Tsujimoto Y, and Taniguchi T (1998) Type 1 interferons are essential mediators of apoptotic death in virally infected cells. Genes Cells 3, 29–37

Taniguchi T, Ogasawara K, Takaoka A, and Tanaka N (2001) IRF family of transcription factors as regulators of host defense. Annu. Rev. Immunol. 19, 623–655

Tapparel C, Hausmann S, Pelet T, Curran J, Kolakofsky D, and Roux L (1997) Inhibition of Sendai virus genome replication due to promoter-increased electivity: a possible role for the accessory C proteins. J. Virol. 71, 9588–9599

Tatsuo H, Ono N, Tanaka K, and Yanagi Y (2000) SLAM(CDw150) is a cellular receptor for measles virus. Nature 406, 893–897

Teng MN, and Collins PL (1999) Altered growth characteristics of recombinant respiratory syncytial viruses, which do not produce NS2 protein. J. Virol. 73, 466–473

Thomas SM, Lamb RA, and Paterson RG (1988) Two mRNAs that differ by two nontemplated nucleotides encode the amino coterminal proteins P and V of the paramyxovirus SV5. Cell 54, 891–902

Tidona CA, Kurz HW, Gelderblom HR, and Darai G (1999) Isolation and molecular characterization of a novel cytopathogenic paramyxovirus from tree shrews. Virology 258, 425–434

Tober C, Seufert M, Schneider H, Billeter MA, Johnston ICD, Niewiesk S, ter Meulen V, and Schneider-Schaulies S (1998) Expression of measles virus V protein is associated with pathogenicity and control of viral RNA synthesis. J. Virol. 72, 8124–8132

Trono D (1995) HIV accessory proteins: Leading roles for the supporting cast. Cell 82, 189–192

Ulane CM, and Horvath CM (2002) Paramyxoviruses SV5 and hPIV2 assemble STAT protein ubiquitin ligase complexes from cellular components. Virology 304, 160–166

Usacheva A, Smith R, Minshall R, Baida G, Seng S, Croz E, and Colamonici O (2001) The WD motif-containing protein receptor for activated protein kinase C (RACK1) is required for recruitment and activation of signal transducer and activator of transcription 1 through the type I interferon receptor. J. Biol. Chem. 276, 22948–22953

Valle M, and Cantell K (1965) The ability of Sendai virus to overcome cellular resistance to vesicular stomatitis, virus I. Ann. Med. Exp. Biol. Fenn. 43, 57–60

Valsamakis A, Schneider H, Auwaerter PG., Kaneshima H, Billeter MA, and Griffin DE (1998) Recombinant measles viruses with mutations in the C, V, or F gene have altered growth phenotypes in vivo. J. Virol. 72, 7754–7761

Vidal S, Curran J, and Kolakofsky D (1990a) A stuttering model for paramyxovirus P mRNA editing. EMBO J. 9, 2017–2022

Vidal S, Curran J, and Kolakofsky D (1990b) Editing of the Sendai virus P/C mRNA by G insertion occurs during mRNA synthesis via a virus-encoded activity. J. Virol. 64, 239–246

Wang LF, Michalski WP, Yu M, Pritchard LI, Crameri G, Shiell B, and Eaton BT (1998) A novel PN/C gene in a new member of the Paramyxovirus family, which causes lethal infection in humans, horses, and other animals. J. Virol. 72, 1482–1490

Wansley EK, and Parks GD (2002) Naturally occurring substitutions in the P/V gene convert the noncytopathic paramyxovirus simian virus 5 into a virus that induces alpha/beta interferon synthesis and cell death. J. Virol. 76, 10109–10121

Whitehead SS, Bukreyev A, Teng MN, Firestone CY, St Claire M, Elkins WR, Collins PL, and Murphy BR (1999) Recombinant respiratory syncytial virus bearing a deletion of either the NS2 or SH gene is attenuated in chimpanzees. J. Virol. 73, 3438–3442

Yanagi Y, Ono N, Tatsuo H, Hashimoto K, and Minagawa H (2002) Measles virus receptor SLAM (CD150) Virology 299, 155–161

Yokosawa N, Yokota S, Kubota T, and Fujii N (2002) C-terminal region of STAT1α is not necessary for its ubiquitination and degradation caused by mumps virus V protein. J. Virol. 17, 12683–12690

Yokota S, Saito H, Kubota T, Yokosawa N, Amano K, and Fujii N (2003) Measles virus suppresses interferon-α signaling pathway: suppression of Jak1 phosphorylation and association of viral accessory proteins, C and V, with interferon-α receptor complex. Virology 305 (in press)

Yonemitsu Y, Kiston C, Ferrari S, Farley R, Griesenbach U, Judd D, Steel R, Scheid P, Zhu J, Jeffery PK, Kato A, Hasan MK, Nagai Y, Fukumura M, Hasegawa M, Geddes DM, and Alton EWFW (2000). Efficient gene transfer to airway epithelium using recombinant Sendai virus. Nat. Biotech. 18, 970–974

Yoshida T, Nagai Y, Yoshii S, Maeno K, Matsumoto T, and Hoshino M (1976) Membrane (M) protein of HVJ (Sendai virus): its role in virus assembly. Virology 71, 143–161

Yoshida T, Nagai Y, Maeno K, Iinuma M, Hamaguchi M, Matsumoto T, Nagayoshi S, and Hoshino M (1979) Studies on the role of M protein in virus assembly using a ts mutant of HVJ (Sendai virus): its role in virus assembly. Virology 71, 143–161

Young DF, Didcock L, Goodbourn S, and Randall RE (2000) Paramyxoviridae use distinct virus-specific mechanisms to circumvent the interferon response. Virology 269, 383–390

Young DF, Chatziandreou N, He B, Goodbourn S, Lamb RA, and Randall RE (2001) Single amino acid substitution in the V protein of simian virus 5 differentiates its ability to block interferon signaling in human and murine cells. J. Virol. 75, 3363–3370

Yu D, Shioda T, Kato A, Hasan MK, Sakai Y, and Nagai Y (1997) Sendai virus-based expression of HIV-1 gp120: reinforcement by the V(−) version. Genes Cells 2, 457–466

Identification and Characterization of Viral Antagonists of Type I Interferon in Negative-Strand RNA Viruses

A. García-Sastre

Department of Microbiology, Mount Sinai School of Medicine,
1 Gustave L. Levy Place, New York, NY 10029, USA
E-mail: adolfo.garcia-sastre@mssm.edu

Abstract Interferons are cytokines secreted in response to viral infections with potent antiviral activity, and they represent a critical component of the innate immune response against viruses. It has now become apparent that many viruses have evolved different mechanisms to counteract the interferon response, allowing their efficient replication and propagation in their hosts. This review discusses how the development of reverse genetics techniques and the increase in our knowledge of the interferon response have led to the discovery of interferon-antagonistic functions of different genes of viruses belonging to the negative-strand RNA virus group. In many cases, these viral genes encode accessory proteins that are not required for viral infectivity but are critical for optimal replication and for virulence in the host.

1 The Interferon System

All viruses must overcome host defense mechanisms to successfully replicate and propagate in their hosts. Among the host innate immune mechanisms involved in antimicrobial defense, the interferon (IFN) system constitutes a powerful antiviral response dedicated to fighting viral infections. IFNs were first identified as cell-secreted substances that interfere with viral replication (Isaacs and Lindenmann 1957). Type I IFNs or IFN-α/βs are a family of cytokines whose synthesis and secretion is in general triggered in response to intracellular viral replication by most cell types. By contrast, type II IFN or IFN-γ is secreted by NK and T cells in response to IL-12 and other immune stimulatory signals (Sen and Ransohoff 1993). All members of the IFN-α/β family share the same receptor, IFNAR, composed of two subunits, IFNAR1 and IFNAR2c. Binding of IFN-α/β to the IFNAR results in activation of two tyrosine kinases, Jak1 and Tyk2, that are associated with the receptor cytoplasmic tails. This leads to tyrosine phosphorylation, heterodimerization, and activation of the latent STAT1 and STAT2 transcription factors, which now associate with IRF-9 and translocate to the nucleus. This trimeric complex, also known as ISGF3, binds to specific DNA sequences, or ISREs, present in the promoter of IFN-α/β-stimulated genes, promoting their transcription. IFN-γ signaling involves a different receptor and leads to the Jak1- and Jak2-mediated activation and dimerization of STAT1. STAT1 homodimers, also known as GAF, bind to GAS elements in the DNA and activate transcription of IFN-γ-inducible genes. Although the primary functions of IFN-α/β and IFN-γ appear to be the induction of a cellular antiviral state and the activation of immune cells and of antibody isotype switching, respectively, both cytokines have roles in innate and adaptive antiviral immune responses (Biron and Sen 2001; Levy and García-Sastre 2001; Stark et al. 1998). Interestingly, a new family of IFNs, IFN-λ, was recently discovered. Like IFN-α/βs, the IFN-λs (IL-28A, IL-28B, and IL-29) are secreted in response to viral infection and activate the formation of the ISGF3 transcription factor. However, the IFN-λs use a different receptor, composed of two subunits, CFR2-12 (now IFN-λR1 or IL-28Rα) and IL-10R2. Although the biological significance of these new IFNs is still unknown, they may represent an IFNAR-independent, ISGF3-dependent host defense antiviral mechanism (Kotenko et al. 2003; Sheppard et al. 2003).

The antiviral action of IFNs is mediated by the transcriptional activation of the IFN-inducible genes. Although more than 100 genes are induced by IFN-α/β (Der et al. 1998), most studies on IFN action have been focused on the antiviral activity of three IFN-inducible genes, the Mx, PKR, and OAS genes. Mx proteins are IFN-inducible GTPases of the dynamin superfamily. Expression of human MxA confers resistance against replication of different viruses, including several families of negative-strand RNA viruses, such as orthomyxoviruses, paramyxoviruses, rhabdoviruses, and bunyaviruses. MxA interferes with the intracellular trafficking of viral nucleocapsids, at least during orthomyxovirus and bunyavirus infections (Haller and Kochs 2002). The double-stranded RNA (dsRNA)-activated protein kinase PKR is an extensively studied serine/threonine kinase with antiviral activity (Williams 1999). PKR activity is regulated not only at the transcriptional level by IFN-α/β but also by direct activation through binding to dsRNA or to specific protein activators such as PACT (Patel and Sen 1998). Activated PKR phosphorylates different substrates, including the translation factor eIF-2α. Phosphorylation of eIF-2α results in inhibition of protein translation. Therefore, PKR activation in virus-infected cells results in a translational block of viral replication. The OAS enzymes are 2′,5′-oligoadenylate synthetases that, like PKR, are transcriptionally induced by IFN-α/β and enzymatically activated by dsRNA. Synthesized 2′,5′-oligoadenylates are coactivators of RNaseL, an enzyme that degrades viral and cellular RNA, stopping viral replication. In addition, eIF-2α- and RNaseL-independent antiviral pathways have been proposed for both PKR and OAS. Moreover, it is clear that in addition to Mx, PKR, and OAS, other IFN-inducible genes participate in the inhibition of viral replication (Samuel 2001).

Inhibition of viral replication by the IFN-α/β system of the host requires the molecular recognition of ongoing viral infection followed by secretion of IFN-α/β. Among the IFN-α/βs, the transcriptional regulation of IFN-β has been more extensively studied. Induction of IFN-β synthesis by viruses is mediated by the activation of three transcription factors, IRF-3 (and/or IRF-7), NF-κB, and AP-1 (Wathelet et al. 1998). Because dsRNA closely recapitulates the induction of IFN-β secretion by viruses, it is assumed that viral dsRNA structures generated as byproducts of viral RNA replication or transcription trigger IFN-β synthesis in infected cells. Recognition of viral dsRNA by the cell results not only in IFN production but also in activation of the PKR and OAS IFN-inducible enzymes (see above). However, other viral products encoded by negative-strand RNA viruses, such as viral nucleocapsids (tenOever et al. 2002) and viral glycoproteins (Zeng et al. 2002a,b) have also been found to induce activation of transcription factors involved in IFN-β synthesis. In addition, it was recently found that extracellular activation of several members of the Toll-like receptor family can induce type I IFN secretion (for a recent review see Akira and Hemmi 2003). Interestingly, binding of dsRNA to TLR3 results in IFN-β secretion (Oshiumi et al. 2003). It is also intriguing that TLR4 interactions with the F protein of respiratory syncytial virus (RSV) appear to be important for innate immunity against RSV (Kurt-Jones et al. 2000), because it has also been found that activation of TLR4 induces IFN-β synthesis (Toshchakov et al. 2002). Although most cells can respond to viral infection by synthesizing IFN-α/β, it has become apparent that plasmacytoid cells are the major source of type I IFN in the host (Cella et al. 1999; Siegal et al. 1999). The viral-host interaction responsible for the induction of the IFN response is then a complex and interesting field of research involving many regulatory mechanisms and cell type specificities.

Because host cells have the ability to quickly respond to viral infections by secreting IFN, which in turn induces an elaborate and powerful antiviral response involving many cellular antiviral genes, efficient viral replication and propagation necessitate viral countermeasures to attenuate the host IFN response (for a recent review see Katze et al. 2002). However, it was not until 1998 that the existence of a viral product required for antagonizing the host IFN response, the nonstructural protein 1 (NS1) of influenza A virus, was demonstrated in a negative-strand RNA virus (García-Sastre et al. 1998). The combination of molecular techniques to study the ability of viral products to inhibit the IFN system with reverse genetics tools to investigate the functional role of such products in the context of virus infections has now resulted in the identification and characterization of viral IFN antagonists among most of the negative-strand RNA viruses. This article reviews the recent findings in this field as well as the potential practical applications of mutant negative-strand RNA viruses with impaired ability to evade the IFN response.

2 Viral Interferon Antagonists Encoded by Negative-Strand RNA Viruses

2.1 The NS1 Protein of Influenza Virus

In 1957 Isaacs and Lindenmann first described IFN as a substance produced after incubation of heat-inactivated influenza A viruses with chicken chorioallantoic membranes (Isaacs and Lindenmann 1957). Interestingly, the ability of influenza A virus to induce efficient IFN secretion was dependent on partial heat inactivation of the virus (56°C for 1 h), because neither live influenza viruses nor virus preparations that were exposed to further inactivation were good inducers of IFN (Isaacs and Burke 1958). These results suggest that IFN induction by influenza viruses depends on viral infection but is inhibited by a viral product whose expression is attenuated by heat inactivation of the virus. With reverse genetics techniques, this viral product was found to be the NS1 of the virus (García-Sastre et al. 1998) (Fig. 1). Growth of a mutant NS1 knock-out influenza A (delNS1) virus was impaired in conventional substrates, such as MDCK cells and 10-day-old embryonated eggs. Remarkably, delNS1 virus regained replication and pathogenicity in $STAT1^{-/-}$ mice, lacking a transactivator required for IFN signaling (García-Sastre et al. 1998). These results suggest that the NS1 functions as an inhibitor of the host IFN antiviral response, facilitating viral replication in IFN-competent substrates and hosts. The IFN antagonist function of the NS1 of influenza A virus was further supported by findings demonstrating that delNS1 virus infection efficiently induces the activation of transcription factors involved in IFN-β synthesis, IRF-3, IRF-7, NF-κB and AP-1, as well as production of IFN-α/β. By contrast, wild-type influenza A virus is a poor inducer of these factors and of IFN-α/β synthesis compared with delNS1 virus (Ludwig et al. 2002; Smith et al. 2001; Talon et al. 2000; Wang et al. 2000). Expression of NS1 by plasmid transfection demonstrated that this viral protein can inhibit IFN production in response to dsRNA in the absence of other viral factors (Talon et al. 2000; Wang et al. 2000). In addition, the ability of influenza viruses to efficiently overcome the IFN response has been correlated with higher levels of NS1 expression (Sekellick et al. 2000).

The NS1 protein of influenza A virus is a protein 202–238 amino acids in length. Naturally occurring mutations in the NS1 gene resulting in in-

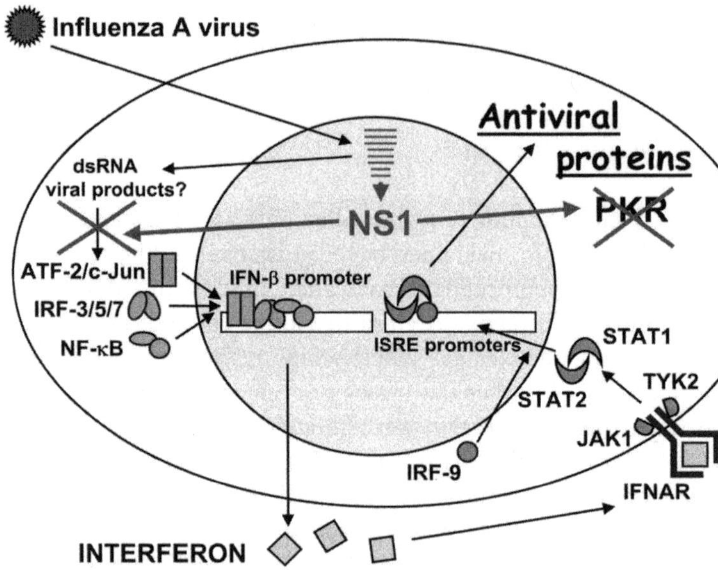

Fig. 1 Inhibition of the type I IFN response by the NS1 protein of influenza A virus. For clarity purposes, not all of the critical components of the IFN response are shown. On viral infection, the eight viral RNA segments are transported to the nucleus. Viral dsRNA generated by transcription, as well as some other uncharacterized viral products, might trigger the activation of ATF-2/c-Jun, IRF, and NF-κB transcription factors, resulting in the production of IFN. Secreted IFN interacts with its receptor, IFNAR, leading to the activation of the Jak1 and Tyk2 kinases and to the tyrosine phosphorylation of STAT1 and STAT2, which dimerize, recruit IRF-9, and activate transcription of genes containing ISRE promoters. This results in synthesis of antiviral proteins, including the translational inhibitor PKR. To counteract this antiviral defense, influenza A virus encodes a viral IFN antagonist, the NS1 protein, that prevents the activation of transcription factors involved in the stimulation of IFN production, as well as the activation of the antiviral protein PKR induced by IFN

fluenza A virus strains with shorter NS1 proteins are also described (Norton et al. 1987). This protein is known to bind different species of RNA, including dsRNA (Hatada and Fukuda 1992). Binding to dsRNA is mediated by the first 73 amino acids and requires protein dimerization (Nemeroff et al. 1995; Qian et al. 1995). Overexpression of the amino-terminal, RNA binding domain of the NS1 was sufficient to inhibit the stimulation of IFN synthesis. By contrast, a mutant NS1 protein defective in its ability to bind dsRNA was impaired in its ability to inhibit IFN production (Wang et al. 2000). Nevertheless, deletion of the carboxy-terminal region of the NS1 leads to recombinant influenza viruses with virulence phenotypes and IFN-inducing properties in between wild-type and delNS1 viruses. The levels of attenuation of these NS1-truncated viruses correlated with the length of the truncation (Egorov et al. 1998; Talon et al. 2000). Strikingly, the pathogenicity in mice of recombinant influenza viruses expressing carboxy-terminally truncated NS1 proteins is restored to levels close to wild-type viruses when a dimerization domain is fused to the amino-terminal domain of the NS1 (Wang et al. 2002). These results suggest that the carboxy-terminal region of the NS1 protein has a structural (dimerization promoting) role in NS1 function. This region might also unspecifically contribute to the inhibition of the IFN response by attenuating cellular mRNA polyadenylation (Noah et al. 2003).

The NS1 of influenza A virus inhibits the IFN system not only at the level of IFN synthesis but also at the level of IFN action. Because synthesis of IFN is most likely not completely prevented during influenza virus infection, the ability of the NS1 to counteract the antiviral action of IFN should also result in increased viral replication in the host. Specifically, it has been shown that the NS1 protein inhibits dsRNA-mediated activation of the IFN-inducible enzyme PKR in vitro (Lu et al. 1995). In fact, infection of cells with NS1 *ts* mutant influenza viruses at the nonpermissive temperature results in uncontrolled activation of PKR (Hatada et al. 1999). Similar observations were made in cells infected with delNS1 virus (Bergmann et al. 2000). The ability of the NS1 protein to counteract the action of PKR most likely accounts for the nonspecific translational enhancing properties of the NS1 (Salvatore et al. 2002). The importance of PKR in the antiviral action of IFN is underscored by the finding that delNS1 virus is pathogenic not only in STAT1$^{-/-}$ mice but also in PKR$^{-/-}$ mice (Bergmann et al. 2000). However, because most laboratory strains of mice lack Mx, the individual contributions of PKR and Mx in inhibiting replication of an influenza virus lacking the NS1 protein are unclear at this moment. The critical role that PKR plays in the antiviral IFN-mediated response is evidenced by the presence of viral encoded PKR inhibitors in several virus families, including RNA and DNA viruses (Gale and Katze 1998). It is also interesting that influenza viruses have evolved a second mechanism of PKR inhibition that is mediated by the activation of a cellular inhibitor of PKR, the p58IPK protein (Lee et al. 1990, 1992).

The mechanism of action responsible for the IFN-antagonistic properties of the NS1 protein is still not completely understood. dsRNA produced during viral infection is believed to stimulate IFN production and PKR activation. Because the IFN antagonist properties of the NS1 correlate with its dsRNA binding activity, it is attractive to speculate that binding of NS1 to dsRNA prevents the activation of these pathways. In this respect, the NS1 protein will be functionally closely related to the E3L protein of vaccinia virus, which also binds dsRNA and inhibits IFN synthesis and PKR activation. However, recent evidence indicating that NS1 function is host specific (see Sect. 3) may indicate the existence of direct protein-protein interactions between the NS1 protein and one or more cellular key components of the IFN response. For instance, a direct interaction between NS1 and PKR has been postulated (Tan and Katze 1998). It might be possible that NS1 interactions with cellular proteins involved in IFN synthesis are cooperatively enhanced by dsRNA binding. The identification of such cellular proteins will provide an invaluable aid in elucidating the molecular mechanism of NS1 function.

Most of the studies on NS1 function have been performed in influenza A virus. Despite major sequence differences between influenza A, B, and C viruses, all three types of viruses encode an NS1 protein. Interestingly, although these NS1 proteins share only limited amino acid sequence identity, NS1s of both influenza A and B viruses have an amino-terminal RNA binding domain and inhibit PKR activation (Wang and Krug 1996). It is therefore conceivable that the NS1 protein of influenza B virus also inhibits IFN production. Interestingly, in addition to PKR, the NS1 protein of influenza B virus also inhibits the function of a second IFN-inducible protein, the ubiquitin-like ISG15 protein (Yuan and Krug 2001). This specific activity of the NS1 protein of influenza B virus is independent of its dsRNA binding activity (Yuan et al. 2002). With respect to influenza C virus, the role of its NS1 protein in antagonizing the IFN response is not known at this moment.

2.2
The ML Protein of Thogotovirus

The orthomyxovirus group includes not only the influenza A, B, and C viruses but also several other segmented negative-strand RNA viruses. Among these viruses, the tick-borne thogotovirus deserves special mention because it lacks an NS segment and therefore does not encode an

NS1 protein. Instead, thogotovirus encodes an interferon antagonist protein from its M segment. In the case of influenza A virus, the M segment directs the synthesis of two viral proteins by alternative splicing, the M1 (matrix) and M2 (ion channel) proteins, which only share the first 9 amino terminal amino acids. By contrast, thogotovirus M segment encodes, also by alternative splicing, the ML and M (matrix) proteins, the ML protein being a 38-amino acid carboxy-terminal extended form of the M protein. Interestingly, a thogotovirus isolate deficient in ML expression as well as recombinant thogotoviruses lacking ML expression were able to efficiently induce IFN in infected cells, whereas ML-expressing wild-type virus strains were poor IFN inducers. Moreover, expression of the ML protein inhibited the activation of the IFN-β promoter by dsRNA, suggesting that, similar to the NS1 protein of influenza A virus, the ML protein of thogotovirus is a virus-encoded IFN antagonist that blocks the transcriptional activation of IFN (Hagmaier et al. 2003).

2.3
V Proteins of Paramyxoviruses

Cells infected with different paramyxoviruses, such as human PIV3, Sendai virus, and SV5, become unable to respond to IFN (Didcock et al. 1999; Hermodsson 1963). In 1998, Yokosawa et al. made the interesting observation that the levels of STAT1, one of the critical components of the ISGF3 complex activated by IFN signaling, become reduced in cells infected with mumps virus, a paramyxovirus belonging to the rubulavirus group (Yokosawa et al. 1998). Subsequent studies using the closely related SV5 virus demonstrated that STAT1 is targeted for proteasome-mediated degradation in SV5-infected cells and that expression of the SV5-encoded protein V is sufficient to mediate this effect (Didcock et al. 1999). Degradation of STAT1 renders SV5- and mumps virus-infected cells unable to respond to IFN. The V proteins of paramyxoviruses are encoded by the P gene, which directs expression of P, V, and D/ W/ I proteins by an mRNA editing mechanism that results in the addition of nontemplated G residues at a specific region of the P gene during mRNA transcription (Table 1). This results in expression of three proteins with identical amino-terminal domains and distinct carboxy-terminal regions. Whereas the P protein is required for RNA transcription and replication, the V proteins encoded by rubulaviruses appear to mediate inhibition of IFN signaling. SV5 (Didcock et al. 1999), mumps virus

Table 1 Proteins encoded by the P gene of paramyxoviruses[a]

Respiroviruses	Sendai virus	P	**V**[b]	W	**C′**	**C**	**Y1**	**Y2**
Rubulaviruses	Mumps virus	P	**V**	I				
Avulaviruses	NDV	P	**V**	W				
Morbilliviruses	Measles virus	P	V	W	C			
Henipaviruses	Nipah virus	P	**V**	**W**	**C**			
Pneumoviruses[c]	RSV	P						
Metapneumoviruses[c]	Avian pneumovirus	P						

[a] Representative examples of each group are shown.

[b] Shown in bold are proteins for which an IFN antagonistic activity has been described.

[c] Pneumoviruses encode IFN antagonist NS1 and NS2 proteins as independent transcriptional units; for the closely related metapneumoviruses, that lack these two proteins, no IFN antagonist has yet been identified.

(Kubota et al. 2001), and SV41 virus (Nishio et al. 2001) V proteins target STAT1 to degradation, allowing for inhibition of both IFN-α/β and IFN-γ signaling. However, human PIV2 V protein targets STAT2 to degradation, specifically inhibiting only IFN-α/β signaling (Nishio et al. 2001; Parisien et al. 2001; Young et al. 2000).

The V protein of the avulavirus Newcastle disease virus (NDV) also appears to have IFN antagonistic properties, because its expression results in inhibition of the antiviral IFN response in chicken embryo fibroblasts, allowing more efficient viral replication of IFN-sensitive viruses (Park et al. 2003). Similarly, stable expression in human cells of the V protein of SV5 increases the replication properties of several viruses, most likely by inhibiting the IFN-mediated antiviral response (Young et al. 2003). Nipah virus, a recently identified paramyxovirus that was responsible for a highly lethal outbreak in humans in Malaysia in 1998–1999 and that is closely related to Hendra virus, also encodes a V protein that inhibits IFN signaling, although its mechanism of action appears to be different from that of other paramyxoviruses (Park et al. 2003; Rodriguez et al. 2002) (see below).

The V protein carboxy-terminal region is highly conserved among paramyxoviruses and consists of a cysteine-rich region that binds two Zn^{+2} atoms (Lamb and Kolakofsky 2001). Expression of just the unique carboxy-terminal V region of mumps virus resulted in decreased levels of STAT1 and in inhibition of IFN signaling (Kubota et al. 2001). Similar results with respect to STAT2 degradation were achieved by expression of the unique carboxy-terminal V region of hPIV2 (Nishio et al. 2001). Moreover, expression of the carboxy-terminal V region of NDV complements viral growth in the presence of an IFN response (Park et al. 2003). With reverse genetics, recombinant hPIV2, NDV, and SV5 expressing a truncated V protein lacking the carboxy-terminal region have been generated (He et al. 2002; Kawano et al. 2001; Mebatsion et al. 2001). These viruses were compromised for growth in IFN-competent substrates and partly regained their ability to replicate at high titers in IFN-deficient substrates. These results are consistent with an essential role of the carboxy-terminal region of V in its IFN-antagonistic functions.

Despite the evidence that the carboxy-terminal V domain of mumps virus and hPIV2 is sufficient to mediate STAT1 and STAT2 degradation, respectively (Kubota et al. 2001; Nishio et al. 2001), the amino-terminal domain of V of SV5 also appears to modulate V activity. Three amino acid differences in the amino-terminal domain of V of an isolate of SV5 virus [CPI(−)] were found to be responsible for the inability of this strain to induce STAT1 degradation (Chatziandreou et al. 2002). Substitution of the V protein of SV5 by the V protein of the CPI(−) strain resulted in a recombinant SV5 virus unable to mediate STAT1 degradation (Wansley and Parks 2002). In contrast to other paramyxovirus V proteins, expression of only the amino-terminal first amino acids of Nipah virus V resulted in inhibition of IFN signaling (Park et al. 2003). Interestingly, the Nipah virus P gene is characterized by a unique amino-terminal region not present in any other paramyxoviruses, with the exception of Hendra virus, that is shared by the P, V, and W proteins of this virus. Thus the W protein of Nipah virus, which shares the same amino-terminal domain as the V protein, also inhibits IFN signaling (Park et al. 2003).

STAT-mediated degradation by the V proteins of rubulaviruses appears to be mainly mediated by the proteasome, because both STAT1 and STAT2 levels are restored in cells expressing SV5 and hPIV2 V proteins by treatment with proteasome inhibitors (Andrejeva et al. 2002). It is interesting that although SV5 and hPIV2 V proteins target different transcription factors for degradation (STAT1 and STAT2, respectively) they require the presence of both STAT1 and STAT2 to specifically target one of these factors for degradation. This suggests that V forms a complex with both STAT1 and STAT2 factors, and that V cannot recognize these factors individually (Parisien et al. 2002). The carboxy-terminal region of the STATs containing their protein-activating tyrosine phosphorylation and functional src homology 2 (SH2) domains were not required for being targeted by the V protein, but their amino terminals were essential (Parisien et al. 2002). Similarly, the carboxy terminal of STAT1 was not needed for its degradation by the mumps virus V protein (Yokosawa et al. 2002). Protein-protein interaction analysis indicated that V and STAT proteins interact physically in vitro and in vivo (Nishio et al. 2002; Parisien et al. 2002). A tryptophan-rich motif in the carboxy-terminal domain of mumps virus V was required for this interaction (Nishio et al. 2002). On the other hand, it has also been reported that mutations in the cysteine residues of the mumps virus V carboxy-terminal domain involved in Zn^{+2} binding abrogate STAT1 binding (Yokosawa et al. 2002).

The SV5 V protein is also known to interact with the damage-specific DNA-binding protein DDB1 (Lin et al. 1998). Interestingly, binding of SV5 V to DDB1 correlates with its ability to target STAT1 for degradation (Andrejeva et al. 2002). SV5 and hPIV2 V proteins copurified from human cells with a complex including STAT1, STAT2, DDB1, and Cul4A, a protein related to a family of cellular ubiquitin ligase complex subunits (Ulane and Horvath 2002). Both DDB1 and Cul4A were required for SV5 V-mediated STAT1 degradation (Andrejeva et al. 2002; Ulane and Horvath 2002), suggesting that V targets STAT1 for degradation by the formation of a complex with STAT1 and STAT2 and cellular components involved in ubiquitin-mediated proteasome degradation, such as DDB1 and Cul4A. Consistent with this hypothesis, ubiquitination of STAT1 has been demonstrated in mumps virus-infected cells (Yokosawa et al. 2002). Intriguingly, mumps virus V protein was recently found to associate with yet another cellular protein, RACK1 (Kubota et al. 2002). RACK1 is known to interact with STAT1 and to function as an adaptor molecule, mediating binding of STAT1 to the IFN-α/β receptor (Usacheva et al. 2001). This binding is required for Jak-mediated phosphorylation of STAT1 in response to IFN. Association of mumps virus V with RACK1 prevented the interaction of STAT1 with RACK1 and with the IFN-α/β receptor (Kubota et al. 2002), and this should result in inhibition of IFN signaling. However, it is unclear whether, as found for DDB1, RACK1 is required for the induction of STAT1 degradation by the V protein. Finally, as already mentioned, the V protein of Nipah virus is unusual in its ability to block IFN signaling. As with the rubulavirus V proteins, Nipah virus V binds to STAT1 and STAT2, but this does not result in STAT degradation. Rather, Nipah virus V inhibits STAT phosphorylation and

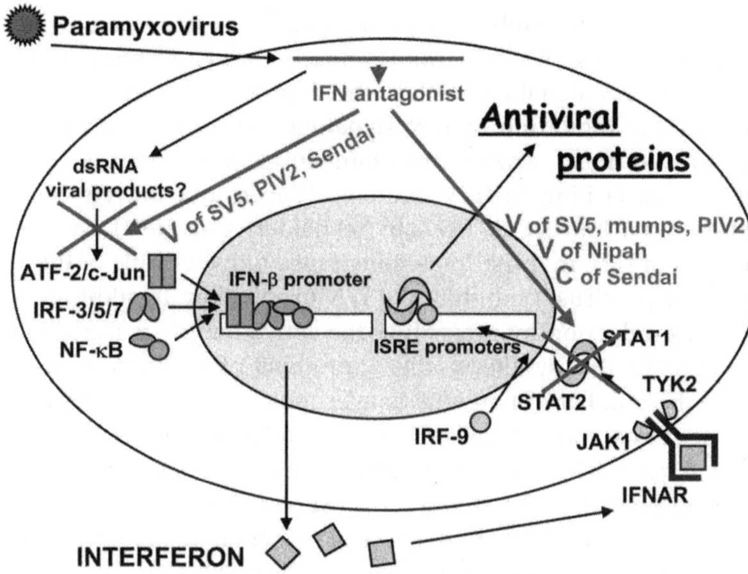

Fig. 2 Inhibition of the type I IFN response by the V and C proteins of paramyxoviruses. On viral infection, the viral genomic RNA is released into the cytoplasm. For the induction of the IFN response, see Fig. 1 legend. The V proteins of the respiroviruses SV5, mumps virus, and PIV2 and of the Nipah virus and the C proteins of Sendai virus inhibit IFN signaling by interactions with STAT1/STAT2, sometimes resulting in degradation of the STATs. In addition, it has recently been shown that the V proteins of several paramyxoviruses are also able to prevent the activation of transcription factors involved in IFN synthesis

translocation to the nucleus (Rodriguez et al. 2002). In contrast to other V proteins, the amino-terminal region of Nipah virus V appears to be responsible for its IFN-antagonistic properties (Park et al. 2003).

Recently, a second role for the paramyxovirus V proteins has been described with respect to their ability to inhibit the IFN system (Fig. 2). A recombinant SV5 virus lacking the carboxy-terminal domain of V not only was defective in its ability to inhibit IFN signaling but also induced higher levels of IFN than wild-type viruses (He et al. 2002). Interestingly, similarly to the NS1 protein of influenza virus, SV5 V protein expression resulted in inhibition of IRF3 and NF-κB activation by dsRNA, resulting in inhibition of IFN production. Expression of only the carboxy-terminal domain of V was able to mediate this inhibition (Poole et al. 2002). Not only the V protein of SV5 but also the V proteins of hPIV2 and of

Sendai virus were able to inhibit IFN production, . The latter paramyxovirus belongs to the respirovirus group, whose V proteins appear to be unable to inhibit IFN signaling. Rather, the C proteins of Sendai virus are responsible for inhibition of IFN signaling (see below). It will be interesting to determine whether recombinant Sendai, measles, and rinderpest viruses lacking their V proteins (Baron and Barrett 2000; Delenda et al. 1997; Kato et al. 1997a,b; Schneider et al. 1997) are also higher inducers of IFN, in order to demonstrate a universal role of the V protein of paramyxoviruses in inhibiting IFN production. In addition to inhibiting IFN production by preventing the activation of transcription factors involved in IFN synthesis, the V proteins of paramyxoviruses might reduce the number of potential danger molecules that trigger IFN production in virus-infected cells, such as dsRNA or nucleocapsids, through their known RNA replication-inhibitory activities (Curran et al. 1991). In this respect, the CPI(−) SV5 virus expressed higher levels of viral RNA and proteins in infected cells because of mutations in its V gene and induced higher levels of IFN (Wansley and Parks 2002).

2.4
C Proteins of Sendai Virus

Among the paramyxoviruses, the respirovirus group includes viruses that encode a V protein, such as Sendai virus and bovine PIV3, and viruses for which there is no biochemical evidence for the presence of a V protein, such as human PIV1 and PIV3. Moreover, the respiroviruses, but not the rubulaviruses, encode at least one C protein from the P gene by alternative use of initiation codons (Table 1). For instance, Sendai virus expresses a total of four C proteins (C′, C, Y1, and Y2) through alternative translation from different start codons (ACG81, AUG114, AUG183, and AUG201) in the same open reading frame. The C proteins therefore have identical carboxy-terminal sequences, and they do not share any amino acid sequence with the P protein (Lamb and Kolakofsky 2001). Interestingly, the C proteins of Sendai virus, and not the V protein, are responsible for inhibition of IFN signaling (Garcin et al. 1999; Gotoh et al. 1999) (Fig. 2). In contrast to wild-type Sendai virus, recombinant Sendai viruses lacking the AUG114-initiated C protein or containing a single amino acid substitution in the C protein are unable to efficiently block IFN action (Garcin et al. 1999). However, only a 4C (C′, C, Y1 and Y2) knockout Sendai virus was completely unable to suppress the induction of IFN-stimulated gene products, indicating that most likely all C proteins have some IFN -antagonistic properties (Gotoh et al. 1999). Expression of individual C proteins in cells has demonstrated that all C proteins, even the smallest Y2 protein, are able to inhibit IFN signaling (Garcin et al. 2000; Kato et al. 2001). Moreover, only the carboxy-terminal half of the C protein is required for this inhibition (Kato et al. 2002). Nevertheless, the larger C′ and C proteins appear to be critical for IFN inhibition in the context of virus infection, because recombinant Sendai viruses expressing the Y1 and Y2 proteins, but lacking expression of the larger C and C′ proteins, are unable to rescue vesicular stomatitis virus (VSV) replication in IFN-treated cells (Garcin et al. 2001).

Sendai virus-infected cells demonstrated low levels of Tyk2 phosphorylation, and this might explain the blockade of IFN-α/β signaling induced by this virus (Komatsu et al. 2000). However, Sendai virus not only inhibits IFN-α/β signaling, but also IFN-γ (Tyk2 independent) signaling (Young et al. 2000). STAT function appears to be modified by Sendai virus infection, because both serine (Young et al. 2000) and tyrosine (Komatsu et al. 2002; Saito et al. 2002) phosphorylation of STAT1 are deregulated by Sendai virus infection because of the C proteins. In addition, STAT1 becomes unstable in murine cells infected with Sendai virus (Garcin et al. 2000).

The mechanism of inhibition of IFN signaling by the C proteins of Sendai virus appears to be a rather complex process. Takeuchi et al. found that Sendai C proteins directly bind to STAT1 (Takeuchi et al. 2001). Although all four C proteins physically interact with STAT1 and inhibit IFN signaling, only the large C′ and C proteins, in an analogous way to the V proteins of rubulaviruses, induce STAT1 ubiquitination and its subsequent proteasome-mediated degradation in murine cells (Garcin et al. 2002). STAT1 degradation might explain the inhibition of IFN signaling in Sendai virus-infected cells. However, C-mediated STAT1 degradation appears to be cell specific and although it is clearly seen in mouse embryo fibroblasts, it does not occur in other cell lines, for example, human HeLa cells, in which C-mediated inhibition of IFN signaling also occurs (Garcin et al. 2002; Komatsu et al. 2000; Young et al. 2000). The mechanism of action of C is further complicated by the observation that, despite the inhibition of STAT1 tyrosine phosphorylation in response to short-term IFN stimulation, tyrosine-phosphorylated STAT1 accumulates in Sendai virus-infected cells or in cells expressing the C proteins (Komatsu et al. 2002; Saito et al. 2002). Recently, Garcin et al

(Garcin et al. 2003) found that although all C proteins have the ability to bind to STAT1 in response to IFN, resulting in STAT1 tyrosine phosphorylation inhibition, the large C′ and C proteins can also interact with STAT1 independently of IFN signaling. This second interaction leads to STAT1 instability and to a paradoxical increase, even in the absence of IFN, in tyrosine-phosphorylated STAT1, which most likely is not functional because of its association with C. Interestingly, Gotoh et al. also recently reported that binding of C to STAT1 facilitates an indirect interaction of C with STAT2, leading to a blockade in tyrosine phosphorylation of STAT2 in response to IFN (Gotoh et al. 2003). In any case, the available evidence indicates that the C proteins of Sendai virus inhibit IFN signaling through binding to STAT1, resulting in a deregulation of the activating STAT1 and STAT2 phosphorylation events accompanied in some instances by instability of STAT1, leading to inhibition of STAT1/STAT2-mediated transcription.

Are other C proteins from other paramyxoviruses involved in inhibiting IFN signaling? Although this question still remains unanswered for most paramyxoviruses, the C protein of Nipah virus was found to be able to promote the replication of NDV in the presence of an IFN response, suggesting that this protein inhibits the host IFN system at some level (Park et al. 2003).

2.5
Nonstructural Proteins of Respiratory Syncytial Viruses

The most divergent paramyxovirus group is the pneumoviruses, which include human RSV. Besides the P protein, the pneumoviruses do not appear to express any additional proteins from the P gene. Moreover, the genome of these viruses contains specific unique transcriptional units encoding the NS1, NS2, and M2 proteins, not present in any other paramyxovirus. RSV has been found to be highly resistant to the antiviral action of IFN, as evidenced by its ability to replicate in cells pretreated with IFN-α/β (Atreya and Kulkarni 1999; Young et al. 2000). With reverse genetics techniques, the IFN-resistance phenotype of bovine RSV was mapped to the NS1 and NS2 genes (Schlender et al. 2000). Recombinant bovine RSVs lacking the NS1, the NS2, or both genes became sensitive to the action of IFN, and their replication was impaired in IFN-competent cells. Moreover, expression of NS1 and NS2 proteins of bovine RSV by recombinant rabies viruses conferred IFN resistance to this otherwise IFN-sensitive virus. IFN resistance was achieved only when both proteins were expressed, suggesting that bovine RSV NS1 and NS2 proteins cooperate to mediate this effect (Schlender et al. 2000). Similarly, the NS1 and NS2 proteins of human RSV are also required for viral resistance against IFN in human cells (Bossert and Conzelmann 2002). The mechanism by which these two viral proteins cooperate to overcome the IFN response is still unknown.

2.6
The M Protein of Vesicular Stomatitis Virus

The ability of VSV to prevent IFN production in infected cells correlates with its ability to inhibit cellular protein expression (Francoeur et al. 1987). The M protein of VSV is a potent inhibitor of cellular mRNA transcription and transport (Black and Lyles 1992; Her et al. 1997). This activity also prevents efficient induction of IFN synthesis (Ferran and Lucas-Lenard 1997). However, additional uncharacterized viral determinants also appear to participate in the differential IFN induction by different strains of VSV (Marcus et al. 1998). Inhibition of cellular mRNA transport by VSV M is mediated by interactions with the nucleoporin Nup98 (von Kobbe et al. 2000). Interestingly, IFN treatment leads to an increase in the levels of Nup98 and Nup96, reversing the M protein-mediated inhibition of cellular gene expression, suggesting that hosts have evolved counteracting mechanisms for the IFN-antagonistic properties of VSV (Enninga et al. 2002).

2.7
The VP35 Protein of Ebola Virus

The negative-strand RNA filoviruses (Marburg and Ebola viruses) are among the most lethal viruses in humans. Endothelial cells infected with Ebola virus were found to lose their ability to respond to dsRNA and to IFN (Harcourt et al. 1998, 1999). Similar results were obtained in human macrophages and peripheral blood mononuclear cells (Gupta et al. 2001). By using a biological screening assay for IFN antagonists based on virus growth complementation it was found that the VP35 protein of Ebola, which carries the same function in RNA replication as the P protein of paramyxoviruses, inhibits the type I IFN system. Inhibition appears to be at the level of induction of IFN synthesis (Basler et al. 2000).

It remains to be determined whether Ebola virus also encodes an inhibitor of IFN signaling.

2.8
NSs Proteins of Bunyaviruses

The three-segmented negative-strand RNA viruses belong to the broad group of the *Bunyaviridae*. These viruses are further subdivided into five genera: *Bunyavirus*, *Hantavirus*, *Nairovirus*, *Phlebovirus*, and *Tospovirus*. By using reverse genetics techniques established for Bunyamwera virus, the bunyavirus prototype, it has been demonstrated that the NSs protein of this virus is required for preventing the induction of IFN-α/β in infected cells (Weber et al. 2002). Thus a recombinant Bunyamwera virus lacking the NSs protein induces activation of NF-κB and IRF-3, resulting in production of IFN, whereas these effects are not observed in wild-type virus-infected cells. Moreover, expression of NSs inhibited the transcriptional induction of IFN by dsRNA. The poorly replicating delNSs Bunyamwera virus regained replication in IFN-deficient substrates, including knockout mice for the IFNAR (Weber et al. 2002). Therefore, the NSs of Bunyamwera virus allows efficient viral replication in the presence of an intact IFN system.

The NSs protein of Bunyamwera virus is encoded by an alternative open reading frame in the small (S) segment, which also encodes the nucleocapsid (N) protein. Rift Valley fever virus, a *Phlebovirus*, also encodes N and NSs proteins from the S segment. The coding strategy is, however, different, and it is based in ambisense expression from nonoverlapping open reading frames (Overton et al. 1987). Analysis of Rift valley fever virus strains with spontaneously acquired mutations in the NSs gene demonstrated that a deletion in the NSs gene resulted in a high-IFN-inducing virus. This virus was attenuated in wild-type mice but virulent in IFNAR$^{-/-}$ mice, suggesting that the NSs of phleboviruses inhibits IFN production, promoting viral replication and pathogenicity in IFN-competent hosts (Bouloy et al. 2001).

The genus *Tospovirus* includes plant-infecting bunyaviruses, and these viruses also express NSs proteins from an ambisense S RNA segment. Interestingly, the NSs of the tospovirus tomato spotted wilt virus inhibits an important antiviral defense mechanism in plants, RNA silencing (Bucher et al. 2003; Takeda et al. 2002). At this moment it is unclear whether the ability of the NSs of tospoviruses to suppress RNA silencing is related to the ability of the NSs of bunyaviruses and phleboviruses to suppress IFN induction.

NSs proteins have not been identified among the hantaviruses and nairoviruses. Nevertheless, different hantaviruses appear to induce different levels of IFN-induced genes in infected endothelial cells. Nonpathogenic hantaviruses induced higher levels of IFN-stimulated genes at early times after infection, suggesting that pathogenic hantaviruses suppress early cellular IFN responses (Geimonen et al. 2002). The viral gene product responsible for this suppression is still unknown.

3 Role of Viral Interferon Antagonists in Virulence and Host Tropism

Through the use of reverse genetics techniques, recombinant negative-strand RNA viruses with altered IFN antagonist genes have been generated. These viruses display an attenuated phenotype in animal models of infection, indicating that these genes are important virulence factors. In some cases, the main reason for attenuation of these recombinant viruses is their inability to inhibit the IFN antiviral response. This can be demonstrated if the attenuated phenotype reverts to a virulent phenotype in mice with defined deficiencies in the IFN response, such as IFNAR$^{-/-}$, STAT1$^{-/-}$, or PKR$^{-/-}$ mice. This was the case for delNS1 influenza A viruses and for NSs-mutated bunyaviruses (Bergmann et al. 2000; Bouloy et al. 2001; García-Sastre et al. 1998; Weber et al. 2002). However, because of the multifunctional characteristics of most of the proteins encoded by negative-strand RNA viruses, the determination of specific mutations that only affect IFN antagonist function might be needed to evaluate the contribution to pathogenesis of virus-encoded IFN inhibitors. For example, the V and C proteins of paramyxoviruses also have roles in RNA replication and viral assembly (Lamb and Kolakofsky 2001), and therefore the attenuation of V and C knockout viruses may represent a complex phenotype not only due to their inability to prevent the IFN antiviral response (Baron and Barrett 2000; Durbin et al. 1999; Escoffier et al. 1999; Kato et al. 1997; Kurotani et al. 1998; Mebatsion et al. 2001; Mrkic et al. 2000; Patterson et al. 2000; Tober et al. 1998; Valsamakis et al. 1998). In the case of influenza A viruses, the NS1 protein has been reported not only to inhibit the IFN response but also to enhance specifically viral mRNA translation and to inhibit cellular mRNA processing (Aragón et al. 2000; Chen et al. 1999).

The ability of negative-strand RNA viruses to overcome the IFN system, and therefore their IFN-antagonist proteins, play an important role in the species tropism of these viruses. SV5 does not replicate well in mice, and this correlates with the fact that SV5 inhibits IFN signaling in human cells but not in murine cells (Didcock et al. 1999). This specificity can be changed by a single amino acid mutation in V (Young et al. 2001). SV5 V interacts with the human STAT1/STAT2 complex, and interestingly the specificity of SV5 V for human STAT1 does not seem to depend on the origin of STAT1, but of STAT2. Thus expression of human STAT2 in mouse cells renders murine STAT1 susceptible to SV5 V-mediated degradation (Parisien et al. 2002). Several other negative-strand RNA viruses have also been shown to encode IFN antagonists that demonstrate preferences for inhibiting the IFN system of their natural hosts. For instance, human RSV NS1 and NS2 proteins are better inhibitors of the IFN system in human cells than in bovine cells, whereas the contrary seems to be the case for bovine RSV NS1 and NS2 proteins (Bossert and Conzelmann 2002). Exchange of the NS1 gene of a mouse-adapted influenza A virus for that of the human pandemic strain of 1918 results in a low-virulence phenotype in mice (Basler et al. 2001). The co-evolution of a virus with its host, resulting in adaptation of viral factors to interact with specific host factors, is most likely responsible for the species specificity of the viral IFN antagonists.

Do highly virulent strains of negative-strand RNA viruses express more efficient IFN antagonist proteins than strains of moderate or low virulence? Because infection with influenza viruses induces to some extent the IFN response (Kim et al. 2002), highly virulent influenza viruses may possess highly efficient anti-IFN NS1 genes. In this respect, the NS1 proteins of the highly lethal 1918 and HK/1997 influenza A virus strains appear to be strong antagonists of the IFN response in human and pig cells, respectively (Geiss et al. 2002; Palese et al. 2002; Seo et al. 2002). Although the NS1 of HK/1997, a virus that jumped directly from birds to humans, allows the virus to efficiently overcome the antiviral state induced by IFN and TNF (Seo et al. 2002), the same NS1 appears to be a poor inhibitor of cytokine induction in human macrophages, and this combination might be responsible for abnormally high levels of proinflammatory cytokines in infected humans, resulting in increased pathogenicity (Cheung et al. 2002). However, the high virulence of both the 1918 and HK/1997 influenza A viruses in humans most likely represents a multifactorial characteristic, in which several viral genes, in addition to the NS1 gene, are involved (Hatta et al. 2001; Katz et al. 2000; Tumpey et al. 2002).

Viral strains from several other negative-strand RNA viruses, such as measles viruses, bunyaviruses, and hantaviruses, have also demonstrated a correlation between virulence and ability to inhibit the IFN response (Bouloy et al. 2001; Geimonen et al. 2002; Naniche et al. 2000). In the case of SV5, the differential ability of viral strains to suppress the IFN response might be associated with their ability to establish persistent infections (Chatziandreou et al. 2002). Future studies will help us to better delineate the relationships between pathogenicity, host and tissue tropism, and the ability of negative-strand RNA viruses to evade the host IFN response.

4 Use of Recombinant Viruses Containing Modified Viral Interferon Antagonists

4.1 Live Attenuated Vaccines

Most viral IFN antagonists of negative-strand RNA viruses appear to be accessory proteins, dispensable for viral infectivity but required for pathogenicity and efficient replication in the host. Therefore, recombinant viruses containing mutations or deletions in their IFN antagonist genes might represent live attenuated vaccine candidates if proven safe and immunogenic (see other chapters of this book). However, highly attenuated viruses, although safe, might be poor immunogens. By evaluating the vaccine potential of a panel of recombinant viruses having different degrees of impairment in their IFN antagonist genes, and therefore different degrees of attenuation, it might be possible to select live vaccine candidates with an optimal window between attenuation and immunogenicity (Talon et al. 2000). Attenuated recombinant viruses might also be used as viral vectors with vaccine or therapeutic purposes (see chapter by von Messling and Cattaneo, this volume).

4.2 Oncolytic Viruses

Several negative-strand RNA viruses, such as measles viruses, VSV, and NDV, have been found to have oncolytic properties, that is, they preferentially replicate in transformed cells, resulting in selective tumor cell killing (Grote et al. 2001; Peng et al. 2001; Schirrmacher et al. 1999; Stojdl et al. 2000). Interestingly, the selective viral replication in tumor cells seems to correlate with an impaired IFN response in most cancer cells (Coffey et al. 1998; Stojdl et al. 2000). It is then likely that recombinant negative-strand RNA viruses with altered IFN antagonist proteins will have a higher specificity for replication in IFN-deficient cancer cells, while sparing IFN-competent normal cells. This was found to be the case for a recombinant influenza A virus lacking the IFN antagonist NS1 protein (Bergmann et al. 2001). Oncolytic negative-strand RNA viruses may be used as therapeutic antitumor agents in combination with other tumor therapies.

5 Conclusions and Perspectives

The use of reverse genetics techniques has completely changed our view of how negative-strand RNA viruses deal with the innate antiviral immune response induced by IFN. It is now clear that most, if not all, negative-strand RNA viruses, despite their compact, small genomes, encode at least one viral product to counteract the action of the host IFN response. Nevertheless, it is still unknown whether some of these viruses, for example, the arenaviruses, express an inhibitor of the IFN response. Viral IFN antagonists are critical virulence factors required for efficient viral replication in the host, and they therefore represent attractive novel targets for antiviral therapy. Although our knowledge of how negative-strand RNA viruses interact with the host IFN response has advanced considerably, we still do not know all the cellular and viral players responsible for these interactions. This represents an area of active research that will have an important impact on understanding viral pathogenicity, not only for negative-strand RNA viruses but also for other viruses (Katze et al. 2002). In addition, these studies may lead to the rational design of improved viral vaccines and viral therapeutic vectors.

Acknowledgements. Work in the author's laboratory is supported by grants from the National Institutes of Health.

References

Akira S, Hemmi H (2003) Recognition of pathogen-associated molecular patterns by TLR family. Immunol. Lett. 85:85–95

Andrejeva J, Poole E, Young DF, Goodbourn S, Randall RE (2002) The p127 subunit (DDB1) of the UV-DNA damage repair binding protein is essential for the targeted degradation of STAT1 by the V protein of the paramyxovirus simian virus 5. J. Virol. 76:11379–11386

Andrejeva J, Young DF, Goodbourn S, Randall RE (2002) Degradation of STAT1 and STAT2 by the V proteins of simian virus 5 and human parainfluenza virus type 2, respectively: consequences for virus replication in the presence of alpha/beta and gamma interferons. J. Virol. 76:2159–2167

Aragón T, de La Luna S, Novoa I, Carrasco L, Ortín J, Nieto A (2000) Eukaryotic translation initiation factor 4GI is a cellular target for NS1 protein, a translational activator of influenza virus. Mol. Cell. Biol. 20:6259–6268

Atreya PL, Kulkarni S (1999) Respiratory syncytial virus strain A2 is resistant to the antiviral effects of type I interferons and human MxA. Virology 261:227–241

Baron MD, Barrett T (2000) Rinderpest viruses lacking the C and V proteins show specific defects in growth and transcription of viral RNAs. J. Virol. 74:2603–2611

Basler CF, Wang X, Mühlberger E, Volchkov V, Paragas J, Klenk H-D, García-Sastre A, Palese P (2000) The Ebola virus VP35 protein functions as a type I interferon antagonist. Proc. Natl. Acad. Sci. (USA) 97:12289–12294

Basler CF, Reid AH, Dybing JK, Janczewski TA, Fanning TG, Zheng H, Salvatore M, Perdue ML, Swayne DE, García-Sastre A, Palese P, Taubenberger JK (2001) Sequence of the 1918 pandemic influenza virus nonstructural gene (NS) segment and characterization of recombinant viruses bearing the 1918 NS genes. Proc. Natl. Acad. Sci. (USA) 98:2746–2751

Bergmann M, García-Sastre A, Carnero E, Pehamberger H, Wolff K, Palese P, Muster T (2000) Influenza virus NS1 protein counteracts PKR-mediated inhibition of replication. J. Virol. 74:6203–6206

Bergmann M, Romirer I, Sachet M, Fleischhacker R, García-Sastre A, Palese P, Wolff K, Pehamberger H, Jakesz R, Muster T (2001) A genetically engineered influenza A virus with ras-dependent oncolytic properties. Cancer Res. 61:8188–8193

Biron CA, Sen GC (2001) Interferons and other cytokines. In: Knipe DM, Howley PM, Griffin DE, Lamb RA, Martin MA, Roizman B, Straus SE (eds) Fields Virology. Lippincott-Raven, Philadelphia, pp 321–351

Black BL, Lyles DS (1992) Vesicular stomatitis virus matrix protein inhibits host cell-directed transcription of target genes in vivo. J. Virol. 66:4058–4064

Bossert B, Conzelmann KK (2002) Respiratory syncytial virus (RSV) nonstructural (NS) proteins as host range determinants: a chimeric bovine RSV with NS genes from human RSV is attenuated in interferon-competent bovine cells. J. Virol. 76:4287–4293

Bouloy M, Janzen C, Vialat P, Khun H, Pavlovic J, Huerre M, Haller O (2001) Genetic evidence for an interferon-antagonistic function of rift valley fever virus nonstructural protein NSs. J. Virol. 75:1371–1377.

Bucher E, Sijen T, De Haan P, Goldbach R, Prins M (2003) Negative-strand tospoviruses and tenuiviruses carry a gene for a suppressor of gene silencing at analogous genomic positions. J. Virol. 77:1329-1336

Cella M, Jarrossay D, Facchetti F, Alebardi O, Nakajima H, Lanzavecchia A, Colonna M (1999) Plasmacytoid monocytes migrate to inflamed lymph nodes and produce large amounts of type I interferon. Nat. Med. 5:919–923

Chatziandreou N, Young D, Andrejeva J, Goodbourn S, Randall RE (2002) Differences in interferon sensitivity and biological properties of two related isolates of simian virus 5: a model for virus persistence. Virology 293:234–242

Chen Z, Li Y, Krug RM (1999) Influenza A virus NS1 protein targets poly(A)-binding protein II of the cellular 3′-end processing machinery. EMBO J. 18:2273–2283

Cheung CY, Poon LL, Lau AS, Luk W, Lau YL, Shortridge KF, Gordon S, Guan Y, Peiris JS (2002) Induction of proinflammatory cytokines in human macrophages by influenza A (H5N1) viruses: a mechanism for the unusual severity of human disease? Lancet 360:1831–1837

Coffey MC, Strong JE, Forsyth PA, Lee PW (1998) Reovirus therapy of tumors with activated Ras pathway. Science 282:1332–1334

Curran J, Boeck R, Kolakofsky D (1991) The Sendai virus P gene expresses both an essential protein and an inhibitor of RNA synthesis by shuffling modules via mRNA editing. EMBO J. 10:3079–3085

Delenda C, Hausmann S, Garcin D, Kolakofsky D (1997) Normal cellular replication of Sendai virus without the trans-frame, nonstructural V protein. Virology 228:55–62

Der SD, Zhou A, Williams BR, Silverman RH (1998) Identification of genes differentially regulated by interferon alpha, beta, or gamma using oligonucleotide arrays. Proc. Natl. Acad. Sci. (USA) 95:15623–15628

Didcock L, Young DF, Goodbourn S, Randall RE (1999) The V protein of simian virus 5 inhibits interferon signalling by targeting STAT1 for proteasome-mediated degradation. J.Virol. 73:9928–9933

Didcock L, Young DF, Goodbourn S, Randall RE (1999) Sendai virus and simian virus 5 block activation of interferon-responsive genes: importance for virus pathogenesis. J. Virol. 73:3125–3133

Durbin AP, McAuliffe JM, Collins PL, Murphy BR (1999) Mutations in the C, D, and V open reading frames of human parainfluenza virus type 3 attenuate replication in rodents and primates. Virology 261:319–330

Egorov A, Brandt S, Sereinig S, Romanova J, Ferko B, Katinger D, Grassauer A, Alexandrova G, Katinger H, Muster T (1998) Transfectant influenza A viruses with long deletions in the NS1 protein grow efficiently in Vero cells. J. Virol. 72:6437–6441

Enninga J, Levy DE, Blobel G, Fontoura BM (2002) Role of nucleoporin induction in releasing an mRNA nuclear export block. Science 295:1523–1525

Escoffier C, Manie S, Vincent S, Muller CP, Billeter M, Gerlier D (1999) Nonstructural C protein is required for efficient measles virus replication in human peripheral blood cells. J. Virol. 73:1695–1698

Ferran MC, Lucas-Lenard JM (1997) The vesicular stomatitis virus matrix protein inhibits transcription from the human beta interferon promoter. J. Virol. 71:371–377

Francoeur AM, Poliquin L, Stanners CP (1987) The isolation of interferon-inducing mutants of vesicular stomatitis virus with altered viral P function for the inhibition of total protein synthesis. Virology 160:236–245

Gale MJ, Katze MG (1998) Molecular mechanisms of interferon resistance mediated by viral-directed inhibition of PKR, the interferon-induced protein kinase. Pharmacol. Ther. 78:29–46

García-Sastre A, Egorov A, Matassov D, Brandt S, Levy DE, Durbin JE, Palese P, Muster T (1998) Influenza A virus lacking the NS1 gene replicates in interferon-deficient systems. Virology 252:324–330

Garcin D, Latorre P, Kolakofsky D (1999) Sendai virus C proteins counteract the interferon-mediated induction of an antiviral state. J. Virol. 73:6559–6565

Garcin D, Curran J, Kolakofsky D (2000) Sendai virus C proteins must interact directly with cellular components to interfere with interferon action. J. Virol. 74:8823–8830

Garcin D, Curran J, Itoh M, Kolakofsky D (2001) Longer and shorter forms of Sendai virus C proteins play different roles in modulating the cellular antiviral response. J. Virol. 75:6800–6807

Garcin D, Marq JB, Strahle L, le Mercier P, Kolakofsky D (2002) All four Sendai Virus C proteins bind Stat1, but only the larger forms also induce its mono-ubiquitination and degradation. Virology 295:256–265

Garcin D, Marq JB, Goodbourn S, Kolakofsky D (2003) The amino-terminal extensions of the longer Sendai virus C proteins modulate pY701-Stat1 and bulk Stat1 levels independently of interferon signaling. J. Virol. 77:2321–2329

Geimonen E, Neff S, Raymond T, Kocer SS, Gavrilovskaya IN, Mackow ER (2002) Pathogenic and nonpathogenic hantaviruses differentially regulate endothelial cell responses. Proc. Natl. Acad. Sci. (USA) 99:13837–13842

Geiss GK, Salvatore M, Tumpey TM, Carter VS, Wang X, Basler CF, Taubenberger JK, Bumgarner RE, Palese P, Katze MG, García-Sastre A (2002) Cellular transcriptional profiling in influenza A virus infected lung epithelial cells: the role of the nonstructural NS1 protein in the evasion of the host innate defense and its potential contribution to pandemic influenza. Proc. Natl. Acad. Sci. (USA) 99:10736–10741

Gotoh B, Takeuchi K, Komatsu T, Yokoo J, Kimura Y, Kurotani A, Kato A, Nagai Y (1999) Knockout of the Sendai virus C gene eliminates the viral ability to prevent the interferon-α/β-mediated responses. FEBS Lett. 459:205–210

Gotoh B, Takeuchi K, Komatsu T, Yokoo J (2003) The STAT2 activation process is a crucial target of Sendai virus C protein for the blockade of alpha interferon signaling. J Virol 77:3360–70

Grote D, Russell SJ, Cornu TI, Cattaneo R, Vile R, Poland GA, Fielding AK (2001) Live attenuated measles virus induces regression of human lymphoma xenografts in immunodeficient mice. Blood 97:3746–3754

Gupta M, Mahanty S, Ahmed R, Rollin PE (2001) Monocyte-derived human macrophages and peripheral blood mononuclear cells infected with ebola virus secrete MIP-1alpha and TNF-alpha and inhibit poly-IC-induced IFN-alpha in vitro. Virology 284:20–25

Hagmaier K, Jennings S, Buse J, Weber F, Kochs G (2003) Novel gene product of thogotovirus segment 6 codes for an interferon antagonist. J. Virol. 77:2747–2752
Haller O, Kochs G (2002) Interferon-induced Mx proteins: dynamin-like GTPases with antiviral activity. Traffic 3:710–717
Harcourt BH, Sanchez A, Offermann MK (1998) Ebola virus inhibits induction of genes by double-stranded RNA in endothelial cells. Virology 252:179–188
Harcourt BH, Sanchez A, Offermann MK (1999) Ebola virus selectively inhibits responses to interferons, but not to interleukin-1beta, in endothelial cells. J. Virol. 73:3491–3496
Hatada E, Fukuda R (1992) Binding of influenza A virus NS1 protein to dsRNA in vitro. J. Gen. Virol. 73:3325–3329
Hatada E, Saito S, Fukuda R (1999) Mutant influenza viruses with a defective NS1 protein cannot block the activation of PKR in infected cells. J. Virol. 73:2425–2433
Hatta M, Gao P, Halfmann P, Kawaoka Y (2001) Molecular basis for high virulence of Hong Kong H5N1 influenza A viruses. Science 293:1840–1842
He B, Paterson RG, Stock N, Durbin JE, Durbin RK, Goodbourn S, Randall RE, Lamb RA (2002) Recovery of paramyxovirus simian virus 5 with a V protein lacking the conserved cysteine-rich domain: the multifunctional V protein blocks both interferon-beta induction and interferon signaling. Virology 303:15–32
Her LS, Lund E, Dahlberg JE (1997) Inhibition of Ran guanosine triphosphatase-dependent nuclear transport by the matrix protein of vesicular stomatitis virus. Science 276:1845–1848
Hermodsson S (1963) Inhibition of interferon by an infection with parainfluenza virus type 3 (PIV-3). Virology 29:333–343
Isaacs A, Lindenmann J (1957) Virus interference. 1. The interferon. Proc. R. Soc. Lond. B. 147:258–267
Isaacs A, Burke DC (1958) Mode of action of interferon. Nature 4642:1073–1076
Kato A, Kiyotani K, Sakai Y, Yoshida T, Nagai Y (1997) The paramyxovirus, Sendai virus, V protein encodes a luxury function required for viral pathogenesis. EMBO J. 16:578–587
Kato A, Kiyotani K, Sakai Y, Yoshida T, Shioda T, Nagai Y (1997) Importance of the cysteine-rich carboxyl-terminal half of V protein for Sendai virus pathogenesis. J. Virol. 71:7266–7272
Kato A, Ohnishi Y, Kohase M, Saito S, Tashiro M, Nagai Y (2001) Y2, the smallest of the Sendai virus C proteins, is fully capable of both counteracting the antiviral action of interferons and inhibiting viral RNA synthesis. J. Virol. 75:3802–3810
Kato A, Ohnishi Y, Hishiyama M, Kohase M, Saito S, Tashiro M, Nagai Y (2002) The amino-terminal half of Sendai virus C protein is not responsible for either counteracting the antiviral action of interferons or down-regulating viral RNA synthesis. J. Virol. 76:7114–7124
Katz JM, Lu X, Tumpey TM, Smith CB, Shaw MW, Subbarao K (2000) Molecular correlates of influenza A H5N1 virus pathogenesis in mice. J. Virol. 74:10807–10810
Katze MG, He Y, Gale M, Jr. (2002) Viruses and interferon: a fight for supremacy. Nat. Rev. Immunol. 2:675–687
Kawano M, Kaito M, Kozuka Y, Komada H, Noda N, Nanba K, Tsurudome M, Ito M, Nishio M, Ito Y (2001) Recovery of infectious human parainfluenza type 2 virus

from cDNA clones and properties of the defective virus without V-specific cysteine-rich domain. Virology 284:99–112

Kim MJ, Latham AG, Krug RM (2002) Human influenza viruses activate an interferon-independent transcription of cellular antiviral genes: outcome with influenza A virus is unique. Proc. Natl. Acad. Sci. (USA) 99:10096–10101

Komatsu T, Takeuchi K, Yokoo J, Tanaka Y, Gotoh B (2000) Sendai virus blocks alpha interferon signaling to signal transducers and activators of transcription. J. Virol. 74:2477–2480

Komatsu T, Takeuchi K, Yokoo J, Gotoh B (2002) Sendai virus C protein impairs both phosphorylation and dephosphorylation processes of Stat1. FEBS Lett. 511:139–144

Kotenko SV, Gallagher G, Baurin VV, Lewis-Antes A, Shen M, Shah NK, Langer JA, Sheikh F, Dickensheets H, Donnelly RP (2003) IFN-λs mediate antiviral protection through a distinct class II cytokine receptor complex. Nat. Immunol. 4:69–77

Kubota T, Yokosawa N, Yokota S, Fujii N (2001) C terminal Cys-rich region of mumps virus structural V protein correlates with block of interferon α and γ signal transduction pathway through decrease of STAT 1-α. Biochem. Biophys. Res. Commun. 283:255–259

Kubota T, Yokosawa N, Yokota S, Fujii N (2002) Association of mumps virus V protein with RACK1 results in dissociation of STAT-1 from the alpha interferon receptor complex. J. Virol. 76:12676–12682

Kurotani A, Kiyotani K, Kato A, Shioda T, Sakai Y, Mizumoto K, Yoshida T, Nagai Y (1998) Sendai virus C proteins are categorically nonessential gene products but silencing their expression severely impairs viral replication and pathogenesis. Genes Cells 3:111–124

Kurt-Jones EA, Popova L, Kwinn L, Haynes LM, Jones LP, Tripp RA, Walsh EE, Freeman MW, Golenbock DT, Anderson LJ, Finberg RW (2000) Pattern recognition receptors TLR4 and CD14 mediate response to respiratory syncytial virus. Nat. Immunol. 1:398–401

Lamb RA, Kolakofsky D (2001) Paramyxoviridae: The viruses and their replication. In: Knipe DM, Howley PM, Griffin DE, Lamb RA, Martin MA, Roizman B, Straus SE (eds) Fields Virology. Lippincott-Raven, Philadelphia, pp 1305–1340

Lee TG, Tomita J, Hovanessian AG, Katze MG (1990) Purification and partial characterization of a cellular inhibitor of the interferon-induced protein kinase of Mr 68,000 from influenza virus-infected cells. Proc. Natl. Acad. Sci. (USA) 87:6208–6212

Lee TG, Tomita J, Hovanessian AG, Katze MG (1992) Characterization and regulation of the 58,000-dalton cellular inhibitor of the interferon-induced, dsRNA-activated protein kinase. J. Biol. Chem. 267:14238–14243

Levy DE, García-Sastre A (2001) The virus battles: IFN induction of the antiviral state and mechanisms of viral evasion. Cytokine Growth Factor Rev. 12:143–156.

Lin GY, Paterson RG, Richardson CD, Lamb RA (1998) The V protein of the paramyxovirus SV5 interacts with damage-specific DNA binding protein. Virology 249:189–200

Lu Y, Wambach M, Katze MG, Krug RM (1995) Binding of the influenza virus NS1 protein to double-stranded RNA inhibits the activation of the protein kinase that phosphorylates the elF-2 translation initiation factor. Virology 214:222–228

Ludwig S, Wang X, Ehrhardt C, Zheng H, Donelan N, Planz O, Pleschka S, García-Sastre A, Heins G, Wolff T (2002) The influenza A virus NS1 protein inhibits activation of Jun N-terminal kinase and AP-1 transcription factors. J. Virol. 76:11166–11171

Marcus PI, Rodriguez LL, Sekellick MJ (1998) Interferon induction as a quasispecies marker of vesicular stomatitis virus populations. J. Virol. 72:542–549

Mebatsion T, Verstegen S, De Vaan LT, Romer-Oberdorfer A, Schrier CC (2001) A recombinant Newcastle disease virus with low-level V protein expression is immunogenic and lacks pathogenicity for chicken embryos. J. Virol. 75:420–428

Mrkic B, Odermatt B, Klein MA, Billeter MA, Pavlovic J, Cattaneo R (2000) Lymphatic dissemination and comparative pathology of recombinant measles viruses in genetically modified mice. J. Virol. 74:1364–1372

Naniche D, Yeh A, Eto D, Manchester M, Friedman RM, Oldstone MB (2000) Evasion of host defenses by measles virus: wild-type measles virus infection interferes with induction of alpha/beta interferon production. J. Virol. 74:7478–7484

Nemeroff ME, Qian XY, Krug RM (1995) The influenza virus NS1 protein forms multimers in vitro and in vivo. Virology 212:422–428

Nishio M, Tsurudome M, Ito M, Kawano M, Komada H, Ito Y (2001) High resistance of human parainfluenza type 2 virus protein-expressing cells to the antiviral and anti-cell proliferative activities of alpha/beta interferons: cysteine-rich V-specific domain is required for high resistance to the interferons. J. Virol. 75:9165–9176

Nishio M, Garcin D, Simonet V, Kolakofsky D (2002) The carboxyl segment of the mumps virus V protein associates with STAT proteins in vitro via a tryptophan-rich motif. Virology 300:92–99

Noah DL, Twu KY, Krug RM (2203) Cellular antiviral processes against influenza A virus are countered at the posttranscriptional level by the viral NS1A protein via its binding to a cellular protein required for the 3' end processing of cellular pre-mRNAs. Virology 307:386–395

Norton GP, Tanaka T, Tobita K, Nakada S, Buonaugurio DA, Greenspan D, Krystal M, Palese P (1987) Infectious influenza A and B virus variants with long carboxyl terminal deletions in the NS1 polypeptides. Virology 156:204–213

Oshiumi H, Matsumoto M, Funami K, Akazawa T, Seya T (2003) TICAM-1, an adaptor molecule that participates in Toll-like receptor 3-mediated interferon-beta induction. Nat. Immunol. 4:161–167

Overton HA, Ihara T, Bishop DH (1987) Identification of the N and NSs proteins coded by the ambisense S RNA of Punta Toro phlebovirus using monospecific antisera raised to baculovirus expressed N and NSs proteins. Virology 157:338–350

Palese P, Basler CF, García-Sastre A (2002) The makings of a killer. Nat. Med. 8:927–928

Parisien JP, Lau JF, Rodriguez JJ, Sullivan BM, Moscona A, Parks GD, Lamb RA, Horvath CM (2001) The V protein of human parainfluenza virus 2 antagonizes type I interferon responses by destabilizing signal transducer and activator of transcription 2. Virology 283:230–239

Parisien JP, Lau JF, Horvath CM (2002) STAT2 acts as a host range determinant for species-specific paramyxovirus interferon antagonism and simian virus 5 replication. J. Virol. 76:6435–6441

Parisien JP, Lau JF, Rodriguez JJ, Ulane CM, Horvath CM (2002) Selective STAT protein degradation induced by paramyxoviruses requires both STAT1 and STAT2 but is independent of alpha/beta interferon signal transduction. J. Virol. 76:4190–4198

Park MS, Shaw ML, Munoz-Jordan J, Cros JF, Nakaya T, Bouvier N, Palese P, García-Sastre A, Basler CF (2003) Newcastle disease virus (NDV)-based assay demonstrates interferon-antagonist activity for the NDV V protein and the Nipah virus V, W, and C proteins. J. Virol. 77:1501–1511

Patel RC, Sen GC (1998) PACT, a protein activator of the interferon-induced protein kinase, PKR. EMBO J. 17:4379–4390

Patterson JB, Thomas D, Lewicki H, Billeter MA, Oldstone MB (2000) V and C proteins of measles virus function as virulence factors in vivo. Virology 267:80–89

Peng KW, Ahmann GJ, Pham L, Greipp PR, Cattaneo R, Russell SJ (2001) Systemic therapy of myeloma xenografts by an attenuated measles virus. Blood 98:2002–2007

Poole E, He B, Lamb RA, Randall RE, Goodbourn S (2002) The V proteins of simian virus 5 and other paramyxoviruses inhibit induction of interferon-beta. Virology 303:33–46

Qian XY, Chien CY, Lu Y, Montelione GT, Krug RM (1995) An amino-terminal polypeptide fragment of the influenza virus NS1 protein possesses specific RNA-binding activity and largely helical backbone structure. RNA 1:948–956

Rodriguez JJ, Parisien JP, Horvath CM (2002) Nipah virus V protein evades alpha and gamma interferons by preventing STAT1 and STAT2 activation and nuclear accumulation. J. Virol. 76:11476–11483

Saito S, Ogino T, Miyajima N, Kato A, Kohase M (2002) Dephosphorylation failure of tyrosine-phosphorylated STAT1 in IFN-stimulated Sendai virus C protein-expressing cells. Virology 293:205–209

Salvatore M, Basler CF, Parisien J-P, Horvath CM, Bourmakina S, Zheng H, Muster T, Palese P, García-Sastre A (2002) Effects of influenza A virus NS1 protein on protein expression: the NS1 protein enhances translation and is not required for shutoff of host protein synthesis. J. Virol. 76:1206–1212

Samuel CE (2001) Antiviral actions of interferons. Clin. Microbiol. Rev. 14:778–809

Schirrmacher V, Haas C, Bonifer R, Ahlert T, Gerhards R, Ertel C (1999) Human tumor cell modification by virus infection: an efficient and safe way to produce cancer vaccine with pleiotropic immune stimulatory properties when using Newcastle disease virus. Gene Ther. 6:63–73

Schlender J, Bossert B, Buchholz U, Conzelmann KK (2000) Bovine respiratory syncytial virus nonstructural proteins NS1 and NS2 cooperatively antagonize alpha/beta interferon-induced antiviral response. J. Virol. 74:8234–8242

Schneider H, Kaelin K, Billeter MA (1997) Recombinant measles viruses defective for RNA editing and V protein synthesis are viable in cultured cells. Virology 227:314–322

Sekellick MJ, Carra SA, Bowman A, Hopkins DA, Marcus PI (2000) Transient resistance of influenza virus to interferon action attributed to random multiple packaging and activity of NS genes. J. Interferon Cytokine Res. 20:963–970

Sen GC, Ransohoff RM (1993) Interferon-induced antiviral actions and their regulation. Adv. Virus Res. 42:57–102

Seo SH, Hoffmann E, Webster RG (2002) Lethal H5N1 influenza viruses escape host anti-viral cytokine responses. Nat. Med. 8:950–954

Sheppard P, Kindsvogel W, Xu W, Henderson K, Schlutsmeyer S, Whitmore TE, Kuestner R, Garrigues U, Birks C, Roraback J, Ostrander C, Dong D, Shin J, Presnell S, Fox B, Haldeman B, Cooper E, Taft D, Gilbert T, Grant FJ, Tackett M, Krivan W, McKnight G, Clegg C, Foster D, Klucher KM (2003) IL-28, IL-29 and their class II cytokine receptor IL-28R. Nat. Immunol. 4:63–68

Siegal FP, Kadowaki N, Shodell M, Fitzgerald-Bocarsly PA, Shah K, Ho S, Antonenko S, Liu YJ (1999) The nature of the principal type 1 interferon-producing cells in human blood. Science 284:1835–1837

Smith EJ, Marié I, Prakash A, García-Sastre A, Levy DE (2001) IRF3 and IRF7 phosphorylation in virus-infected cells does not require double-stranded RNA-dependent protein kinase R or IκB kinase but is blocked by vaccinia virus E3L protein. J. Biol. Chem. 276:8951–8957

Stark GR, Kerr IM, Williams BR, Silverman RH, Schreiber RD (1998) How cells respond to interferons. Annu. Rev. Biochem. 67:227–264

Stojdl DF, Lichty B, Knowles S, Marius R, Atkins H, Sonenberg N, Bell JC (2000) Exploiting tumor-specific defects in the interferon pathway with a previously unknown oncolytic virus. Nat. Med. 6:821–825

Takeda A, Sugiyama K, Nagano H, Mori M, Kaido M, Mise K, Tsuda S, Okuno T (2002) Identification of a novel RNA silencing suppressor, NSs protein of Tomato spotted wilt virus. FEBS Lett. 532:75–79

Takeuchi K, Komatsu T, Yokoo J, Kato A, Shioda T, Nagai Y, Gotoh B (2001) Sendai virus C protein physically associates with Stat1. Genes Cells 6:545–557

Talon J, Horvath CM, Polley R, Basler CF, Muster T, Palese P, García-Sastre A (2000) Activation of interferon regulatory factor 3 is inhibited by the influenza A virus NS1 protein. J. Virol. 74:7989–7996

Talon J, Salvatore M, O'Neill RE, Nakaya Y, Zheng H, Muster T, García-Sastre A, Palese P (2000) Influenza A and B viruses expressing altered NS1 proteins: a vaccine approach. Proc. Natl. Acad. Sci. (USA) 97:4309–4314

Tan SL, Katze MG (1998) Biochemical and genetic evidence for complex formation between the influenza A virus NS1 protein and the interferon-induced PKR protein kinase. J. Interferon Cytokine Res. 18:757–766

tenOever BR, Servant MJ, Grandvaux N, Lin R, Hiscott J (2002) Recognition of the measles virus nucleocapsid as a mechanism of IRF-3 activation. J. Virol. 76:3659–3669

Tober C, Seufert M, Schneider H, Billeter MA, Johnston IC, Niewiesk S, ter Meulen V, Schneider-Schaulies S (1998) Expression of measles virus V protein is associated with pathogenicity and control of viral RNA synthesis. J. Virol. 72:8124–8132

Toshchakov V, Jones BW, Perera PY, Thomas K, Cody MJ, Zhang S, Williams BR, Major J, Hamilton TA, Fenton MJ, Vogel SN (2002) TLR4, but not TLR2, mediates IFN-beta-induced STAT1alpha/beta-dependent gene expression in macrophages. Nat. Immunol. 3:392–398

Tumpey TM, García-Sastre A, Mikulasova A, Taubenberger JK, Swayne DE, Palese P, Basler CF (2002) Existing antivirals are effective against influenza viruses with genes from the 1918 pandemic virus. Proc. Natl. Acad. Sci. (USA) 99:13849–13854

Ulane CM, Horvath CM (2002) Paramyxoviruses SV5 and hPIV2 assemble STAT protein ubiquitin ligase complexes from cellular components. Virology 304:160–166

Usacheva A, Smith R, Minshall R, Baida G, Seng S, Croze E, Colamonici O (2001) The WD motif-containing protein receptor for activated protein kinase C (RACK1) is required for recruitment and activation of signal transducer and activator of transcription 1 through the type I interferon receptor. J. Biol. Chem. 276:22948–22953

Valsamakis A, Schneider H, Auwaerter PG, Kaneshima H, Billeter MA, Griffin DE (1998) Recombinant measles viruses with mutations in the C, V, or F gene have altered growth phenotypes in vivo. J. Virol. 72:7754–7761

von Kobbe C, van Deursen JM, Rodrigues JP, Sitterlin D, Bachi A, Wu X, Wilm M, Carmo-Fonseca M, Izaurralde E (2000) Vesicular stomatitis virus matrix protein inhibits host cell gene expression by targeting the nucleoporin Nup98. Mol. Cell 6:1243–1252

Wang W, Krug RM (1996) The RNA-binding and effector domains of the viral NS1 protein are conserved to different extents among influenza A and B viruses. Virology 223:41–50

Wang X, Li M, Zheng H, Muster T, Palese P, Beg AA, García-Sastre A (2000) Influenza A virus NS1 protein prevents the activation of NF-κB and induction of type I IFN. J. Virol. 74:11566–11573

Wang X, Basler CF, Williams BRG, Silverman RH, Palese P, García-Sastre A (2002) Functional replacement of the carboxy-terminal two thirds of the influenza A virus NS1 protein with short heterologous dimerization domains. J. Virol. 76:12951–12962

Wansley EK, Parks GD (2002) Naturally occurring substitutions in the P/V gene convert the noncytopathic paramyxovirus simian virus 5 into a virus that induces alpha/beta interferon synthesis and cell death. J. Virol. 76:10109–10121

Wathelet MG, Lin CH, Parekh BS, Ronco LV, Howley PM, Maniatis T (1998) Virus infection induces the assembly of coordinately activated transcription factors on the IFN-beta enhancer in vivo. Mol. Cell 1:507–518

Weber F, Bridgen A, Fazakerley JK, Streitenfeld H, Kessler N, Randall RE, Elliott RM (2002) Bunyamwera bunyavirus nonstructural protein NSs counteracts the induction of alpha/beta interferon. J. Virol. 76:7949–7955

Williams BR (1999) PKR; a sentinel kinase for cellular stress. Oncogene 18:6112–6120

Yokosawa N, Kubota T, Fujii N (1998) Poor induction of interferon-induced 2′,5′-oligoadenylate synthetase (2–5 AS) in cells persistently infected with mumps virus is caused by decrease of STAT-1α. Arch. Virol. 143:1985–1992

Yokosawa N, Yokota S, Kubota T, Fujii N (2002) C-terminal region of STAT-1α is not necessary for its ubiquitination and degradation caused by mumps virus V protein. J Virol 76:12683–12690

Young DF, Didcock L, Goodbourn S, Randall RE (2000) Paramyxoviridae use distinct virus-specific mechanisms to circumvent the interferon response. Virology 269:383–390

Young DF, Chatziandreou N, He B, Goodbourn S, Lamb RA, Randall RE (2001) Single amino acid substitution in the V protein of simian virus 5 differentiates its ability to block interferon signaling in human and murine cells. J. Virol. 75:3363–3370

Young DF, Andrejeva L, Livingstone A, Goodbourn S, Lamb RA, Collins PL, Elliott RM, Randall RE (2003) Virus replication in engineered human cells that do not respond to interferons. J. Virol. 77:2174–2181

Yuan W, Krug RM (2001) Influenza B virus NS1 protein inhibits conjugation of the interferon (IFN)-induced ubiquitin-like ISG15 protein. EMBO J. 20:362–371

Yuan W, Aramini JM, Montelione GT, Krug RM (2002) Structural basis for ubiquitin-like ISG 15 protein binding to the NS1 protein of influenza B virus: a protein-protein interaction function that is not shared by the corresponding N-terminal domain of the NS1 protein of influenza A virus. Virology 304:291–301

Zeng J, Fournier P, Schirrmacher V (2002a) Stimulation of human natural interferon-alpha response via paramyxovirus hemagglutinin lectin-cell interaction. J. Mol. Med. 80:443–451

Zeng J, Fournier P, Schirrmacher V (2002b) Induction of interferon-alpha and tumor necrosis factor-related apoptosis-inducing ligand in human blood mononuclear cells by hemagglutinin-neuraminidase but not F protein of Newcastle disease virus. Virology 297:19-30

Toward Novel Vaccines and Therapies Based on Negative-Strand RNA Viruses

V. von Messling · R. Cattaneo
Molecular Medicine Program, Mayo Foundation, 200 1st Street SW, Rochester, MN 55905, USA
E-mail: Cattaneo.Roberto@mayo.edu

Abstract The study of negative-strand RNA viruses has suggested new strategies to produce more attenuated viruses. Reverse genetics has allowed the implementation of the strategies, and new or improved monovalent vaccines are being developed. In addition, recombinant viruses expressing foreign proteins or epitopes have been produced with the aim of developing multivalent vaccines capable of stimulating humoral and cellular immune responses against more than one pathogen. Finally, recombinant viruses that selectively enter cells expressing tumor markers or the HIV envelope protein have been engineered and shown to lyse target cells. Preclinical and clinical trials of improved and multivalent vaccines and therapeutic (oncolytic) viruses are ongoing.

1 Introduction

Negative-strand RNA viruses are important agents of disease, including age-old diseases of humans and animals as well as newly recognized emerging diseases. Human deaths attributable to measles, influenza, and the parainfluenza viruses worldwide, especially in children, are numbered over a million annually (Clements and Cutts 1995). An emerging paramyxovirus named Nipah has recently claimed more than hundred human lives and has led to the slaughter of almost all porcine livestock in Malaysia (Chua et al. 1999). Similarly, a pathogenic influenza strain has claimed several human lives in Hong Kong, and the possibility of pandemics has repeatedly induced Hong Kong authorities to preventively eliminate millions of chickens (Hatta et al. 2001).

There are many pathogenic negative-strand RNA viruses and too few effective vaccines, and those vaccines (against measles, mumps, and influenza) are not affordable for certain nations or are not optimally effective in all settings. Reverse genetics technology for segmented and nonsegmented negative-strand RNA viruses has been developed to a level of sophistication (see chapter by Neumann and Kawaoka and chapter by Conzelmann, this volume) allowing the design not only of improved vaccines against the homologous pathogens but also of multivalent vaccines inducing humoral and cellular immune response against other pathogens. Moreover, this technology is allowing us to revisit cytoreductive virotherapy, a field of experimentation that developed in the 1950s but was superseded by chemo- and radiotherapy protocols for treatment of cancer.

In this chapter, we briefly discuss those recent advances in understanding of the replication and assembly of negative-strand RNA viruses that are of paramount importance for the development of improved live vaccines. We also review the available data on the genetic stability of recombinant viruses from the perspective of the interplay between the accumulation of mutations and of selection. The focus then shifts to the core of the matter: recent studies developing recombinant, improved monovalent as well as multivalent vaccines. Finally, the paradigms on which a new field of experimentation aiming at the production of therapeutic viruses is developing are critically evaluated.

1.1
Negative-Strand RNA Virus Vectors: Genetic and Structural Constraints

Of the seven families of negative-strand RNA viruses three (*Arenaviridae*, *Bunyaviridae*, and *Orthomyxoviridae*) have segmented genomes and the other four (*Bornaviridae*, *Filoviridae*, *Paramyxoviridae*, and *Rhabdoviridae*) have nonsegmented genomes. We first discuss the parameters influencing the stability of nonsegmented genomes and the issues that must be considered when the modification of these genomes is planned.

The lack of proofreading by the polymerases of RNA viruses implies higher mutation rates compared with DNA viruses. Nevertheless, the mutation rate is only one factor influencing genome evolution. The stringency of selective constraints determines how fast mutants are selected. Recombinant *Rhabdoviridae* expressing additional genes have been found to be surprisingly stable (Schnell et al. 1996). However, there is no doubt that mutants of any natural or recombinant virus involving stop codons or other genetic alterations within the added gene(s) arise continuously, and when subjected to strong selective pressure these mutants can be rapidly selected (Quinones-Kochs et al. 2001; Rager et al. 2002; Schnell et al. 1998). Therefore, it appears that under standard tissue culture conditions selective constraints on recombinant viruses that encode selectively neutral genes are minor.

The observed stability of added genes probably also is due to the relaxed constraints on genome length in negative-strand RNA viruses. Their particles have a helical nucleocapsid and a pleomorphic envelope, a combination that tolerates considerable changes in genome length. This is in contrast to viral particles of icosahedral symmetry. Early experiments based on particle sedimentation and ultraviolet inactivation indicated that the pleomorphic particles of parainfluenza viruses often incorporate more than one genome (Dahlberg and Simon 1969; Hosaka et al. 1966), and more recent studies indicate that polyploid particles are as infective as standard viruses (Rager et al. 2002), consistent with the stable maintenance of large segments of foreign genetic materials by recombinant viruses.

The lack of a significant nuclear phase in most nonsegmented negative-strand RNA viruses is another favorable aspect from the perspective of vector development because it insulates these infectious agents from extensive genetic interference with cellular processes. Finally, the fact that genomes of negative-strand RNA viruses do not recombine is also of relevance for vector development.

1.2 Engineering Nonsegmented Negative-Sense RNA Viruses

Shortly after the establishment of systems to generate recombinant nonsegmented negative-sense RNA viruses entirely from cDNA, it was demonstrated that foreign genes can be expressed from an additional transcription unit. The foreign gene must be flanked by virus-specific gene-start and gene-end motifs, which direct transcription by the viral RNA-dependent RNA polymerase (Bukreyev et al. 1996; Mebatsion and Conzelmann 1996). The location of this transcription unit determines the amount of protein that is expressed, because of attenuation of transcription at gene boundaries inherent in the transcriptional mode of the viral RNA polymerase (Cattaneo et al. 1987; Iverson and Rose 1981). Alternative methods for the expression of foreign sequences include bicistronic transcription units (Sect. 3.2) or fusion proteins (Sect. 3.1).

The recovery of viruses with up to three additional genes, each in a separate transcription unit at different positions in the genome, has been reported. The resulting increase in size of the viral genome can reach as much as 40% (Skiadopoulos et al. 2002; L. Hangartner, T. Cornu and M. Billeter, personal communication). Depending on the virus and the characteristics of the additional gene(s), the introductions lead to a 3- to 20-fold reduction of virus yield in vitro (Haglund et al. 2000; Sakai et al. 1999). The stability of the additional gene seems to be strongly dependent on potential adverse effects of its encoded protein on virus replication. Whereas reporter genes like chloramphenicol acetyl transferase (CAT) or enhanced green fluorescent protein (eGFP) are efficiently expressed over 20–30 passages, other proteins that interfere with virus replication are rapidly eliminated by the introduction of point mutations or the addition or deletion of single nucleotides (Quinones-Kochs et al. 2001; Rager et al. 2002; Wertz et al. 2002).

A different issue that is relevant in this context is the expression of foreign proteins intended to be incorporated in the viral envelope to improve immunological recognition (Sect. 3.2) or to retarget cell entry (Sect. 4.3).

1.3 Engineering Segmented Negative-Strand RNA Viruses

The engineering of segmented negative-strand RNA viruses must consider genetic reassortment. An additional gene presented as vRNA flanked by the appropriate 5′ and 3′ terminal bases is initially packaged into viral particles (Luytjes et al. 1989) but lost quickly during subsequent rounds of replication if it does not perform an essential viral function. To avoid rapid loss of expression, the nonessential reading frame must be coupled to a viral gene whose maintenance is essential for virus viability, typically by engineering bicistronic transcripts. Bicistronic transcripts have been obtained with internal ribosomal entry sites (IRES) or a short self-cleaving protease sequence.

By inserting additional sequences downstream of the open reading frame (ORF) of the neuraminidase (NA) gene it was shown that influenza A virus can tolerate an insertion of at least 1 kb in that gene (Garcia-Sastre et al. 1994b). In a different experimental approach using an IRES, an additional ORF was introduced upstream of the NA ORF. By this means conservation of certain characteristics of the first ORF is necessary to ensure NA synthesis, and therefore inserts tend to be more stable (Garcia-Sastre et al. 1994a). As an alternative system for the tandem expression of genes the self-cleaving protease 2A of the foot and mouth disease virus has been inserted in place of the IRES between the additional gene and NA. Again, in this bicistronic transcription unit conservation of the upstream (foreign) ORF is required to express the coupled viral ORF (Percy et al. 1994). These systems, in combination with the recently established highly efficient system to generate influenza A virus entirely from cDNA (Fodor et al. 1999; Neumann et al. 1999), have extended the versatility of the reverse genetics system for influenza A and allow the rapid generation of recombinant viruses that express additional genes.

2 Improved Vaccine Viruses

Live attenuated vaccines have traditionally been generated by repeated passages of virulent virus strains through various cultured cells as well as eggs. Alternatively, closely related viruses that have a different species as original host have been used as vaccines. For certain viruses, the genomes of the attenuated vaccine viruses and their direct wild-type ancestors are available and have been completely sequenced; for other viruses, these matched pairs are not available. The systematic characterization of these genes and then of the single point mutations responsible for attenuation began once reverse genetics systems became available, taking advantage of those matched pairs.

2.1 Identification and Importation of Attenuating Mutations

A cold-adapted (cp) human parainfluenza virus-3 (hPIV3 cp45) that is a promising vaccine candidate was found to have 15 differences compared with its wild-type parent. Systematic genetic analysis of the importance of these mutations revealed that three amino acid substitutions in the polymerase (L) gene are crucial for attenuation. Comprehensive studies in rodents and primates indicated that not only these three residues contribute to a temperature-sensitive phenotype and cause a 3–5 times decrease in replication in the upper and lower respiratory tract of hamsters (Skiadopoulos et al. 1998).

This work set the stage for the next development, the importation of mutations responsible for attenuation from one virus species to the other. This is a distinct advantage of reverse genetics technology for vaccine production. In this experimental paradigm, amino acid exchanges shown to affect the efficiency of a replicative process in one virus may be transferred to the homologous protein of a related virus, even one that is relatively divergent. Proof of principle for the importation of attenuating mutations from related virus species was established again in the hPIV system, by importing into the hPIV polymerase a mutation characterized in the corresponding protein of respiratory syncytial virus (RSV). The high conservation of certain segments of L proteins among nonsegmented negative-strand RNA viruses was the rationale for introducing a phenylalanine-to-leucine mutation conferring temperature sensitivity in RSV. Introduction of that mutation resulted in a temperature-sensitive phenotype in a recombinant hPIV and caused a 10- to 100-fold decrease in replication in hamsters when combined with the three previously identified mutations of the cp45 live attenuated vaccine candidate. A fivefold decrease in replication was observed in chimpanzees, and a moderate- to high-level antibody response was characterized (Skiadopoulos et al. 1999a).

An alternative approach to attenuation, gene order rearrangement, was pioneered with vesicular stomatitis virus (VSV). In this system the linear order of the first four genes on the genome was systematically altered, and the effects of gene rearrangement on replication in vitro and on pathogenesis in a natural host were monitored (Flanagan et al. 2001; Wertz et al. 1998). Recombinant VSV strains with different attenuation levels have been shown to be genetically stable and to protect animals from disease. Application of similar principles to human nonsegmented negative-strand RNA viruses may now be envisaged.

2.2 Glycoprotein Exchange

Glycoprotein exchange involves the use of one viral backbone to carry and express the antigenic determinants of another virus. This can be useful for vaccine design, in which one attenuated viral backbone can be used to express the antigenic determinants of a second, wild-type virus for which vaccine strains do not exist, thus expediting vaccine development. In addition, glycoprotein exchange can be attenuating, likely because of possible suboptimal interaction of the heterologous proteins during virus replication and assembly and within the virions. This method is illustrated with the PIV system, where the glycoproteins, or segments thereof, were exchanged between an attenuated hPIV-3 backbone and wild-type hPIV-1 or hPIV-2, for which attenuated strains or a reverse genetics system did not exist at the time. A recombinant hPIV-3 with the fusion (F) and hemagglutinin/neuraminidase (HN) glycoproteins of hPIV-2, whose cytoplasmic tails were replaced with their hPIV-3 counterparts, was generated, and replicated efficiently in vitro. Hamsters, African green monkeys, and chimpanzees immunized with the recombinant virus were protected against challenge with hPIV-2, despite the attenuation for replication in the respiratory tract that was observed in vivo (Tao et al. 2000b). A similar approach was taken by introducing the F and HN proteins of hPIV-1 into the live attenuated hPIV-3 cp45 background. In vivo, the chimeric virus was about fivefold more restricted for replication in the respiratory tract of hamsters than the attenuated hPIV-3 cp45 parental virus (Skiadopoulos et al. 1999b). Interestingly, hamsters immunized with the chimeric virus not only were resistant to hPIV-1 challenge but showed moderate resistance to challenge with hPIV-3 as well, even though only the internal proteins of that virus were presented (Tao et al. 1999). This protective effect likely was mediated by cytotoxic T cells specific to the internal hPIV-3 proteins, and it waned within the space of several months. Nonetheless, a protective immune response against hPIV-1 could be induced with this virus even in animals with preexisting immunity against hPIV-3 (Tao et al. 2000a).

In another series of experiments, attenuation was sought by engineering viruses combining the replication apparatus of an animal virus with the envelope of its human homolog. In particular, the F and HN genes of bovine (b)PIV-3 and hPIV-3 were introduced into the respective heterologous systems, and the replication efficiency in the upper and lower respiratory tract of rhesus macaques was assessed. The replication efficiency of hPIV-3 carrying the bovine glycoproteins was restricted to a level similar to bPIV-3, whereas that of bPIV-3 with the human glycoproteins lay between the two parental viruses. This suggests that host range determinants are not exclusively located on the glycoproteins, even though these clearly play an important role in tropism (Schmidt et al. 2000). The importance of this approach for human vaccines was that it combined the backbone of bPIV-3, a virus that is attenuated in primates because of a natural host range restriction, with the major antigenic determinants of hPIV-3, thus providing a novel vaccine against hPIV-3.

A similar cross-species approach was taken with a recombinant bovine RSV (bRSV), in which the F gene alone or in combination with the attachment (G) gene were replaced with the respective proteins of two different human strains. The viruses displayed intermediate growth characteristics in human and bovine cells, but their replication was highly restricted in chimpanzees compared with human (h)RSV. Animals immunized with the recombinant viruses were not significantly protected against challenge with wild-type hRSV, suggesting that the magnitude of growth restriction was too great to provide adequate antigen expression and that the replication efficiency of this chimeric vaccine candidate in primates should be improved by the import of additional hRSV genes (Buchholz et al. 2000).

Thus immunization with a closely related animal virus that is nonpathogenic for humans, or the use of such an animal virus as a recombinant vector to bear the antigenic determinants of the relevant human pathogen, provides "Jennerian" and "modified Jennerian" means of vaccination, respectively. In general, glycoproteins representing the major neutralization antigens are the relevant proteins for inducing a long-lasting protective response. This was illustrated by the hPIV-3-1 virus described above, where the internal genes provided only transient protection. Also, the replacement of the RV G protein with that of VSV led to a loss of protection against challenge with a pathogenic RV strain, suggesting that in this system the expression of internal proteins alone also is not sufficient to elicit an effective immune response (Foley et al. 2000).

2.3
Ablation of Proteins

A third attenuation method that we will discuss only briefly is the silencing of the expression of accessory proteins essential for efficient virus replication and pathogenesis in an organism but dispensable for virus replication in cultured cells. Such "accessory proteins" often work by antagonizing host defense factors (see chapters by Nagai and Kato and García-Sastre, this volume). Silencing or, when possible, deleting the genes coding for these proteins is an attractive strategy because this may improve the safety and efficacy of the resulting vaccine. Moreover, it is practical because recombinant viruses lacking these proteins frequently grow to high titers in cultured cells, a feature necessary for efficient preparation of vaccines.

The expression of the V and C proteins of several paramyxoviruses has been silenced by various means, and studies with C- and/or V-silenced viruses have given important insights in the mechanisms of actions of theses proteins (Kato et al. 1997; Kurotani et al. 1998; chapter by Nagai and Kato, this volume). The question about the applicability of C or V "knock-out" mutations for the production of more attenuated vaccines has been asked not only in the Sendai system (Kano et al. 2002) but also in most detail once again in the hPIV-3 system. It was concluded that the simultaneous ablation of the C and V proteins provides an intermediate level of attenuation in the lower respiratory tract but that the C ablation mutant is overattenuated (Durbin et al. 1999). Because the V and C proteins of other viruses may have sets of functions that are only partially overlapping, it is important to verify the consequences of protein silencing for each protein and virus individually.

In addition to these mutants, involving genes that might have been anticipated to be nonessential in vitro, viruses with mutations in proteins that had been expected to be essential for replication have also been obtained. These mutations include, for example, the ablation of two glycoproteins in hRSV (Karron et al. 1997; Techaarpornkul et al. 2001) or expression silencing of the influenza M2 ion channel (Takeda et al. 2002; Watanabe et al. 2001). In contrast to the experience with mutants in the accessory proteins, it appears likely that these mutants may be too attenuated to produce a standard vaccine. Nevertheless, such mutants may become the vectors of choice for the vaccination of immunosuppressed individuals.

3
Multivalent and Vectored Monovalent Vaccine Viruses

As a logical extension of the work on obtaining improved vaccines against individual viruses, the exploration of the potential of negative-strand RNA viruses as multivalent vaccines has begun. To this end, different approaches have been used. First, the incorporation of immunogenic epitopes into viral proteins has been sought, by using strategies allowing preservation of the function of the "carrier" proteins. Second, expression of the protein of interest from an additional ORF in the viral genome has been achieved. The common goal is to generate a protective immune response against other pathogens. Because the replicating, recombinant viruses express the gene of interest in the target cells, the corresponding protein is presented in the immunogenic context of a viral infection that may act as an adjuvant.

A live attenuated vaccine candidate not only must be satisfactorily attenuated but also must maintain sufficiently efficient replication in cultured cells for production and in humans to be immunogenic. Therefore, a balance between attenuation and immunogenicity is required. In the previous section, we discussed attempts at developing viruses with different levels of attenuation. In the case of multivalent vaccines, attenuation may be inherent with the insertion of any additional gene and maintaining efficient replication becomes the issue.

3.1
Introduction of Foreign Epitopes into Viral Proteins

One approach toward the generation of recombinant vaccines is the incorporation of a T or B cell epitope of interest into a viral glycoprotein while maintaining its functionality. The influenza virus system is best characterized for this approach. The introduction of a cytotoxic T cell epitope of lymphocytic choriomeningitis virus (LCMV) into the NA stalk of influenza A virus led to the protection of mice against LCMV challenge for at least 4 months after vaccination (Castrucci et al. 1994). A more commonly used display site is the antigenic site B, residing in the head of the influenza A HA protein. Several epitopes have been displayed in this position. Proof of principle for this display technology was brought initially with expression of the hybrid protein and injection of purified material. Mice immunized with a chimeric protein, in which the V3 loop of human immunodeficiency virus-1 (HIV-1) *env* was introduced, developed high levels of antibodies against HA and the inserted epitope, even though only nanograms of protein were injected (Kalyan et al. 1994).

With the possibility of generating recombinant influenza viruses, this approach was tested in the viral context. A virus carrying the chimeric HA protein with the V3 loop of HIV-1 elicited anti-HIV antibodies and cytotoxic T cells against the epitope in vaccinated mice (Li et al. 1993a). Vaccination of mice using a recombinant influenza A virus with the neutralizing HIV-1 gp41 epitope ELDKWA in the HA protein also resulted in the production of antibodies and antibody-secreting cells not only in the upper respiratory tract, which is the site of viral replication, but also in blood, spleen, and the genital tract (Ferko et al. 1998). A vaccination strategy combining a recombinant influenza virus that expressed the V3 loop of HIV-1 with a recombinant vaccinia virus expressing the complete *env* protein led to an enhanced $CD8^+$ T cell response in mice compared with vaccination with only one of the viruses (Gonzalo et al. 1999).

The presentation of epitopes of interest positioned in an antigenic site of a viral protein within the viral context has also proven to be effective for nonviral antigens for which B or T cell epitopes have been characterized. Mice vaccinated with an influenza virus expressing a B cell epitope of the *Plasmodium falciparum* circumsporozoite protein, followed by a booster with a recombinant vaccinia virus expressing the complete protein, developed a strong $CD8^+$ T cell response (Miyahira et al. 1998). Similar results were obtained by using the same strategy with the rodent-specific parasite *P. yoelii*, protecting mice against sporozoite-induced malaria (Li et al. 1993b).

A systematic approach was taken with peptide 10, one of the two characterized linear B cell epitopes in the outer membrane protein F of *Pseudomonas aeruginosa.*. Influenza viruses that carried penta- to dodecameric peptides in the HA protein were generated, and their ability to elicit an immune response in mice was characterized. The 8- and 11-residue inserts led to the generation of high antibody titers (Gilleland et al. 1997). All these studies concur in showing that one or more arms of the immune system can be mobilized to extend protection against additional pathogens. Nevertheless, for vaccination purposes the use of complete proteins of pathogens may be more efficient. Therefore, the expression of additional proteins from recombinant viruses was sought.

3.2 Expression of Additional Proteins and Glycoprotein Replacement

3.2.1 Rhabdovirus-Based Vectors

Among nonsegmented RNA viruses, the proven high efficiency of foreign protein incorporation in the envelope, the tolerance of the particles in accepting larger genomes, and the low seroprevalence predestined the rhabdoviruses VSV and RV to be the vector systems initially used for expression of heterologous antigenic proteins. In particular, it was demonstrated that a recombinant RV in which the endogenous G protein was replaced by a chimeric HIV-1 *env* protein with the cytoplasmic tail of RV G replicated exclusively in CD4-expressing cells and could be neutralized by anti-HIV sera (Mebatsion and Conzelmann, 1996). Similar viruses that expressed the *env* protein of HIV-1 strains with different tropism infected various blood cells and entered CD4-expressing cells, in contrast to wild-type RV but similar to HIV-1, by a pH-independent pathway (Foley et al. 2002).

Another approach involved the introduction in the RV genome of the HIV *env* genes from strains with different tropism. These genes were embedded in additional transcription units. Mice inoculated with the recombinant viruses mounted a strong humoral immune response against *env* and high levels of neutralizing antibodies against HIV after a single boost with recombinant HIV-1 gp120 protein (Schnell et al. 2000). Furthermore, a strong cytotoxic T-lymphocyte response was induced, able to cross-kill target cells expressing *env* from heterologous HIV-1 strains (McGettigan et al. 2001).

The bovine rhabdovirus VSV was developed in parallel with RV as a vaccine virus for HIV-1. Recombinant VSVs expressing the HIV *gag* with or without *env* protein were generated. Production of HIV-like particles in addition to VSV particles was observed after expression of *gag* in this system, and *env* was specifically incorporated into the HIV-like particles (Haglund et al. 2000).

Another new concept in vaccination, that of envelope exchange for boosting, was proven in the VSV system. Because recombinant viruses expressing foreign antigens are bound to induce high-titer neutralizing antibodies directed against the vector envelope glycoprotein and these antibodies can prevent reinfection and boosting with the same vector, it was hypothesized that efficient boosting could be achieved by replacing the envelope with G proteins from other serotypes. Indeed, mice immunized with VSVs expressing HIV *env* in addition to the G protein of the Indiana strain could be boosted efficiently by using other recombinants carrying the New Jersey or Chandipura G protein in its place (Rose et al. 2000). The immunization of Rhesus macaques with this strategy, using viruses that express the HIV *env* and *gag* protein, proved to be effective to protect animals from a challenge with pathogenic SHIV (Rose et al. 2001). This vaccine strategy is currently one of the most promising novel developments in the field of AIDS prevention. It is directly analogous to using vaccinia virus as a vector, but VSV offers more flexibility in making different serotype-based viruses, in attenuating, and in the likely absence of antagonists of host immunity. Its efficiency is approached only by repeated vaccinations with high amounts of a DNA-based vaccine followed by boosting with a recombinant modified vaccinia virus Ankara expressing multiple HIV proteins (Amara et al. 2001) and matched by a combined plasmid DNA/replication-incompetent adenovirus/*gag* vaccine protocol (Shiver et al. 2002).

Another avenue of experimentation that was developed with the VSV system is the incorporation of foreign proteins into the envelope: Many foreign glycoproteins expressed from recombinant VSV are incorporated efficiently into particles (Schnell et al. 1996). However, certain proteins may interfere with virus particle assembly, and it is not possible to predict whether a foreign protein will be well tolerated. The display of foreign proteins on VSV particles may improve recognition and enhance the immune response. It was shown that the influenza A glycoproteins

HA and NA expressed from additional transcription units are efficiently incorporated into viral particles (Kretzschmar et al. 1997). Mice immunized intranasally with the recombinant VSV expressing influenza HA produced high levels of neutralizing antibodies and were protected against challenge with a lethal dose of virus (Roberts et al. 1998). Because some VSV-related pathogenicity was observed in this study, the effectiveness of viruses with a truncation in the G protein cytoplasmic domain or without G protein was examined. Both viruses were entirely nonpathogenic and provided complete protection from lethal influenza virus challenge. Furthermore, the virus without G protein had the added advantage of not inducing neutralizing antibodies to the vector itself (Roberts et al. 1999).

A similar approach was pursued with the aim of generating a vaccine against hRSV. Recombinant VSV that expressed either the F or the G protein of hRSV from additional transcription units were produced. Both proteins were transported to the cell surface and incorporated into VSV particles, and hRSV F protein-expressing viruses were shown to fuse cells at neutral pH, reflecting the functionality of the incorporated protein (Kahn et al. 1999). Mice immunized with those VSV recombinants elicited hRSV-specific antibodies in serum as well as neutralizing antibodies to hRSV and were completely protected against RSV challenge. In a second study, nonpropagating VSVs that lack the VSV G gene and were transcomplemented with either of the RSV F and G glycoproteins were tested for their ability to induce protective immunity in the mouse model. Specific antibodies against RSV, but no neutralizing antibodies, were detected only in the animals immunized with the virus transcomplemented with the RSV F protein, which were protected against RSV challenge (Kahn et al. 2001).

Recombinant VSV expressing the MV hemagglutinin (H) protein were generated to explore the possibility that these viruses may circumvent the strong maternal immunity present in infants from vaccinated or previously infected mothers. The virus induced high levels of neutralizing antibodies in cotton rats in the presence of passively transferred human or cotton rat anti-MV antibodies and protected them from subsequent challenge with MV. Because MV H is not a functional part of the VSV envelope, MV-specific antibodies only slightly inhibited replication of the recombinant virus in vitro (Schlereth et al. 2000).

In summary, VSV- and RV-based vectors have proven to be exceptionally effective in introducing immune responses against additional viral pathogens, possibly because of an "adjuvant-like" effect of viral infection, based in part on the co-incorporation of certain proteins in the viral envelope. The development of rhabdovirus-based vectors as HIV vaccines is being pursued energetically. It may, however, be less likely that these vectors will be used as vaccines for measles, influenza, or other human diseases for which a vaccine is available. Neither VSV nor RV represents a needed vaccine, which reduces their desirability as a vector for the general population.

3.2.2
Multivalent Vaccine Vectors Based on Paramyxovirus

Vectors based on paramyxoviruses have also been constructed for the expression of foreign glycoproteins and vaccine development. A safe and effective live attenuated MV vaccine is available for protection of children, and PIV and RSV vaccines are currently being vigorously developed. In parallel, the construction of multivalent vaccines based on hPIV-3, a representative of the *Respirovirus* genus, has been sought.

By introducing the hPIV-2 HN protein in an additional transcription unit into a recombinant hPIV-3 with the glycoproteins of hPIV-1 it was demonstrated that this virus tolerates insertion of additional genes into its genome. The resulting virus induced resistance against challenge with both hPIV-1 and hPIV-2 in hamsters vaccinated intranasally (Tao et al. 2001). This demonstrated that an attenuated hPIV-3 backbone can be used to express the antigenic determinants of heterologous PIVs as gene substitutions (hPIV-1) and as added genes (hPIV-2). The introduction of the MV H gene as additional transcription unit into different locations in the hPIV3wt genome, or of an attenuated hPIV-3 (rcp45L) genome, caused a modest further attenuation of the recombinant viruses in vitro. Hamsters vaccinated with the recombinant viruses generated a strong immune response against hPIV-3 and MV. Viruses with the H gene at upstream positions (high expression levels) induced about 400-fold more MV-neutralizing antibodies than the virus with MV H at a downstream position (lower expression levels) (Durbin et al. 2000).

A further attenuated virus that carried the bPIV-3 nucleoprotein (N) instead of its own, and additionally expressed the MV H protein, proved to be attenuated for replication but was still able to elicit a strong immune response against MV and hPIV-3 in monkeys immunized intranasally (Skiadopoulos et al. 2001). Moreover, a recombinant virus expressing all of the above HN proteins (hPIV-1, -2, and -3 and MV H) was constructed with a hPIV-3 backbone, and its pathogenicity and immunogenicity in the respiratory tract of hamsters were characterized. The study clearly indicates that the type, number, and level of expression of the foreign proteins produced do influence the efficiency of virus production and of the immune response. Therefore, the number of foreign proteins to be expressed by potential multivalent vaccines may be limited, but nonetheless it appears to be possible to make effective bi- and trivalent vaccine viruses based on a hPIV backbone.

Bovine RSV, a *Pneumovirus* genus member, was also developed as a multivalent vaccine vector. The tolerance of bRSV toward coexpressed heterologous glycoproteins was explored by replacing the F and G protein individually or in concert with the F and/or HN protein of bPIV-3. Viruses in which the G protein was replaced with the HN protein or both glycoproteins were exchanged could be recovered, whereas viruses in which only the F proteins were exchanged were not viable. Because RSV requires only its F protein for replication in vitro, it seemed likely that the expression of an additional, heterologous glycoprotein would be tolerated, but the recovery of a fully functional virus with completely exchanged envelope is remarkable (Stope et al. 2001). The efficacy of these recombinant viruses as veterinary vaccines can now be tested.

MV is a *Morbillivirus* genus member of interest for human vaccination because of its ability to elicit life-long immunity. The available extremely safe live attenuated MV vaccine is an ideal base for the development of multivalent vaccines. A recombinant MV that expressed the small surface antigen of hepatitis B virus (HBsAg) produced humoral immune responses against HBsAg as well as MV. The appropriately glycosylated HBsAg was uniformly expressed after 10 passages, indicating a sufficient stability of the transgene (Singh et al. 1999). Recombinant MV expressing either the mumps HN or F or the SIV *gag*, *pol*, or *env* proteins from an additional transcription unit have also been characterized. The expression of the respective additional proteins, with the exception of mumps F, was maintained for up to 20 passages (Wang et al. 2001), but the effectiveness of those viruses as vaccines has not been tested in animals yet. In the MV system, an envelope exchange protocol was also developed. Replacement of the F and H glycoproteins with the VSV G protein was possible, but virus particle assembly was slightly more efficient in the presence of the F protein cytoplasmic tail. Mice immunized with a recombinant MV expressing the VSV G protein instead of its endogenous envelope were protected against challenge with lethal doses of wild-type VSV (Spielhofer et al. 1998).

Finally, the potential of the other *Respirovirus* Sendai virus (SeV) as a vaccine vector was also assessed. SeV is highly virulent for mice, and therefore the study of virus-host interactions in this homologous small-animal model is straightforward. HIV-1 *env* was introduced into an attenuated SeV mutant with enhanced gene expression due to the functional inactivation of the nonessential V gene. The protein was expressed stably at high levels, was readily purified from the culture supernatant, and maintained its functionality and immunological properties (Yu et al. 1997). By employing a prime/boost regimen, the efficacy of a SeV vector in protecting macaques from AIDS was tested in combination with a naked DNA vaccine priming. A single boost with SeV expressing HIV-1 *gag* efficiently generated a secondary immune response in macaques primed with *env*- and *nef*-deleted proviral DNA and protected from acute CD4 depletion after challenge with intravenous SHIV89.6PD, a highly virulent strain. The animals also showed greatly reduced peak viral loads compared with controls. Vaccination with the DNA alone or SeV *gag* alone was not enough to confer the consistent protection from CD4 cell depletion, although it led to efficient secondary CD8 T-cell responses, indicating the value of the prime/boost strategy (Matano et al. 2001). In summary, paramyxoviruses of different subfamilies are currently being developed as multivalent vaccine vectors. The different characteristics of these viruses may result in a number of different vector applications. In particular, paramyxovirus-based vaccines with the potential of being rapidly developed for clinical trials include MV or SeV expressing HIV antigens and pediatric vaccines based on hPIV backbones.

4
Cytoreductive Virotherapy

The cytoreductive potential of certain negative-strand RNA viruses was characterized as early as the 1950s, but virotherapy was superseded when chemo- and radiotherapy proved effective and easier to control. Recently, the resistance of many cancers to those conventional therapies and the development of reverse genetics systems have revitalized the virotherapy field (Ring 2002; Russell 1994). Current developments in this field include clinical trials with natural and recombinant virus strains and much-needed investigations of the mechanisms of oncolysis. Here we review the developments based on natural or recombinant negative-strand RNA viruses that are being evaluated as vaccines boosting immunity against cancerous cells or as vectors capable of eliciting the elimination of infected cells by cytolysis.

4.1 Newcastle Disease Virus

The cytoreductive potential of Newcastle disease virus (NDV), an avian paramyxovirus, has been thoroughly investigated. A large variety of human tumor cells were shown to support efficient NDV replication independently of tumor cell proliferation. Such a property makes NDV a suitable agent for immunotherapy: stimulation of the immune response against the tumor. In this experimental therapy freshly isolated and gamma-irradiated patient-derived tumor cells are modified by NDV infection with the aim of eliciting an immune response against the tumor after immunization with the protein preparation, and the therapy does not involve exposure of patients to replicating virus. For the apathogenic nonlytic strain NDV-Ulster, used in clinical vaccine trials, selective replication in tumor cells compared with corresponding normal cells was characterized (Schirrmacher et al. 1999). Clinical protocols were based on the vaccination of patients with lysate of allogeneic and autologous tumor cells previously infected with live NDV, which resulted in a marked increase in short- and long-term survival rates in patients with various tumors, with no adverse effects (Batliwalla et al. 1998; Ockert et al. 1996). It appeared that immunomodulation with autologous or allogeneic tumor cell vaccines is mainly due to cytokine induction, whereas tumor-specific humoral or cellular responses are not detectable in the peripheral blood of patients (Zorn et al. 1997).

In addition, the benefits of direct application of replication-competent NDV have increasingly been explored during recent years, leading to promising results for the virotherapy in initial clinical trials. NDV strains with more or less extensive cytopathic effect (in short lytic or nonlytic) strains were compared with regard to their antitumor effects after local or systemic application. When tumor cells were preinfected in vitro, it appeared that the non-lytic strain showed stronger antitumor activity, whereas only the lytic strains were effective for the treatment of preestablished human melanomas by intratumoral injection of NDV. Intra- or peritumoral application of NDV or NDV-infected tumor cells showed more pronounced antitumoral activity than systemic application (Schirrmacher et al. 2001). With the use of human carcinoma xenografts in athymic mice, it was shown that NDV administered either locally or systemically is an effective antitumoral therapy in this system and that replication competence was necessary for maximal effect. Furthermore, it was demonstrated that multiple NDV doses are more effective than a single dose (Phuangsab et al. 2001). Together, these data indicate that both a direct cytopathic effect of NDV and the indirect activation of an antitumoral immune response contribute to oncolysis.

Clinical phase I trials based on the above experiments have been performed. In these trials, it was observed that repeated intravenous administration of up to 100 times 10^9 plaque-forming units of the NDV strain PV701 to patients with treatment-resistant solid tumors was well tolerated; furthermore, in a few cases NDV treatment may have resulted in progression-free survival. Adverse effects were observed: They consisted of flulike symptoms that decreased with each subsequent dose, tumor site-specific adverse effects, and acute dosing reactions, but there was no cumulative toxicity. Thus virotherapy with NDV will be explored further (Pecora et al. 2002).

4.2
Vesicular Stomatitis Virus

VSV has been identified as another candidate vector for cytoreductive therapy, because it is minimally pathogenic for humans and extremely sensitive to the antiviral effects of interferon (IFN). It was demonstrated that it selectively induces the cytolysis of numerous transformed human cell lines in vitro and potently inhibits the growth of p53-null C6 glioblastoma tumors in vivo without replicating in normal tissue. Inhibition of VSV replication is due to the presence of the double-stranded RNA-activated protein kinase PKR and a functional IFN system in primary human cells (Balachandran and Barber, 2000; Dummer et al. 2001; Stojdl et al. 2000). VSV replicates selectively in tumor cells that are defective in p53 function or transformed with *myc*. Cells expressing activated *ras* are also susceptible to viral cytolysis in vivo. VSV was oncolytic in the absence of any significant cytotoxic T-lymphocyte response, and VSV infection resulted in significant inhibition of tumor growth after intravenous administration in immunocompetent hosts (Balachandran et al. 2001).

To enhance the VSV oncolytic activity, recombinant viruses expressing thymidine kinase (TK) or the cytokine interleukin 4 (IL-4) were produced and their antitumoral activity was characterized in the same system. In vitro, high levels of biologically active TK or IL-4 were detectable, and the viruses replicated at similar levels as the parental virus. After direct intratumoral inoculation, VSVs expressing either TK or IL-4 were considerably more oncolytic in syngeneic breast or melanoma tumors in mice than the parental virus. In addition, VSV expressing IL-4 or TK, but not GFP, was more effective against metastatic disease after intravenous administration. This demonstrates the validity of the approach based on recombinant VSV expressing conditionally toxic or immunostimulating proteins for the treatment of malignant diseases (Fernandez et al. 2002).

The target of cytoreductive virotherapy has recently been extended from cancer to infectious diseases. Two groups (Mebatsion et al. 1997; Schnell et al. 1997) have presented an original approach for targeting and eliminating HIV-infected cells, and thus toward the treatment of AIDS, relying on *Rhabdovirus* vectors. These authors hypothesized that it may be possible to replace the envelope protein of the two rhabdoviruses, VSV and RV, with CD4 and CXCR4, the receptor and coreceptor of HIV-1. In particular, it was envisaged that the recombinant viruses may selectively infect cells expressing the HIV-1 glycoprotein *env*. To test this hypothesis in the VSV system, the HIV-1 receptor and coreceptor were expressed from an additional transcription unit, whereas for RV these proteins were provided *in trans* by an expression plasmid (Mebatsion et al. 1997; Schnell et al. 1997). Indeed, the viruses were unable to infect normal cells but replicated efficiently in HIV-1-infected cells that were killed because of the lytic character of a VSV infection. The recombinant VSV led to a reduction of the HIV titer by 10^4 in HIV-1-infected cells that were superinfected with this virus (Schnell et al. 1997). In both the VSV and RV systems, the cytoplasmic tails of the foreign glycoproteins were replaced by that of the respective G protein to ensure efficient incorporation into viral particles. Targeting one virus to cells infected by another could result in decrease of viral load and thus control disease progression. This strategy may be applicable not only for HIV but also for other chronic viral diseases.

4.3
Measles Virus

Infection with wild-type strains of MV causes considerable morbidity and mortality, which result from the profound immune suppression that accompanies infection. However, measles can be effectively prevented by vaccination. The live attenuated vaccine strain Edmonston is one of the safest and most successful vaccines in use and is recommended even for vaccination of HIV-infected children other than those severely immunocompromised (Moss et al. 1999). The oncolytic properties of MV make it particularly suitable for development as a cytoreductive agent designed to eliminate cancerous cells. There have been several reports of regression of Hodgkin and non-Hodgkin lymphoma after natural MV infection (Bluming and Ziegler 1971; Taqi et al. 1981). Moreover, the MV vaccine strain replicates considerably more efficiently in PBMC of myeloma patients than in PBMC of healthy controls. Finally, it was shown that intratumoral injection of unmodified MV induces the regression of large established human lymphoma or myeloma xenografts in immunocompromised mice and that intravenous administration of MV also results in considerable slowing of tumor progression (Grote et al. 2001; Peng et al. 2001).

A major issue in regard to the application of MV-based therapeutics is the high prevalence of antibodies in the population. Even if in mice the antitumoral effect still occurred in the presence of passively transferred anti-MV antibodies, the effects of anti-MV antibodies and the strong immunosuppressive properties of the virus must be examined further before a therapeutic application is possible. Possible measures to counteract the effects of antibodies are plasmaphoresis or envelope exchange, as demonstrated with VSV. Measures counteracting immunosuppression may include weakening the interactions with its receptor, the immune cell-specific protein SLAM (CD150), and expression of IL-12, a cytokine that is downregulated in MV infections.

The demonstration that MV can enter cells through targeted receptors, for example, via proteins overexpressed in certain types of cancer like the EGF receptor or carcinoembryogenic antigen (CEA), has broadened the scope of MV-based oncolytic therapy (Hammond et al. 2001; Schneider et al. 2000). To redirect cell entry, replicating MVs have been engineered in which the gene for the standard attachment (H) protein is replaced by one with an additional specificity domain, namely, EGF or single-chain antibody against CEA, appended to the ectodomain. In all protein hybrids a flexible linker inserted between the H ectodomain and the specificity determinant reduces structural constraints on the protein hybrid. This genetic "single-chain antibody display" technology appears to be broadly applicable: Recombinant MV that enter cells through CD20, a lymphoma-specific protein, or CD38, a myeloma specific protein, have been developed (Bucheit et al. 2003; Peng et al. 2003). Thus it is possible to use specificity domains as diverse as growth factors or single-chain antibodies to confer entry through targeted receptors.

However, all the above viruses also maintain the ability to enter cells via the natural receptors, the immune cell-specific protein (CD150) and the ubiquitous protein CD46 (MCP). Several residues supporting the interactions of MV H with one or the other of these natural receptors have been identified recently, and recombinant viruses that efficiently enter cells through only one of the natural receptors have been produced (Vongpunsawad et al. 2004). Even if exclusive targeting to a predetermined receptor may be difficult, the available viruses will make it possible to address in vivo questions regarding virus dissemination and pathogenesis/therapeutic effects. Moreover, even preferential targeting of certain tumor markers may improve the specificity and efficacy of recombinant MV in oncolytic therapy.

4.4 Other Viruses, Other Applications

Certain strains of the segmented negative-strand RNA virus influenza A were also considered as antitumoral agents in the 1960s and 1970s. More recently, a protocol based on the nesting of cytotoxic T cell epitopes of tumor-associated antigens (TAA) within the HA or NA open reading frame has been developed. Initially, it was shown that recombinant influenza viruses expressing a model TAA mediated the regression of established pulmonary metastases in mice through the induction of cytotoxic T-cell responses (Restifo et al. 1998). This study was expanded by using survival as the end point of the assay. Animals with a high tumor burden showed extended survival times when treated with a recombinant influenza virus expressing a TAA, but they finally succumbed, presenting a small number of large tumors in the lungs, which no longer expressed the TAAs. On the other hand, mice with a lower tumor burden showed complete tumor regression and survival for more than 6 months, and were protected against tumor challenge, when treated with the recombinant virus (Zheng et al. 2000).

The above experimental systems have been developed with the aim of eliminating cancerous or HIV-infected cells. On the other hand, a paramyxovirus has been developed for the purpose of delivering therapeutic genes and thus for the correction of metabolic diseases. Toward this aim, genes of interest have been introduced into an additional transcription unit in the SeV genome. With viruses that express β-galactosidase, high efficiency of gene transfer into lung epithelial cells of different species was characterized in vivo (Yonemitsu et al. 2000). Moreover, the intramuscular injection of a recombinant SeV expressing the human insulin-like growth factor induced significant myofiber regeneration. Transgene expression was observed up to 1 month after injection into the rat anterior tibialis muscle (Shiotani et al. 2001). The attractive features of this system are entry efficiency and high expression levels combined with virtually no preexisting antiviral immunity in the general population. However, the issue of the immune response antagonizing long-term expression from a nonintegrating, replication-competent virus must be resolved if chronic diseases like cystic fibrosis are targeted. Vectors based on replicating viruses may be easier to apply to cytoreductive therapy: In these protocols, a strong immune response is expected to synergize with viral replication in eliciting a long-term effect.

5
Conclusions and Perspectives

More than 200 years ago, Jenner showed that inoculation of humans with a nonpathogenic animal virus (cowpox or, later, vaccinia virus) conferred protective immunity against a related human pathogen (variola virus, or smallpox). Because the animal virus was derived from cattle (Latin, *vacca*), the process was named vaccination. Fifty years ago, with the advent of methods for in vitro virus cultivation, vaccines against several additional human viruses were produced by attenuation of these viruses by passage on cells derived from other species or on eggs. Today, detailed knowledge of the biology and pathology of viruses allows us not only to devise rational approaches to virus attenuation but also to extend the scope of vaccination to more than one pathogen. Reverse genetics has been instrumental in testing these new concepts, initially with positive-strand RNA viruses and DNA viruses. The urgency of implementation of new vaccination strategies has received an additional boost by the specter of the use of viruses as biological warfare or bioterrorism agents.

Negative-strand RNA viruses have waited for reverse genetics much longer, but now their biological characteristics predestine them to lead exciting developments in vaccination against multiple pathogens, including HIV, and possibly even in therapeutic clinical interventions against different types of cancer or certain infectious diseases. The versatility of the envelope structure that is not constrained by icosahedral symmetry allows the encapsidation of genomes of vastly different lengths. In *Paramyxoviridae* the distribution of receptor attachment and membrane fusion functions on two proteins facilitates engineering of targeted viruses. In general, the simplicity of the viral envelope and its tolerance for the incorporation of foreign proteins qualify these viruses as preferred vectors.

An exciting prospect is the transformation of negative-strand RNA viruses into targeted therapeutics. This aim is currently being pursued with different families of viruses, and again the variable cargo capacity and the relatively straightforward engineering have allowed a swift pace of progress. In particular, the principle of using alternative receptors for cell entry has been established with a paramyxovirus: It was shown that recombinant measles viruses displaying different single-chain antibodies as an extension of their attachment protein selectively enter cells through the targeted receptor. Moreover, a few negative-strand RNA viruses do replicate preferentially in transformed cells; it is now essential to elucidate the basis for this attenuation in order to rationally pursue oncolytic virotherapy.

Acknowledgements. We thank Peter Collins for careful review of the manuscript and Steve Russell for discussions. This work was supported by grants of the Mayo, Siebens, and Eisenberg foundations, and the NIH (CA 90636). The salary of V. von Messling was provided by a career development award of the German Research Foundation (DFG; Emmy Noether-Program).

References

Amara, R. R., Villinger, F., Altman, J. D., Lydy, S. L., O'Neil, S. P., Staprans, S. I., Montefiori, D. C., Xu, Y., Herndon, J. G., Wyatt, L. S., Candido, M. A., Kozyr, N. L., Earl, P. L., Smith, J. M., Ma, H. L., Grimm, B. D., Hulsey, M. L., Miller, J.,

McClure, H. M., McNicholl, J. M., Moss, B., and Robinson, H. L. (2001). Control of a mucosal challenge and prevention of AIDS by a multiprotein DNA/MVA vaccine. Science 292(5514), 69–74

Balachandran, S., and Barber, G. N. (2000). Vesicular stomatitis virus (VSV) therapy of tumors. IUBMB Life 50(2), 135–8

Balachandran, S., Porosnicu, M., and Barber, G. N. (2001). Oncolytic activity of vesicular stomatitis virus is effective against tumors exhibiting aberrant p53, Ras, or myc function and involves the induction of apoptosis. J Virol 75(7), 3474–9

Batliwalla, F. M., Bateman, B. A., Serrano, D., Murray, D., Macphail, S., Maino, V. C., Ansel, J. C., Gregersen, P. K., and Armstrong, C. A. (1998). A 15-year follow-up of AJCC stage III malignant melanoma patients treated postsurgically with Newcastle disease virus (NDV) oncolysate and determination of alterations in the CD8 T cell repertoire. Mol Med 4(12), 783–94

Bluming, A. Z., and Ziegler, J. L. (1971). Regression of Burkitt's lymphoma in association with measles infection. Lancet 2(7715), 105–6

Bucheit, A. D., Kumar, S., Grote, D., Lin, Y., von Messling, V., Cattaneo, R., and Fielding, A. K. (2003). An oncolytic measles virus engineered to enter cells through the CD20 antigen. Mol Ther 7(1), 62–72

Buchholz, U. J., Granzow, H., Schuldt, K., Whitehead, S. S., Murphy, B. R., and Collins, P. L. (2000). Chimeric bovine respiratory syncytial virus with glycoprotein gene substitutions from human respiratory syncytial virus (HRSV): effects on host range and evaluation as a live-attenuated HRSV vaccine. J Virol 74(3), 1187–99

Bukreyev, A., Camargo, E., and Collins, P. L. (1996). Recovery of infectious respiratory syncytial virus expressing an additional, foreign gene. J Virol 70(10), 6634–41

Castrucci, M. R., Hou, S., Doherty, P. C., and Kawaoka, Y. (1994). Protection against lethal lymphocytic choriomeningitis virus (LCMV) infection by immunization of mice with an influenza virus containing an LCMV epitope recognized by cytotoxic T lymphocytes. J Virol 68(6), 3486–90

Cattaneo, R., Rebmann, G., Schmid, A., Baczko, K., ter Meulen, V., and Billeter, M. A. (1987). Altered transcription of a defective measles virus genome derived from a diseased human brain. EMBO J 6(3), 681–8

Chua, K. B., Goh, K. J., Wong, K. T., Kamarulzaman, A., Tan, P. S., Ksiazek, T. G., Zaki, S. R., Paul, G., Lam, S. K., and Tan, C. T. (1999). Fatal encephalitis due to Nipah virus among pig-farmers in Malaysia. Lancet 354(9186), 1257–9

Clements, C. J., and Cutts, F. T. (1995). The epidemiology of measles: thirty years of vaccination. Curr Top Microbiol Immunol 191, 13–33

Dahlberg, J. E., and Simon, E. H. (1969). Physical and genetic studies of Newcastle disease virus: evidence for multiploid particles. Virology 38(4), 666–78

Dummer, R., Dobbeling, U., Geertsen, R., Willers, J., Burg, G., and Pavlovic, J. (2001). Interferon resistance of cutaneous T-cell lymphoma-derived clonal T-helper 2 cells allows selective viral replication. Blood 97(2), 523–7

Durbin, A. P., McAuliffe, J. M., Collins, P. L., and Murphy, B. R. (1999). Mutations in the C, D, and V open reading frames of human parainfluenza virus type 3 attenuate replication in rodents and primates. Virology 261(2), 319–30

Durbin, A. P., Skiadopoulos, M. H., McAuliffe, J. M., Riggs, J. M., Surman, S. R., Collins, P. L., and Murphy, B. R. (2000). Human parainfluenza virus type 3 (PIV3) expressing the hemagglutinin protein of measles virus provides a potential method for immunization against measles virus and PIV3 in early infancy. J Virol 74(15), 6821–31

Ferko, B., Katinger, D., Grassauer, A., Egorov, A., Romanova, J., Niebler, B., Katinger, H., and Muster, T. (1998). Chimeric influenza virus replicating predominantly in the murine upper respiratory tract induces local immune responses against human immunodeficiency virus type 1 in the genital tract. J Infect Dis 178(5), 1359–68

Fernandez, M., Porosnicu, M., Markovic, D., and Barber, G. N. (2002). Genetically engineered vesicular stomatitis virus in gene therapy: application for treatment of malignant disease. J Virol 76(2), 895–904

Flanagan, E. B., Zamparo, J. M., Ball, L. A., Rodriguez, L. L., and Wertz, G. W. (2001). Rearrangement of the genes of vesicular stomatitis virus eliminates clinical disease in the natural host: new strategy for vaccine development. J Virol 75(13), 6107–14

Fodor, E., Devenish, L., Engelhardt, O. G., Palese, P., Brownlee, G. G., and Garcia-Sastre, A. (1999). Rescue of influenza A virus from recombinant DNA. J Virol 73(11), 9679–82

Foley, H. D., McGettigan, J. P., Siler, C. A., Dietzschold, B., and Schnell, M. J. (2000). A recombinant rabies virus expressing vesicular stomatitis virus glycoprotein fails to protect against rabies virus infection. Proc Natl Acad Sci USA 97(26), 14680–5

Foley, H. D., Otero, M., Orenstein, J. M., Pomerantz, R. J., and Schnell, M. J. (2002). Rhabdovirus-based vectors with human immunodeficiency virus type 1 (HIV-1) envelopes display HIV-1-like tropism and target human dendritic cells. J Virol 76(1), 19–31

Garcia-Sastre, A., Muster, T., Barclay, W. S., Percy, N., and Palese, P. (1994a). Use of a mammalian internal ribosomal entry site element for expression of a foreign protein by a transfectant influenza virus. J Virol 68(10), 6254–61

Garcia-Sastre, A., Percy, N., Barclay, W., and Palese, P. (1994b). Introduction of foreign sequences into the genome of influenza A virus. Dev Biol Stand 82, 237–46

Gilleland, H. E., Jr., Gilleland, L. B., Staczek, J., Harty, R. N., Garcia-Sastre, A., Engelhardt, O. G., and Palese, P. (1997). Chimeric influenza viruses incorporating epitopes of outer membrane protein F as a vaccine against pulmonary infection with *Pseudomonas aeruginosa*. Behring Inst Mitt(98), 291–301

Gonzalo, R. M., Rodriguez, D., Garcia-Sastre, A., Rodriguez, J. R., Palese, P., and Esteban, M. (1999). Enhanced CD8+ T cell response to HIV-1 env by combined immunization with influenza and vaccinia virus recombinants. Vaccine 17(7–8), 887–92

Grote, D., Russell, S. J., Cornu, T. I., Cattaneo, R., Vile, R., Poland, G. A., and Fielding, A. K. (2001). Live attenuated measles virus induces regression of human lymphoma xenografts in immunodeficient mice. Blood 97(12), 3746–54

Haglund, K., Forman, J., Krausslich, H. G., and Rose, J. K. (2000). Expression of human immunodeficiency virus type 1 Gag protein precursor and envelope proteins from a vesicular stomatitis virus recombinant: high-level production of virus-like particles containing HIV envelope. Virology 268(1), 112–21

Hammond, A. L., Plemper, R. K., Zhang, J., Schneider, U., Russell, S. J., and Cattaneo, R. (2001). Single-chain antibody displayed on a recombinant measles virus confers entry through the tumor-associated carcinoembryonic antigen. J Virol 75(5), 2087–96

Hatta, M., Gao, P., Halfmann, P., and Kawaoka, Y. (2001). Molecular basis for high virulence of Hong Kong H5N1 influenza A viruses. Science 293(5536), 1840–2

Hosaka, Y., Kitano, H., and Ikeguchi, S. (1966). Studies on the pleomorphism of HVJ virons. Virology 29(2), 205–21

Iverson, L. E., and Rose, J. K. (1981). Localized attenuation and discontinuous synthesis during vesicular stomatitis virus transcription. Cell 23(2), 477–84

Kahn, J. S., Roberts, A., Weibel, C., Buonocore, L., and Rose, J. K. (2001). Replication-competent or attenuated, nonpropagating vesicular stomatitis viruses expressing respiratory syncytial virus (RSV) antigens protect mice against RSV challenge. J Virol 75(22), 11079–87

Kahn, J. S., Schnell, M. J., Buonocore, L., and Rose, J. K. (1999). Recombinant vesicular stomatitis virus expressing respiratory syncytial virus (RSV) glycoproteins: RSV fusion protein can mediate infection and cell fusion. Virology 254(1), 81–91

Kalyan, N. K., Lee, S. G., Wilhelm, J., Pisano, M. R., Hum, W. T., Hsiao, C. L., Davis, A. R., Eichberg, J. W., Robert-Guroff, M., and Hung, P. P. (1994). Immunogenicity of recombinant influenza virus haemagglutinin carrying peptides from the envelope protein of human immunodeficiency virus type 1. Vaccine 12(8), 753–60

Kano, M., Matano, T., Kato, A., Nakamura, H., Takeda, A., Suzaki, Y., Ami, Y., Terao, K., and Nagai, Y. (2002). Primary replication of a recombinant Sendai virus vector in macaques. J Gen Virol 83(Pt 6), 1377–86

Karron, R. A., Buonagurio, D. A., Georgiu, A. F., Whitehead, S. S., Adamus, J. E., Clements-Mann, M. L., Harris, D. O., Randolph, V. B., Udem, S. A., Murphy, B. R., and Sidhu, M. S. (1997). Respiratory syncytial virus (RSV) SH and G proteins are not essential for viral replication in vitro: clinical evaluation and molecular characterization of a cold-passaged, attenuated RSV subgroup B mutant. Proc Natl Acad Sci USA 94(25), 13961–6

Kato, A., Kiyotani, K., Sakai, Y., Yoshida, T., and Nagai, Y. (1997). The paramyxovirus, Sendai virus, V protein encodes a luxury function required for viral pathogenesis. EMBO J 16(3), 578–587

Kretzschmar, E., Buonocore, L., Schnell, M. J., and Rose, J. K. (1997). High-efficiency incorporation of functional influenza virus glycoproteins into recombinant vesicular stomatitis viruses. J Virol 71(8), 5982–9

Kurotani, A., Kiyotani, K., Kato, A., Shioda, T., Sakai, Y., Mizumoto, K., Yoshida, T., and Nagai, Y. (1998). Sendai virus C proteins are categorically nonessential gene products but silencing their expression severely impairs viral replication and pathogenesis. Genes Cells 3(2), 111–124

Li, S., Polonis, V., Isobe, H., Zaghouani, H., Guinea, R., Moran, T., Bona, C., and Palese, P. (1993a). Chimeric influenza virus induces neutralizing antibodies and cytotoxic T cells against human immunodeficiency virus type 1. J Virol 67(11), 6659–66

Li, S., Rodrigues, M., Rodriguez, D., Rodriguez, J. R., Esteban, M., Palese, P., Nussenzweig, R. S., and Zavala, F. (1993b). Priming with recombinant influenza virus followed by administration of recombinant vaccinia virus induces CD8+ T-cell-mediated protective immunity against malaria. Proc Natl Acad Sci USA 90(11), 5214–8

Luytjes, W., Krystal, M., Enami, M., Pavin, J. D., and Palese, P. (1989). Amplification, expression, and packaging of foreign gene by influenza virus. Cell 59(6), 1107–13

Matano, T., Kano, M., Nakamura, H., Takeda, A., and Nagai, Y. (2001). Rapid appearance of secondary immune responses and protection from acute CD4 depletion after a highly pathogenic immunodeficiency virus challenge in macaques vaccinated with a DNA prime/Sendai virus vector boost regimen. J Virol 75(23), 11891–6

McGettigan, J. P., Foley, H. D., Belyakov, I. M., Berzofsky, J. A., Pomerantz, R. J., and Schnell, M. J. (2001). Rabies virus-based vectors expressing human immunodeficiency virus type 1 (HIV-1) envelope protein induce a strong, cross-reactive cytotoxic T-lymphocyte response against envelope proteins from different HIV-1 isolates. J Virol 75(9), 4430–4

Mebatsion, T., and Conzelmann, K. K. (1996). Specific infection of CD4+ target cells by recombinant rabies virus pseudotypes carrying the HIV-1 envelope spike protein. Proc Natl Acad Sci USA 93(21), 11366–70

Mebatsion, T., Finke, S., Weiland, F., and Conzelmann, K. K. (1997). A CXCR4/CD4 pseudotype rhabdovirus that selectively infects HIV-1 envelope protein-expressing cells. Cell 90(5), 841–7

Miyahira, Y., Garcia-Sastre, A., Rodriguez, D., Rodriguez, J. R., Murata, K., Tsuji, M., Palese, P., Esteban, M., Zavala, F., and Nussenzweig, R. S. (1998). Recombinant viruses expressing a human malaria antigen can elicit potentially protective immune CD8+ responses in mice. Proc Natl Acad Sci USA 95(7), 3954–9

Moss, W. J., Cutts, F., and Griffin, D. E. (1999). Implications of the human immunodeficiency virus epidemic for control and eradication of measles. Clin Infect Dis 29(1), 106–12

Neumann, G., Watanabe, T., Ito, H., Watanabe, S., Goto, H., Gao, P., Hughes, M., Perez, D. R., Donis, R., Hoffmann, E., Hobom, G., and Kawaoka, Y. (1999). Generation of influenza A viruses entirely from cloned cDNAs. Proc Natl Acad Sci USA 96(16), 9345–50

Ockert, D., Schirrmacher, V., Beck, N., Stoelben, E., Ahlert, T., Flechtenmacher, J., Hagmuller, E., Buchcik, R., Nagel, M., and Saeger, H. D. (1996). Newcastle disease virus-infected intact autologous tumor cell vaccine for adjuvant active specific immunotherapy of resected colorectal carcinoma. Clin Cancer Res 2(1), 21–8

Pecora, A. L., Rizvi, N., Cohen, G. I., Meropol, N. J., Sterman, D., Marshall, J. L., Goldberg, S., Gross, P., O'Neil, J. D., Groene, W. S., Roberts, M. S., Rabin, H., Bamat, M. K., and Lorence, R. M. (2002). Phase I trial of intravenous administration of PV701, an oncolytic virus, in patients with advanced solid cancers. J Clin Oncol 20(9), 2251–66

Peng, K.-W., Donovan, K. A., Schneider, U., Cattaneo, R., Lust, J. A., and Russell, S. J. (2003). Oncolytic measles virus displaying a single chain antibody against CD38, a myeloma cell marker. Blood 101(7), 2557–62

Peng, K. W., Ahmann, G. J., Pham, L., Greipp, P. R., Cattaneo, R., and Russell, S. J. (2001). Systemic therapy of myeloma xenografts by an attenuated measles virus. Blood 98(7), 2002–7

Percy, N., Barclay, W. S., Garcia-Sastre, A., and Palese, P. (1994). Expression of a foreign protein by influenza A virus. J Virol 68(7), 4486–92

Phuangsab, A., Lorence, R. M., Reichard, K. W., Peeples, M. E., and Walter, R. J. (2001). Newcastle disease virus therapy of human tumor xenografts: antitumor effects of local or systemic administration. Cancer Lett 172(1), 27–36

Quinones-Kochs, M. I., Schnell, M. J., Buonocore, L., and Rose, J. K. (2001). Mechanisms of loss of foreign gene expression in recombinant vesicular stomatitis viruses. Virology 287(2), 427–35

Rager, M., Vongpunsawad, S., Duprex, W. P., and Cattaneo, R. (2002). Polyploid measles virus with hexameric genome length. EMBO J 21(10), 2364–72

Restifo, N. P., Surman, D. R., Zheng, H., Palese, P., Rosenberg, S. A., and Garcia-Sastre, A. (1998). Transfectant influenza A viruses are effective recombinant immunogens in the treatment of experimental cancer. Virology 249(1), 89–97

Ring, C. J. (2002). Cytolytic viruses as potential anti-cancer agents. J Gen Virol 83 (Pt 3), 491–502

Roberts, A., Buonocore, L., Price, R., Forman, J., and Rose, J. K. (1999). Attenuated vesicular stomatitis viruses as vaccine vectors. J Virol 73(5), 3723–32

Roberts, A., Kretzschmar, E., Perkins, A. S., Forman, J., Price, R., Buonocore, L., Kawaoka, Y., and Rose, J. K. (1998). Vaccination with a recombinant vesicular stomatitis virus expressing an influenza virus hemagglutinin provides complete protection from influenza virus challenge. J Virol 72(6), 4704–11

Rose, N. F., Marx, P. A., Luckay, A., Nixon, D. F., Moretto, W. J., Donahoe, S. M., Montefiori, D., Roberts, A., Buonocore, L., and Rose, J. K. (2001). An effective AIDS vaccine based on live attenuated vesicular stomatitis virus recombinants. Cell 106(5), 539–49

Rose, N. F., Roberts, A., Buonocore, L., and Rose, J. K. (2000). Glycoprotein exchange vectors based on vesicular stomatitis virus allow effective boosting and generation of neutralizing antibodies to a primary isolate of human immunodeficiency virus type 1. J Virol 74(23), 10903–10

Russell, S. J. (1994). Replicating vectors for cancer therapy: a question of strategy. Semin Cancer Biol 5(6), 437–43

Sakai, Y., Kiyotani, K., Fukumura, M., Asakawa, M., Kato, A., Shioda, T., Yoshida, T., Tanaka, A., Hasegawa, M., and Nagai, Y. (1999). Accommodation of foreign genes into the Sendai virus genome: sizes of inserted genes and viral replication. FEBS Lett 456(2), 221–6

Schirrmacher, V., Griesbach, A., and Ahlert, T. (2001). Antitumor effects of Newcastle Disease Virus in vivo: local versus systemic effects. Int J Oncol 18(5), 945–52

Schirrmacher, V., Haas, C., Bonifer, R., Ahlert, T., Gerhards, R., and Ertel, C. (1999). Human tumor cell modification by virus infection: an efficient and safe way to produce cancer vaccine with pleiotropic immune stimulatory properties when using Newcastle disease virus. Gene Ther 6(1), 63–73

Schlereth, B., Rose, J. K., Buonocore, L., ter Meulen, V., and Niewiesk, S. (2000). Successful vaccine-induced seroconversion by single-dose immunization in the presence of measles virus-specific maternal antibodies. J Virol 74(10), 4652–7

Schmidt, A. C., McAuliffe, J. M., Huang, A., Surman, S. R., Bailly, J. E., Elkins, W. R., Collins, P. L., Murphy, B. R., and Skiadopoulos, M. H. (2000). Bovine parainfluenza virus type 3 (BPIV3) fusion and hemagglutinin- neuraminidase glycoproteins make an important contribution to the restricted replication of BPIV3 in primates. J Virol 74(19), 8922–9

Schneider, U., Bullough, F., Vongpunsawad, S., Russell, S. J., and Cattaneo, R. (2000). Recombinant measles viruses efficiently entering cells through targeted receptors. J Virol 74(21), 9928–36

Schnell, M. J., Buonocore, L., Boritz, E., Ghosh, H. P., Chernish, R., and Rose, J. K. (1998). Requirement for a non-specific glycoprotein cytoplasmic domain sequence to drive efficient budding of vesicular stomatitis virus. EMBO J 17(5), 1289–96

Schnell, M. J., Buonocore, L., Kretzschmar, E., Johnson, E., and Rose, J. K. (1996). Foreign glycoproteins expressed from recombinant vesicular stomatitis viruses are incorporated efficiently into virus particles. Proc Natl Acad Sci USA 93(21), 11359–65

Schnell, M. J., Foley, H. D., Siler, C. A., McGettigan, J. P., Dietzschold, B., and Pomerantz, R. J. (2000). Recombinant rabies virus as potential live-viral vaccines for HIV-1. Proc Natl Acad Sci USA 97(7), 3544–9

Schnell, M. J., Johnson, J. E., Buonocore, L., and Rose, J. K. (1997). Construction of a novel virus that targets HIV-1-infected cells and controls HIV-1 infection. Cell 90(5), 849–57

Shiotani, A., Fukumura, M., Maeda, M., Hou, X., Inoue, M., Kanamori, T., Komaba, S., Washizawa, K., Fujikawa, S., Yamamoto, T., Kadono, C., Watabe, K., Fukuda, H., Saito, K., Sakai, Y., Nagai, Y., Kanzaki, J., and Hasegawa, M. (2001). Skeletal muscle regeneration after insulin-like growth factor I gene transfer by recombinant Sendai virus vector. Gene Ther 8(14), 1043–50

Shiver, J. W., Fu, T. M., Chen, L., Casimiro, D. R., Davies, M. E., Evans, R. K., Zhang, Z. Q., Simon, A. J., Trigona, W. L., Dubey, S. A., Huang, L., Harris, V. A., Long, R. S., Liang, X., Handt, L., Schleif, W. A., Zhu, L., Freed, D. C., Persaud, N. V., Guan, L., Punt, K. S., Tang, A., Chen, M., Wilson, K. A., Collins, K. B., Heidecker, G. J., Fernandez, V. R., Perry, H. C., Joyce, J. G., Grimm, K. M., Cook, J. C., Keller, P. M., Kresock, D. S., Mach, H., Troutman, R. D., Isopi, L. A., Williams, D. M., Xu, Z., Bohannon, K. E., Volkin, D. B., Montefiori, D. C., Miura, A., Krivulka, G. R., Lifton, M. A., Kuroda, M. J., Schmitz, J. E., Letvin, N. L., Caulfield, M. J., Bett, A. J., Youil, R., Kaslow, D. C., and Emini, E. A. (2002). Replication-incompetent adenoviral vaccine vector elicits effective anti-immunodeficiency-virus immunity. Nature 415(6869), 331–5

Singh, M., Cattaneo, R., and Billeter, M. A. (1999). A recombinant measles virus expressing hepatitis B virus surface antigen induces humoral immune responses in genetically modified mice. J Virol 73(6), 4823–8

Skiadopoulos, M. H., Durbin, A. P., Tatem, J. M., Wu, S. L., Paschalis, M., Tao, T., Collins, P. L., and Murphy, B. R. (1998). Three amino acid substitutions in the L protein of the human parainfluenza virus type 3 cp45 live attenuated vaccine candidate contribute to its temperature-sensitive and attenuation phenotypes. J Virol 72(3), 1762–8

Skiadopoulos, M. H., Surman, S., Tatem, J. M., Paschalis, M., Wu, S. L., Udem, S. A., Durbin, A. P., Collins, P. L., and Murphy, B. R. (1999a). Identification of mutations contributing to the temperature-sensitive, cold-adapted, and attenuation phenotypes of the live-attenuated cold- passage 45 (cp45) human parainfluenza virus 3 candidate vaccine. J Virol 73(2), 1374–81

Skiadopoulos, M. H., Surman, S. R., Riggs, J. M., Collins, P. L., and Murphy, B. R. (2001). A chimeric human-bovine parainfluenza virus type 3 expressing measles virus hemagglutinin is attenuated for replication but is still immunogenic in rhesus monkeys. J Virol 75(21), 10498–504

Skiadopoulos, M. H., Surman, S. R., Riggs, J. M., Orvell, C., Collins, P. L., and Murphy, B. R. (2002). Evaluation of the replication and immunogenicity of recombinant human parainfluenza virus type 3 vectors expressing up to three foreign glycoproteins. Virology 297(1), 136–52

Skiadopoulos, M. H., Tao, T., Surman, S. R., Collins, P. L., and Murphy, B. R. (1999b). Generation of a parainfluenza virus type 1 vaccine candidate by replacing the HN and F glycoproteins of the live-attenuated PIV3 cp45 vaccine virus with their PIV1 counterparts. Vaccine 18(5–6), 503–10

Spielhofer, P., Bachi, T., Fehr, T., Christiansen, G., Cattaneo, R., Kaelin, K., Billeter, M. A., and Naim, H. Y. (1998). Chimeric measles viruses with a foreign envelope. J Virol 72(3), 2150–9

Stojdl, D. F., Lichty, B., Knowles, S., Marius, R., Atkins, H., Sonenberg, N., and Bell, J. C. (2000). Exploiting tumor-specific defects in the interferon pathway with a previously unknown oncolytic virus. Nat Med 6(7), 821–5

Stope, M. B., Karger, A., Schmidt, U., and Buchholz, U. J. (2001). Chimeric bovine respiratory syncytial virus with attachment and fusion glycoproteins replaced by bovine parainfluenza virus type 3 hemagglutinin-neuraminidase and fusion proteins. J Virol 75(19), 9367–77

Takeda, M., Pekosz, A., Shuck, K., Pinto, L. H., and Lamb, R. A. (2002). Influenza A virus M2 ion channel activity is essential for efficient replication in tissue culture. J Virol 76(3), 1391–9

Tao, T., Davoodi, F., Cho, C. J., Skiadopoulos, M. H., Durbin, A. P., Collins, P. L., and Murphy, B. R. (2000a). A live attenuated recombinant chimeric parainfluenza virus (PIV) candidate vaccine containing the hemagglutinin-neuraminidase and fusion glycoproteins of PIV1 and the remaining proteins from PIV3 induces resistance to PIV1 even in animals immune to PIV3. Vaccine 18(14), 1359–66

Tao, T., Skiadopoulos, M. H., Davoodi, F., Riggs, J. M., Collins, P. L., and Murphy, B. R. (2000b). Replacement of the ectodomains of the hemagglutinin-neuraminidase and fusion glycoproteins of recombinant parainfluenza virus type 3 (PIV3) with their counterparts from PIV2 yields attenuated PIV2 vaccine candidates. J Virol 74(14), 6448–58

Tao, T., Skiadopoulos, M. H., Davoodi, F., Surman, S. R., Collins, P. L., and Murphy, B. R. (2001). Construction of a live-attenuated bivalent vaccine virus against human parainfluenza virus (PIV) types 1 and 2 using a recombinant PIV3 backbone. Vaccine 19(27), 3620–31

Tao, T., Skiadopoulos, M. H., Durbin, A. P., Davoodi, F., Collins, P. L., and Murphy, B. R. (1999). A live attenuated chimeric recombinant parainfluenza virus (PIV) encoding the internal proteins of PIV type 3 and the surface glycoproteins of PIV type 1 induces complete resistance to PIV1 challenge and partial resistance to PIV3 challenge. Vaccine 17(9–10), 1100–8

Taqi, A. M., Abdurrahman, M. B., Yakubu, A. M., and Fleming, A. F. (1981). Regression of Hodgkin's disease after measles. Lancet 1(8229), 1112

Techaarpornkul, S., Barretto, N., and Peeples, M. E. (2001). Functional analysis of recombinant respiratory syncytial virus deletion mutants lacking the small hydrophobic and/or attachment glycoprotein gene. J Virol 75(15), 6825–34

Vongpunsawad, S., Oezgun, N., Braun, W. and Cattaneo, R. (2004). Selectively receptor-blind measles viruses: identification of residues necessary for SLAM- or CD46-induced fusion and their localization on a new hemagglutinin structural model. J Virol, 78(1), 1

Wang, Z., Hangartner, L., Cornu, T. I., Martin, L. R., Zuniga, A., Billeter, M. A., and Naim, H. Y. (2001). Recombinant measles viruses expressing heterologous antigens of mumps and simian immunodeficiency viruses. Vaccine 19(17–19), 2329–36

Watanabe, T., Watanabe, S., Ito, H., Kida, H., and Kawaoka, Y. (2001). Influenza A virus can undergo multiple cycles of replication without M2 ion channel activity. J Virol 75(12), 5656–62

Wertz, G. W., Moudy, R., and Ball, L. A. (2002). Adding genes to the RNA genome of vesicular stomatitis virus: positional effects on stability of expression. J Virol 76(15), 7642–50

Wertz, G. W., Perepelitsa, V. P., and Ball, L. A. (1998). Gene rearrangement attenuates expression and lethality of a nonsegmented negative strand RNA virus. Proc Natl Acad Sci USA 95(7), 3501–6

Yonemitsu, Y., Kitson, C., Ferrari, S., Farley, R., Griesenbach, U., Judd, D., Steel, R., Scheid, P., Zhu, J., Jeffery, P. K., Kato, A., Hasan, M. K., Nagai, Y., Masaki, I., Fukumura, M., Hasegawa, M., Geddes, D. M., and Alton, E. W. (2000). Efficient gene transfer to airway epithelium using recombinant Sendai virus. Nat Biotechnol 18(9), 970–3

Yu, D., Shioda, T., Kato, A., Hasan, M. K., Sakai, Y., and Nagai, Y. (1997). Sendai virus-based expression of HIV-1 gp120: reinforcement by the V(−) version. Genes Cells 2(7), 457–66

Zheng, H., Palese, P., and Garcia-Sastre, A. (2000). Antitumor properties of influenza virus vectors. Cancer Res 60(24), 6972–6

Zorn, U., Duensing, S., Langkopf, F., Anastassiou, G., Kirchner, H., Hadam, M., Knuver-Hopf, J., and Atzpodien, J. (1997). Active specific immunotherapy of renal cell carcinoma: cellular and humoral immune responses. Cancer Biother Radiopharm 12(3), 157–65

Influenza Vaccines Generated by Reverse Genetics

K. Subbarao · J. M. Katz

Influenza Branch, Centers for Disease Control and Prevention,
Mailstop G-16, 1600 Clifton Road, Atlanta, GA 30333, USA
E-mail: JKatz@cdc.gov

Abstract Influenza viruses cause annual epidemics and occasional pandemics of acute respiratory disease. Vaccination is the primary means to prevent and control the disease. However, influenza viruses undergo continual antigenic variation, which requires the annual reformulation of trivalent influenza vaccines, making influenza unique among pathogens for which vaccines have been developed. The segmented nature of the influenza virus genome allows for the traditional reassortment between two viruses in a coinfected cell. This technique has long been used to generate strains for the preparation of either inactivated or live attenuated influenza vaccines. Recent advancements in reverse genetics techniques now make it possible to generate influenza viruses entirely from cloned plasmid DNA by cotransfection of appropriate cells with 8 or 12 plasmids encoding the influenza virion sense RNA and/or mRNA. Once regulatory issues have been addressed, this technology will enable the routine and rapid generation of strains for either inactivated or live attenuated influenza vaccine. In addition, the technology offers the potential for new vaccine strategies based on the generation of genetically engineered donors attenuated through directed mutation of one or more internal genes. Reverse genetics techniques are also proving to be important for the development of pandemic influenza vaccines, because the technology provides a means to modify genes to remove virulence determinants found in highly pathogenic avian strains. The future of influenza prevention and control lies in the application of this powerful technology for the generation of safe and more effective influenza vaccines.

1 Introduction

Vaccination is the primary means for the prevention of human disease caused by influenza viruses. The currently licensed trivalent inactivated vaccine (TIV) has changed little since its development over 50 years ago. Newer, improved vaccines are desirable; live attenuated influenza vaccines (LAIV) have been studied extensively, and are now licensed in the U.S. Any influenza vaccine must keep pace with antigenic drift by continual reformulation to reflect changes in the major surface glycoproteins of influenza A and B viruses. The segmented nature of the influenza virus genome has in the past allowed for the generation of candidate vaccine strains for either the conventional TIV or LAIV with traditional reassortment techniques. However, the technique of reverse genetics, the generation of negative-sense RNA viruses from cloned cDNAs, has now reached a level of sophistication that enables the generation of virtually any influenza virus, including candidate vaccine strains for either TIV or LAIV. The recently improved reverse genetics technology allows not only the rapid generation of vaccine strain but the ability to manipulate genes for further attenuation and safety of vaccines. In this chapter, we review influenza as a disease, the variability of influenza viruses and the ramifications for the development of effective vaccines, the development and characterization of LAIV, and the application of reverse genetics to the development of vaccines against both epidemic and pandemic influenza.

2
Influenza Viruses and the Disease

Influenza A and B viruses are the cause of annual epidemics of acute respiratory disease among humans. Influenza A viruses also cause occasional pandemics of influenza that are characterized by the spread, in a serologically naïve human population, of a novel virus that has acquired one or both surface glycoproteins from an animal influenza virus. Influenza viruses are transmitted among humans primarily by small-particle aerosol generated by infected persons during coughing and sneezing. Influenza infection is characterized by the abrupt onset of systemic symptoms including fever, myalgia, headache, and malaise and respiratory symptoms including sore throat, rhinitis, and cough. Acute illness may last 3–4 days, although malaise and cough may persist for 1–2 weeks. Pulmonary complications due to primary viral pneumonia are associated with high mortality and are rarer than secondary bacterial pneumonia, which typically occurs in the early convalescent stages of illness. Other rare complications of influenza include myocarditis, pericarditis, and neurological sequelae (Nicholson 1998).

Influenza A and B viruses are negative-sense RNA viruses belonging to the family *Orthomyxoviridae*. Influenza A and B virus genomes consist of eight segments of RNA encoding nine structural and two nonstructural proteins (Wright and Webster 2001; Chen et al. 2001). Hemagglutinin (HA) and neuraminidase (NA) are transmembrane glycoproteins that protrude from the lipid envelope on the viral surface. HA mediates the attachment and penetration of the virus into a susceptible host cell and is the major target of antibodies that neutralize the infectivity of the virus. NA is a sialidase that mediates the release of progeny virus from host cell membrane and is also a target of antibodies. A third transmembrane surface protein, M2, is an ion channel found in influenza A viruses and is essential for virus uncoating. Another transmembrane protein, NB, encoded by RNA segment 6 is found in influenza B viruses. The matrix (M1) protein is the major structural protein that underlies the lipid membrane and interacts with multiple viral components. The ribonucleoprotein (RNP) complex is formed by the nucleoprotein and three polymerase genes, PB2, PB1, and PA. A nonstructural gene encodes two proteins: NS1, which is found only in the virus-infected cells, and NEP (also known as NS2), which is a minor virion component involved in nuclear export of RNP. An additional small nonstructural protein derived from the +1 reading frame of the PB1 gene has recently been identified (Chen et al. 2001).

3 Burden of Disease

Infection rates for epidemic influenza are highest among children. However, serious illness, hospitalization, and death are greatest among persons 65 years of age and over, children less than 2 years of age, or persons of any age who have certain chronic medical conditions. Epidemic influenza is responsible for an average of 114,000 hospitalizations and 20,000 deaths per year in the United States (Centers for Disease Control and Prevention 2002). Influenza-related deaths are primarily a result of pneumonia or exacerbation of cardiopulmonary disorders or other chronic diseases. More than 90% of influenza-related deaths occur in older adults. It has been estimated that the next influenza pandemic will result in up to 730,000 hospitalizations and 207,000 deaths in the U.S. alone (Meltzer et al. 1999). Demographics of previous pandemics indicate that a higher proportion of deaths occur in persons <65 years of age in the initial pandemic period (Simonsen et al. 1998).

4 Antigenic Variability

Antigenic variability is the hallmark of influenza viruses. Two mechanisms of variation exist. The first, termed antigenic drift, occurs in both influenza A and B viruses and is a result of the accumulation of point mutations in the viral HA and NA genes and selection of variants with amino acid mutations that enable the virus to escape neutralization by antibody acquired from previous infections or immunization. Antigenic drift leads to the emergence of new epidemic strains and is the force that drives the continual evolution of influenza viruses in humans. The second form of variation, termed antigenic shift, occurs only in influenza A viruses, of which there are 15 HA (H1–H15) and 9 NA (N1–N9) subtypes. All these subtypes exist in aquatic bird species (Webster et al. 1992). Mammals including humans, swine, horses, and some sea mammals may act as hosts for a limited number of influenza A subtypes. Antigenic shift occurs when a virus with a novel HA, with or without other accompanying genes derived from an avian or animal source, appears in humans. Antigenic shift may result in a pandemic if the novel virus is capable of spreading from person to person within a serologically naïve and susceptible population. Only H1N1, H2N2, and H3N2 viruses have emerged in humans as pandemic strains. Although highly pathogenic avian H5N1 viruses infected humans in Hong Kong in 1997, they were unable to transmit efficiently among humans. Currently, influenza A H1N1 and H3N2 viruses and influenza B viruses cocirculate in humans. During the 2001–2002 influenza season in the northern hemisphere, H1N2 reassortant viruses were also isolated from individuals with influenza-like illness (Gregory et al. 2002).

5
Vaccine Strain Selection

The continual variation that influenza viruses undergo makes them unique among pathogens for which vaccines have been developed. The continual emergence of variants of influenza A and B viruses necessitates the annual reformulation of the TIV. Identification of new variants with epidemic potential is based on global virological and epidemiological surveillance conducted by the international World Health Organization (WHO) network of laboratories. The global surveillance network identifies and characterizes influenza viruses isolated from humans during influenza seasons in the northern and southern hemispheres, as well as unusual viruses with pandemic potential (Cox et al. 1994). Viruses with variant HA molecules are first identified by antigenic analyses using reference viruses and antisera. The nucleotide sequence of the HA genes is analyzed to identify amino acid mutations in key antigenic sites and provide important information on the molecular evolution of influenza viruses. Retrospective analysis has determined that variants responsible for major epidemics typically have at least three or four mutations in at least two of the five antigenic sites located on the HA1 domain that forms the globular head of the HA molecule (Cox and Bender 1995). If a variant is detected that also reacts poorly with human sera from individuals vaccinated with the existing vaccine, and epidemiological surveillance confirms geographic spread, this variant becomes a candidate for the new season's vaccine. Currently, vaccine strain selection occurs biannually; vaccine strain selection for the northern hemisphere occurs in February, whereas selection of vaccine strains for the southern hemisphere occurs in September. Once an influenza A wild-type variant is selected, a high growth reassortant is generated and further characterized to confirm antigenic identity with the parental wild-type strain. The realization that novel epidemic strains may emerge from China or southern Asia has increased awareness of the importance of heightened surveillance in this area (Cox et al. 1994). Likewise, enhanced surveillance in southern Asia is important for pandemic preparedness because the region has been implicated as an epicenter for the emergence of future pandemic strains (Webster et al. 1992).

6 Current Influenza Vaccines

Vaccination is the primary method for the prevention of influenza and its complications in the community. Inactivated trivalent influenza vaccines contain three components, H1N1 and H3N2 influenza A strains and an influenza B strain. Both influenza A components are high-growth reassortant viruses that typically contain the HA and NA genes from the wild-type epidemic strain and the six internal genes from a laboratory-adapted parent strain A/Puerto Rico/8/34 that confers the property of high yield in embryonated eggs. Annual vaccination with the currently licensed inactivated trivalent vaccine is recommended for persons at increased risk for complications from influenza. These includes persons 65 years of age and over, residents of nursing homes and other chronic-care facilities, persons of any age with chronic pulmonary, cardiovascular, metabolic, or immunosuppressive disorders, children and adolescents who are receiving long-term aspirin therapy, and women who will be in the second or third trimester of pregnancy during the influenza season (Centers for Disease Control and Prevention 2002). Vaccination is also recommended for health care workers and others who provide care to persons in high-risk groups. Because the prevalence of high-risk conditions is increased in persons aged 50–64 years, annual influenza vaccination is also recommended for this age group. Influenza vaccination is encouraged in children at 6–23 months of age because this age group is now recognized to have an increased risk for hospitalization due to influenza (Izurieta et al 2000; Neuzil et al. 2000); the recommendation for vaccination of this age group in the U.S. is currently under consideration.

The licensed inactivated vaccine is administered by intramuscular injection and induces immunity to infection in 70–90% of healthy adults <65 years of age when there is a good antigenic match between vaccine and circulating virus strains (Palache 1997; Bridges et al. 2000). Immunity induced by inactivated influenza vaccines is based primarily on the induction of neutralizing serum antibody directed against the HA of the vaccine strain. The serum antibody induced by inactivated vaccine is strain specific, and antigenic differences between the vaccine and circulating strain may reduce the efficacy of the vaccine. Although the majority of children and young adults develop levels of postvaccination HI antibody considered to be protective (titers$\geq$40), older adults and persons with chronic diseases develop lower anti-HA antibody responses after vaccination. In elderly individuals 65 years of age and over, the inactivated influenza vaccine is only 30–50% effective in preventing influenza illness (Govaert et al. 1994; Patriarca et al. 1985). Nevertheless, the inactivated vaccine is 50–80% effective in preventing influenza-related hospitalizations and deaths in this population (Patriarca et al. 1985; Nichol et al. 1998). An additional limitation of the intramuscularly delivered inactivated vaccine is the relatively poor induction of local IgA antibody responses in respiratory secretions and cell-mediated immunity, which are thought to play an important role in protection or recovery from infection, respectively.

7 Live Attenuated Influenza Vaccines

The development of safe, stable, and effective live attenuated influenza vaccines has long been a goal of vaccine researchers. Theoretically, live vaccines delivered intranasally and designed to replicate to a limited extent in the upper respiratory tract should induce immunity similar to that induced by natural infection. Thus such vaccines have the potential to provide immunity superior to that induced by conventional inactivated vaccines. The need to continually update vaccines with the HA and NA genes of currently circulating influenza strains necessitated the development of master (or donor) strains that conferred the attenuation phenotype based on the constellation of internal genes. The segmented viral genome of influenza allows for the preparation of a vaccine strain by reassortment between a master attenuated strain, conferring the attenuation phenotype within the six internal genes and the wild-type circulating strain providing the antigenically relevant HA and NA genes. Such a virus is termed a 6:2 reassortant. Three main approaches have been used in the development of live attenuated influenza master strains.

7.1 Host-Range Vaccines

For influenza A viruses, the availability of animal viruses possessing highly related but host-specific internal genes offered the possibility of developing a master strain from a nonhuman virus that was naturally attenuated in humans. The use of such host-range variants overcame the need for derivation of the attenuation phenotype by laborious in vitro passaging. A number of avian influenza A viruses were found to exhibit restricted replication in nonhuman primates (Murphy et al. 1982a). One of these, A/Mallard/New York/6750/78, was used to develop attenuated avian-human reassortant viruses. Reassortants that possessed the six internal genes from the avian virus and the HA and NA genes from a circulating human influenza H3N2 virus exhibited restricted replication in the upper respiratory tract of nonhuman primates and humans and were safe, genetically stable, and immunogenic in seronegative adults (Clements et al. 1992). Although multiple genes contributed to the attenuation phenotype, the avian PB2 gene conferred a host range restriction phenotype that restricts replication in mammalian cells (Clements et al. 1992; Subbarao et al. 1993b). Because additional clinical trials established that some avian-human reassortants possessed some residual virulence in infants and children, the use of host-range restriction for the development of live attenuated influenza vaccines has not been pursued (Steinhoff et al. 1991).

7.2 Temperature-Sensitive Vaccines

Growth of influenza virus in the presence of the mutagen 5-fluorouracil was used to predictably generate mutants that exhibited restricted growth at 37°C in vitro, that is, a temperature-sensitive (*ts*) phenotype, and restricted replication in the lower respiratory tract of animals at 37–38°C. In the cooler temperature (32–33°C) of the upper respiratory tract, the *ts* mutants replicated and induced antibody that protected mice from lethal infection with wild-type virus (Mills and Chanock 1971). Two master strains were developed and used to generate reassortant vaccine strains. However, *ts* reassortants proved to be genetically unstable. Loss of the *ts* phenotype and reversion to virulence was observed in clinical trials, and the further development of such vaccines was not pursued (Tolpin et al. 1981).

7.3 Cold-Adapted Vaccines

Adaptation of human influenza viruses for growth at less than optimal temperatures results in decreased virulence (Maassab 1967). This strategy has been used by both U.S. and Russian investigators to generate highly stable master strains of influenza A and B viruses with an attenuated cold-adapted (*ca*) phenotype. The *ca* vaccines are the most promising candidates for LAIV.

7.3.1 U.S. *ca* Vaccines

Influenza A (A/Ann Arbor/6/60) and B (B/Ann Arbor/1/66) master *ca* strains currently used to generate LAIV in the U.S. were isolated on chick kidney cells and were adapted for growth at 25°C by a total of 32 stepwise passages at successively lower temperatures. These donor strains replicate efficiently at 25°C and 33°C but fail to replicate efficiently at temperatures above 37°C, and thus they have not only a *ca* phenotype but also a *ts* phenotype (Maassab 1968). The *ca* virus was substantially attenuated (*att* phenotype) for growth in ferrets and rodents. In contrast to the wild-type virus from which it was derived, the *ca* virus did not replicate in the warmer temperatures found in the lungs of ferrets. Furthermore, the *ca* virus was also attenuated in the upper respiratory tract, replicating to lower titers and for a shorter period than the wild-type virus (Maassab et al. 1982). The *ca* and *ts* phenotypes clearly correlated with loss of virulence in ferrets. Recombinant vaccine viruses that possess six internal genes from the donor *ca* strain and the HA and NA genes from the wild-type epidemic strain are generated by traditional reassortment techniques. The resulting 6:2 reassortants possess the attenuation phenotypes of the donor strain and the desired antigenic properties of a given epidemic variant. Such recombinants retain the attenuated growth phenotype of the *ca* donor in both animals and humans (Maassab et al. 1982; Murphy and Coelingh 2002), indicating that the *ca* donor internal genes confer these properties (see below). The *ca* reassortant vaccine strains are at least 1,000-fold less infectious than wild-type viruses in seronegative individuals, and seronegative hosts are 10-fold more susceptible to infection with *ca* vaccine strains than are seropositive individuals. Doses of 10^7 50% tissue culture infectious doses ($TCID_{50}$) are typically safe and immunogenic in all age groups.

7.3.2
Russian *ca* Vaccines

Cold-adapted master strains for attenuated influenza vaccines were also developed and licensed for use in Russia. Influenza vaccine prevention programs in children and working-age adults in the former USSR and Russia have been based almost exclusively on the use of *ca* vaccines. Two master influenza A virus *ca* strains were derived from A/Leningrad/134/57 (H2N2) that had been passaged in embryonated hens' eggs at least 20 times before attenuation. Attenuation was achieved by 17 or 47 further passages in eggs at reduced temperatures, primarily at 25–26°C, resulting in the two *ca* donor strains, A/Leningrad/137/17/57 and A/Leningrad/137/47/57, respectively. These *ca* viruses were also temperature sensitive, failing to replicate efficiently in eggs at 39°C (Kendal et al. 1981). The former virus is licensed in Russia for the preparation of reassortant *ca* vaccines for use in working-age adults, whereas the latter, more attenuated donor strain is licensed for use in the preparation of *ca* vaccines for children 3–14 years of age. Live attenuated vaccines based on these donor strains and influenza B virus donor strain B/USSR/60/69 have been widely used in Russia for the annual vaccination of working-age adults and children (Kendal 1997; Rudenko and Alexandrova 2001). The infectivity of an H1N1 *ca* recombinant generated with the A/Leningrad/134/17/57 *ca* donor was shown to be similar to a recombinant produced with the A/Ann Arbor/6/60 in both seronegative and seropositive individuals (Nicholson et al. 1987).

8
Genetic Basis of Attenuation and Stability of *ca* Vaccines

The genetic basis of the attenuation phenotypes of *ca* vaccines has been investigated by comparing the nucleotide sequences of the entire genomes of cloned *ca* donor strains with those of the wild-type viruses. Using a wild-type virus that had been passaged multiple times in eggs at 34°C for comparison, 11 amino acid substitutions and 2 non-coding region changes in six gene products were identified in the *ca* A/Ann Arbor/6/60 donor strain (Table 1) (Cox et al. 1988). Further sequence com-

Table 1 Amino acid sequence comparison of influenza A *wt* and *ca* master donor strains

Gene product	A/Ann Arbor/6/60[a]			A/Leningrad/137/57[b]			
	Amino acid			Amino acid			
	Residue	*wt*	*ca*	Residue	*wt*	*ca* 17	*ca* 47
PB2	**265**[c]	N	S	478	V	L	L
				490	S	–	R
PB1	**391**	K	E	265	L	N	N
	457	E	D	317	M	–	I
	581	E	G	591	V	I	I
	661	A	T				
PA	613	K	E	28	L	P	P
	715	L	P	341	V	L	L
NP	23	T	N	341	l	–	I
	34	N	G				
M1	–			15	I	V	V
M2	86	A	S	86	A	T	T
NS1	153	A	T	–			
NS2	–			100	M	I	I

[a] From Cox et al. 1988 and Herlocher et al. 1996.

[b] From Klimov et al. 1992.

[c] Amino acids in bold are those associated with *ts* phenotype as defined in Jin et al. 2003.

parison of the *ca* donor strain with several closely related A/Ann Arbor/6/60 wild-type stains representing different passage levels identified seven coding changes in five gene products (Table 1) (Herlocher et al. 1996). A similar level of divergence was observed between the Russian wild type and *ca* donor strain A/Leningrad/137/17/57, which varied from wild-type virus by eight amino acids in six gene products. The A/Leningrad/137/47/57 donor strain contained an additional two amino acid substitutions, one in PB1 and one in PB2, and a third novel substitution in NP (Table 1) Klimov et al. 1992). The use of single gene reassortants possessing only a single gene from the *ca* A/Ann Arbor virus established that the PA gene specified the *ca* phenotype, whereas the PB1 and PB2 genes each specified the *ts* phenotype and all three of these genes contributed independently to the *att* phenotype. Although the M gene was initially also implicated in the *att* phenotype, further studies determined that this was the effect of the gene constellation of a particular 6:2 *ca* reassortant virus (Snyder et al. 1988; Subbarao et al. 1992). Recent studies to establish the genetic determinants of the *ts* phenotype of the U.S. master donor virus, the property expected to contribute to the attenuation of the vaccine in humans, have been refined by the application of reverse genetics. The *ts* phenotype of *ca* A/Ann Arbor was mapped to four major loci in the PB2, PB1, and NP genes of the *ca* A/Ann Arbor virus with contribution of an additional mutation in PB1 (Table 1) (Jin et al. 2003). With traditional single gene reassortment techniques, in the A/Leningrad/137/17/57 *ca* donor strain, the PA and PB2 genes each specified the *ca* phenotype whereas the PB1 and PB2 genes each specified the *ts* phenotype. Of the single gene reassortants possessing a *ca* and/or *ts* phenotype, only those possessing either the PB1 or the PB2 gene were attenuated in ferrets (Klimov et al. 2001). The molecular bases for the *ca, ts,* and *att* phenotypes of the U.S. and Russian influenza B virus *ca* donors are at present less well defined.

An essential property of a live attenuated vaccine is the genetic stability of the attenuation phenotype conferred by the internal genes of the *ca* donor strain. Thus studies examining the genetic stability of the recombinant vaccine strains after replication in study volunteers have become an important part of the clinical evaluation of the *ca* vaccines. Numerous studies conducted with different H3N2 and H1N1 A/Ann Arbor/6/60 6/2 vaccine reassortant strains have established that the *ca* and *ts* phenotypes are highly stable after passage of the attenuated virus in humans (Murphy and Coelingh 2002). In recent studies, the genetic stability of such reassortants was confirmed by genotypic analysis of influenza A and B vaccine strains recovered from children, although further sequence analysis of the polymerase genes was required to confirm the presence of the attenuating mutations (Cha et al. 2002). Reassortant vaccines prepared from the A/Leningrad/17/47 and A/Leningrad/47/57 *ca* donor strains have also confirmed the genetic stability of the *ca* reassortant vaccine strains recovered from vaccinated children and adults (Klimov et al. 1995). Partial sequence analysis of all six internal genes of H1N1 and H3N2 reassortant vaccine viruses confirmed that mutations in the PB1, PB2, and PA genes associated with the *ca* or *ts* phenotypes were preserved. Minor variability in the other internal genes was attributed to heterogeneity in these genes within the *ca* donor strain or heterogeneity in the reassortant vaccine strain itself (Klimov et al. 1996). The high degree of phenotypic stability exhibited by the *ca* reassortant vaccine strains is presumably due to the polygenic nature of attenuation. There is little evidence of transmission of LAIV from vaccinated individuals, presumably because the low levels of virus shed after vaccination are below the threshold required for efficient transmission. The minimal induction of symptoms such as sneezing and coughing that would aid in transmission of the virus likely also limits the spread of the vaccine strains among nonvaccinated individuals.

9
Safety, Immunogenicity, and Efficacy of LAIV

Since 1995, the trivalent investigational LAIV prepared from the U.S. influenza A and B virus *ca* donors has been administered to over 10,000 children and adults. Overall, the vaccines were well tolerated, although the number of reports of runny nose or sore throat were higher in those administered LAIV compared with those who received intranasal placebo (Mendelman et al. 2001). Similarly, a high level of safety in both pediatric and adult populations also has been reported for the Russian LAIV [29, 43, 44]. The immunogenicity of LAIV depends on the strain of virus used, the age of the individual, and the presence of preexisting antibody as a result of prior influenza infection or vaccination or maternally acquired antibodies in the case of infants less than 6 months of age. Although one dose of LAIV induces long-lasting HA-specific serum IgG and mucosal IgA antibody responses in seronegative children, two doses of LAIV are optimal for both younger and older infants and children

(Johnson et al. 1986; Murphy et al. 1982b). In children and adults, serum hemagglutination inhibition (HI) antibody titers induced by LAIV are lower than those induced by inactivated influenza vaccine (Edwards et al. 1994; Rudenko et al. 1993; Clements and Murphy 1986). In adults, including elderly adults aged 65 years, LAIV boosts mucosal IgA responses more frequently than inactivated vaccine (Rudenko et al. 2001; Beyer et al. 2002). Overall, LAIV is less immunogenic in elderly adults because of the limited replication of the vaccine virus, which is associated with the presence of increased levels of serum and local antibodies (Powers et al 1992). In a 2-year study in healthy preschool children, the overall efficacy of the trivalent *ca* LAIV in preventing culture-confirmed illness was 92% (95% CI: 88, 94) (Belshe et al. 1998; Belshe et al. 2000). This included the provision of excellent heterotypic protection against an antigenic drift variant (A/Sydney/5/97) that was not contained in the vaccine (Beyer et al. 2002). In a 5-year comparative study in persons aged 1–65 years, the efficacy of a bivalent LAIV was similar to that of trivalent inactivated vaccine in preventing culture-confirmed H3N2 illness and serologically proven H1N1 infection (Edwards et al. 1994). More recently, trivalent LAIV and inactivated vaccine were shown to be of similar efficacy in preventing laboratory-confirmed H1N1 illness after experimental challenge in healthy adults (Treanor et al. 2000).

10
Advantages and Disadvantages of *ca* Live Attenuated Influenza Vaccines

Compared with inactivated influenza vaccine, the *ca* LAIV induces superior local IgA antibody responses in the respiratory tract. Although mucosal antibody is typically less durable than serum IgG, it nevertheless offers protection at the site of infection. Theoretically, LAIV should also induce superior cell-mediated immunity, although data, and particularly direct comparisons with inactivated vaccines, are limited. Although LAIV induce lower serum antibody responses compared with inactivated vaccines, the protective efficacy is as good if not better than that offered by inactivated vaccines, suggesting that enhanced mucosal and cellular responses overcome the limitations of systemic antibody induction by LAIV. The concept that mucosal antibody responses may be of broader specificity is borne out by the heterotypic efficacy of LAIV demonstrated in one study (Treanor et al. 2000). The needle-free delivery of LAIV is a definite advantage for vaccine acceptability, particularly for children.

Genetic stability studies performed to date have confirmed the stability of the *ca*, *ts*, and *att* phenotypes. However, additional studies are needed to better understand the gene constellation effects observed in different reassortant vaccine strains. The potential for transmission of vaccine strains to unvaccinated contacts has also been considered a disadvantage of LAIV. However, studies conducted in young children have found only rare evidence of such transmission (Murphy and Coelingh, 2002; Gruber et al. 1993; Karron et al. 1995). Nevertheless, the potential for transmission to immunocompromised individuals must be addressed if LAIV is to be used widely in the general population. Another concern is the potential for reassortment between a *ca* vaccine strain and wild-type influenza A virus. Although all the available evidence suggests that reassortment with currently circulating human H1N1 or H3N2, or H1N2 viruses would result in an attenuated strain unable to transmit efficiently among humans, the outcome of reassortment with a novel virus of swine or avian origin is less clear.

11 The Application of Reverse Genetics to Influenza Vaccine Development

The viruses that are included in the licensed trivalent formalin-inactivated influenza vaccine are updated annually to keep pace with antigenic drift among circulating influenza viruses. The influenza A (H1N1 and H3N2) components of the vaccine are high-growth reassortant viruses that derive their surface glycoprotein genes from circulating wild-type (*wt*) viruses and internal protein genes from a vaccine donor strain influenza A/Puerto Rico/8/34 (PR8), a gene constellation that confers high yield in embryonated eggs (Kilbourne 1969) . These high growth 6:2 reassortant viruses are generated by genetic reassortment annually or as needed for inclusion in the vaccine. Significant research efforts are being made to develop better influenza vaccines than those that are currently licensed and vaccines to protect humans against pandemic strains of influenza (Palese and Garcia-Sastre 2002). Ever since techniques were developed to rescue individual influenza virus genes or complete influenza viruses, investigators have sought ways to apply reverse genetics techniques to vaccine development by generating rationally designed vaccines with specific features that can be incorporated at will.

11.1 The First Generation: Ribonucleoprotein Transfection

The first generation of reverse genetics techniques, referred to as ribonucleoprotein (RNP) transfection methods, used a helper virus to rescue a single cDNA-derived influenza virus gene (Luytjes et al 1989; Enami and Palese 1991) . The cDNA-derived gene was transcribed in vitro in the presence of purified RNP to generate a RNP-RNA complex that was transfected into cells infected with a helper virus. Reassortant viruses bearing the cDNA-derived gene segment were selected from the helper virus progeny (Fig. 1). The success of this method depended on the selection system.

RNP transfection was applied to generate 6:2 reassortant human influenza vaccine strains with the HA and NA genes derived from circulating *wt* viruses in a background of internal protein genes of the A/Ann Arbor/6/60 (AA) *ca* vaccine virus.

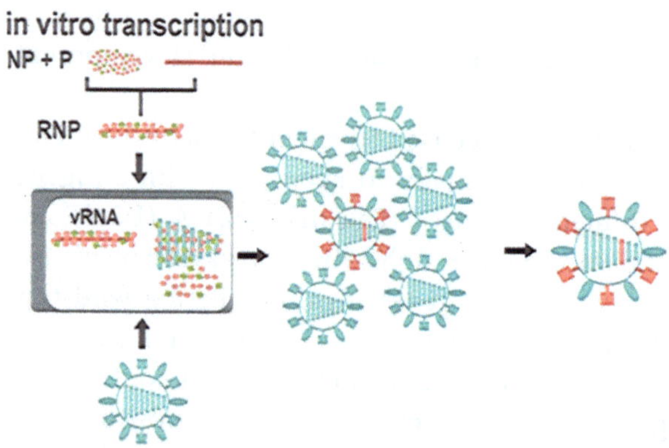

Fig. 1 First generation reverse genetics: ribonucleoprotein transfection. In vitro synthesized vRNA is mixed with NP and polymerase proteins purified from virus and the resulting vRNP complex is transfected into eukaryotic cells. The remaining seven vRNPs are provided by a helper virus and the desired recombinant virus is selected using a selection system such as antibody, host-range, temperature or drug sensitivity. (Reproduced with permission from Neumann and Kawaoka, 2002, John Wiley and Sons Ltd)

11.2
The Second Generation: Plasmid-Based Reverse Genetics

RNP-based transfection methods had significant limitations because they were dependent on a strong selection system and required laborious purification of functional RNP complexes (Luytjes et al 1989). In 1999, two laboratories independently developed plasmid-based reverse genetics techniques to rescue influenza A viruses (Fodor et al. 1999; Neumann et al. 1999). These methods were based on cotransfection of appropriate cells (293-T, Vero, or a mixture of 293-T and MDCK cells) with 12 to 17 plasmids, 8 encoding virion sense RNA (vRNA) under the control of a human Pol I promoter (Neumann and Kawaoka 2002) and 4–9 plasmids encoding messenger RNAs (mRNA) for different influenza proteins under the control of a Pol II promoter (Fig. 2). These methods did not require a helper virus and therefore did not require a selection system. The method was further modified and simplified by engineering

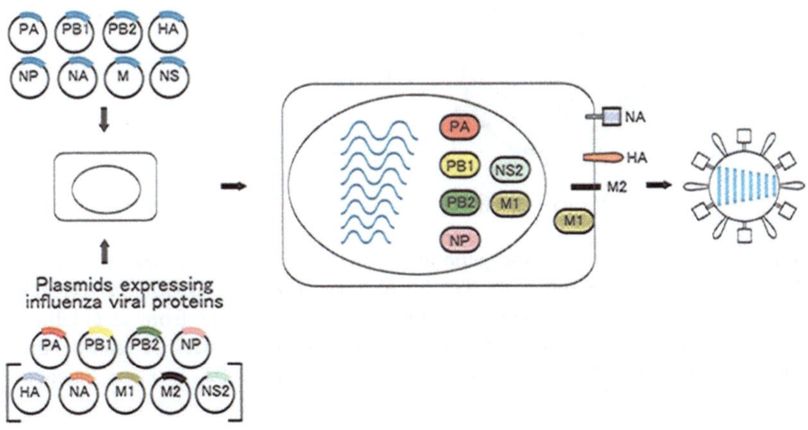

Fig. 2 Second generation reverse genetics: generation of influenza virus entirely from cloned cDNAs. Eight plasmids each encoding a region for a viral segment flanked by RNA polymerase I promoter and terminator sequences, are transfected into eukaryotic cells for vRNA synthesis. Nine plasmids designed for expression of viral structural proteins or alternatively, only four plasmids encoding the polymerases and NP required for replication and transcription of the vRNA, are provided by cotransfection of cells with these protein expressing constructs. (Reproduced with permission from Neumann and Kawaoka, 2002, John Wiley and Sons Ltd)

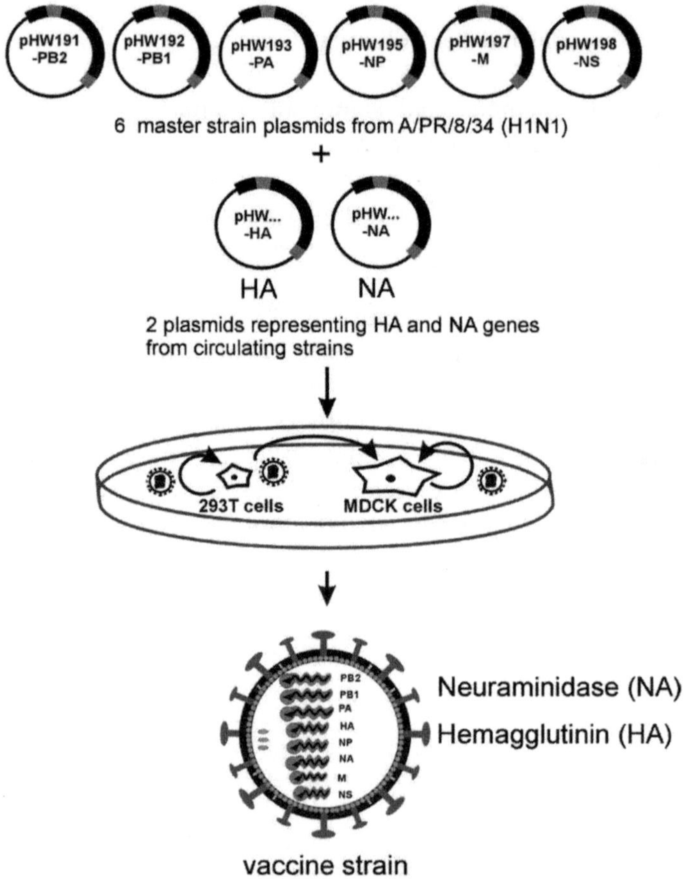

Fig. 3 Eight plasmid system for the generation of reassortant influenza A viruses. Eukaryotic cells are transfected with six plasmids representing A/PR/8/34 (H1N1) master strain and two plasmids encoding the HA and NA genes from circulating wildtype viruses. No selection system is required to isolate the desired reassortant virus. (Reproduced with permission from Hoffmann et al. 2002, Elsevier Science Ltd)

the influenza virus gene segments into a bidirectional or ambisense vector with Pol I and Pol II promoters flanking each gene segment (Hoffmann et al. 2000a; Hoffmann et al. 2000b). This modification reduced the number of plasmids required in the transfections from 12 or 17 to 8 (Fig. 3). The 8- and 12-plasmid systems have been applied to the generation of 6:2 reassortant viruses for use as vaccine seed viruses (Schickli et al. 2001; Hoffmann et al. 2002a). When a new vaccine strain is identified, the HA and NA genes can be rapidly cloned into the appropriate vector and cotransfected with the six internal protein genes of the vaccine donor strain (PR8) to generate new vaccine reassortant viruses. The use of plasmid-based reverse genetics will allow more rapid generation of reassortant viruses without the need for a helper virus or antibody selection.

Influenza B viruses have also been rescued recently by plasmid-based reverse genetics (Hoffmann et al. 2002b). Thus technology will soon be available to routinely generate all three components of trivalent influenza virus vaccines.

12 Attenuation Resulting from Mutations in a Single Gene Segment

12.1 Attenuating Mutations in the PB2 Gene

RNP transfection methods were used to generate a candidate vaccine virus with *ts* mutations inserted by site-directed mutagenesis into the PB2 gene of an influenza A virus, reasoning that a single attenuating gene could be transferred to different *wt* virus backgrounds to generate vaccine viruses (Subbarao et al. 1993a; Subbarao 1995). Previous preclinical and clinical evaluation of *ts* mutants generated by chemical mutagenesis as vaccine candidates established that a *ts* mutant virus with a shut-off temperature for virus replication of 38°C in vitro would provide a satisfactory level of attenuation in humans. The combination of three *ts* mutations in the PB2 gene at amino acids 112, 265, and 556 or at 112, 265, and 658 resulted in *ts* mutant viruses with 37°C shut-off temperatures. The three mutations were encoded by five nucleotide substitutions, providing some confidence in the likelihood of genetic stability and lack of reversion of the mutations. The *ts* mutant virus with the triple-mutant PB2 gene (112+265+556) was attenuated in hamsters and was immunogenic and protected animals from subsequent *wt* virus challenge. However, the shut-off temperature increased when the triple-mutant PB2 gene was transferred by reassortment into another genetic background, indicating that the temperature sensitivity conferred by the triple-mutant PB2 gene was partly determined by the gene constellation of the virus (Murphy et al. 1997). It is likely that the incorporation of additional *ts* or non-*ts* attenuating mutations would result in a satisfactory vaccine candidate, but this approach was not pursued further. A similar strategy was applied to introduce *ts* mutations into the P2 gene by alanine-scanning mutagenesis (Parkin et al. 1996, 1997).

12.2
Altered NS1 Proteins

Among the several functions of the NS1 protein of influenza viruses (reviewed in Salvatore et al. 2002), this protein is a virally encoded inhibitor of interferon-mediated antiviral responses. Influenza A viruses with deletions in the NS1 protein generated by RNP transfection were able to replicate in substrates deficient in the interferon pathway Garcia-Sastre et al. 1998; Talon et al. 2000). Laboratory variants of influenza B viruses have been described with deletions in the NS1 gene (Tanaka et al. 1984; Tobita et al. 1990). Influenza A viruses with targeted deletions and laboratory variants of influenza B viruses with deletions in the NS1 were evaluated as potential vaccine candidates (Talon et al. 2000). Influenza A viruses lacking the NS1 gene (delNS1) or with a154-nt deletion in the NS1 gene resulting in a NS1 protein containing only the first 99 amino acids of the *wt* NS1 (the *wt* gene encodes a proteins of 230 amino acids) were compromised in their ability to replicate to high titer in embryonated eggs. The NS1-99 virus was able to replicate to high titer in 6-day-old embryonated eggs. Both viruses were attenuated in BALB/c mice. At an immunizing dose of 10^6 pfu, both viruses (delNS1 and NS1-99) elicited ELISA antibody responses and protected mice from subsequent challenge with 100 and 5,000 LD_{50} of *wt* PR8 virus (Talon et al. 2000). Reduction in the immunizing dose of the NS1-99 virus to 3.3×10^4 pfu reduced the protection to 78%, and a similar reduction in the immunizing dose of delNS1 resulted in complete loss of protection against lethal challenge (Talon et al. 2000). The influenza B viruses with deletions in the NS1 gene were immunogenic in mice; vaccine efficacy was demonstrated by reduction in lung virus titers because a lethal virus challenge model was not available (Talon et al. 2000).

12.3
NS Mutant Replication-Incompetent Viruslike Particles

Plasmid-based reverse genetics has been used to generate replication-incompetent viruslike particles (VLPs) (Neumann et al. 2000). Replication-incompetent VLPs lacking the entire NS gene (NS-gene deficient VLPs) or lacking only the NS2 gene (NS2 knockout VLPs) were generated by deleting the splice donor site in the polI plasmid encoding the eight gene segment of influenza A/WSN/33 and inserting additional stop codons after the NS1 open reading frame (Watanabe et al. 2002). Cells infected with the VLPs did not produce progeny virus. Although 16 HA units of NS2 knockout VLPs administered intranasally to BALB/c mice did not induce detectable HI antibody titers, immunized mice exhibited a >100-fold reduction in pulmonary replication compared with nonimmunized control mice and were protected from subsequent lethal challenge with *wt* virus. In contrast, the NS gene-deficient VLPs did not induce protection from lung virus replication or from lethal challenge with *wt* virus. The authors proposed that NS2 knockout VLPs could serve as useful vaccine vectors for expression of foreign proteins or epitopes (Watanabe et al. 2002).

12.4 M2-Deficient Influenza A Viruses

A chimeric mutant influenza A virus in which the transmembrane domain of the M2 gene was replaced with that of the HA gene was generated by plasmid-based reverse genetics using 17 plasmids (Watanabe et al. 2001). This virus replicated well in vitro but failed to replicate in the nasal turbinates or lungs of mice; this degree of attenuation may make this virus an attractive vaccine candidate because the mechanism of attenuation is likely to be genetically stable (Watanabe et al. 2001). However, further studies are needed to determine whether the virus replicates sufficiently to stimulate a protective immune response, particularly in view of a recent report that viruses with a deletion of the transmembrane domain appeared to be debilitated in growth in tissue culture to an extent similar to that of influenza viruses grown in the presence of amantadine (Takeda et al. 2002).

13 Attenuation Resulting from a General Approach

Other methods of genetically engineering attenuation of influenza viruses that may be more broadly applicable than the gene-specific attenuation strategies outlined above have been proposed as vaccine design strategies. An example of a more general approach is the replacement of the noncoding ends of the HA or NA gene of an influenza A virus with the corresponding noncoding region of the NS gene of an influenza B virus, which resulted in significant attenuation (Muster et al. 1991)[83].

13.1 Application of Reverse Genetics to Generation of Pandemic Vaccines

In 1997 and 1999 human infections by avian influenza A H5N1 and H9N2 viruses were documented in Hong Kong and China (Subbarao et al. 1998; Subbarao and Katz 2000; Peiris et al. 1999; Guo et al. 1999). These events have raised concerns about the potential of avian influenza viruses of certain subtypes for pandemic spread and have led to efforts to develop vaccines to protect humans from avian influenza A viruses. Reverse genetics techniques are essential for this process if any of the genes need to be modified to remove virulence determinants. For example, a 6:2 reassortant H5N1 virus bearing the internal protein genes of the AA *ca* vaccine donor virus was generated by RNP transfection (Li et al. 1999); the H5 HA gene was genetically modified to remove the multibasic amino acid motif associated with pathogenicity for chickens. The H5N1/AA *ca* virus displayed *ts*, *ca*, and attenuation phenotypes specified by the AA *ca* virus genes and was no longer highly pathogenic for chickens (Li et al. 1999). The immunogenicity and protective efficacy of intravenously administered virus were variable in chickens (Li et al. 1999). Efficacy was not evaluated in a mammalian model.

Plasmid-based reverse genetics techniques using 12 plasmids were used to generate an H5N1/PR8 reassortant virus with a similarly modified H5 HA (Subbarao et al. 2003). The removal of the multibasic amino acid motif in the HA gene and the new genotype of the H5N1/PR8 transfectant virus attenuated the virus for chickens and mice without altering the antigenicity of the HA. Because the virus is no longer lethal for embryonated eggs and is not highly pathogenic or infectious for chickens, it should be possible to manufacture an egg-based vaccine in a biosafety level-2 production facility without posing a biological threat to animal agriculture and the environment. A formalin-inactivated vaccine prepared from the H5N1/PR8 transfectant virus was immunogenic and protected mice from subsequent challenge with *wt* viruses from the homologous and heterologous antigenic subgroups; the protection provided by this vaccine was similar to that provided by a nonlethal *wt* H5N1 live virus infection and a formalin-inactivated vaccine prepared from the *wt* H5N1 virus. The major concerns posed by other H5 vaccine candidates, such as an incomplete match between the HA and NA of the vaccine and those of the HK H5N1 viruses and biosafety concerns with regard to safety of personnel involved in vaccine production as well as the potential threat to agriculture and the environment, are addressed by the H5N1/PR8 vaccine reassortant (Subbarao et al. 2003).

An eight-plasmid system described for the generation of seed viruses that derive internal protein genes from PR8 was used to generate 6:2 reassortant viruses with surface glycoprotein genes from human influenza A H1N1 and H3N2 viruses or avian H5N1, H6N1, and H9N2 viruses (Hoffmann et al. 2002a). The authors demonstrated that vaccine seed viruses could be generated rapidly and reproducibly by cloning the HA and NA genes of different viruses while using the same plasmids for the PR8 internal protein genes (Hoffmann et al. 2002a). Rapid cloning of genes from a variety of influenza viruses into the plasmid backbone for the eight-plasmid system can be facilitated by the use of a set of universal primers (Hoffmann et al.2001)[90].

13.2 Advantages of and Concerns About Vaccines Generated by Reverse Genetics

The advantages of the use of reverse genetics for the generation of vaccine reassortant viruses are the rapidity with which the reassortants can be generated (Hoffmann et al. 2002a), the ability to genetically modify genes as needed to remove virulence markers (Subbarao et al. 2003), or potentially, to introduce attenuating mutations, and the fact that neither helper viruses nor selection pressures are required to select reassortant viruses with the desired genotype.

The primary concerns that will have to be addressed before vaccines for human use are generated by reverse genetics are potential regulatory concerns, including the cell substrate used for transfections (293-T, MDCK, and Vero cells) and the transfection reagents used. In plasmid-based reverse genetics, mammalian cells must be used for transcription because vRNA is transcribed by human pol I polymerase. Therefore, transfections will have to be carried out in mammalian cell substrates that are qualified for use in humans. As is true for any vaccine licensed for use in humans, all materials used in cell culture, plasmid purification, and transfection must be recorded. Until the rates and risk of gene rearrangements are established, more complete genotype and sequence analysis may be required of transfectant viruses than is required of vaccine seed viruses generated by reassortment. Although the reassortant viruses in currently licensed influenza vaccines are generated each time the vaccine composition is updated, years of experience with the methodology have established that the outcomes are reproducible. Regulatory authorities will have to determine the extent of evaluation required of vaccines generated by reverse genetics. When other approaches to attenuate viruses are explored, it will be necessary to establish the reproducibility of attenuation in different gene constellations (Murphy et al. 1997).

14 Summary

It is now technically feasible to use plasmid-based reverse genetics techniques to generate conventional influenza vaccine reassortants; in the future the technique will permit the introduction of genetic changes designed to improve current vaccines. The actual implementation of the technology will occur when regulatory concerns are addressed. Reverse genetics has made it possible to explore several new approaches to the generation of live attenuated influenza viruses. Although preliminary data appear promising, further studies are needed to determine the feasibility of generating vaccines with these approaches. The reproducibility of the attenuation phenotype conferred by abolishing M2 ion channel activity or deleting the NS1 or NS2 proteins must be established in different genetic backgrounds of influenza viruses. The promise of the technology lies in the fact that several attenuating mechanisms can be combined in a rationally designed influenza virus vaccine.

References

Belshe RB, Gruber WC, Mendelman PM, et al. (2000) Efficacy of vaccination with live attenuated, cold-adapted, trivalent, intranasal influenza virus vaccine against a variant (A/Sydney) not contained in the vaccine. J Pediatr 136:168–175

Belshe RB, Mendelman PM, Treanor J, et al. (1998) The efficacy of live attenuated, cold-adapted, trivalent, intranasal influenzavirus vaccine in children. N Eng J Med 338:1405–1412

Beyer WEP, Palache AM, de Jong JC, Osterhaus ADME (2002) Cold-adapted influenza vaccine versus inactivated vaccine: systematic vaccine reactions, local and systemic antibody response, and vaccine efficacy A meta-analysis. Vaccine 20:1340–1353

Bridges CB, Thompson WW Meltzer et al. (2000) Effectiveness and cost benefit of influenza vaccination of healthy working adults: a randomized controlled trial. JAMA 284:1655–1663

Centers for Disease Control and Prevention (2002) Prevention and control of influenza. Recommendations of the Advisory Committee on Immunization Practices (ACIP). Morb. Mort. Wkly Rep 51/No.RR-3:1-31

Cha T-A, Kao K, Zhao J, et al. (2002) Genotypic stability of cold-adapted influenza virus vaccine in an efficacy clinical trial. J Clin Micro 38:839–845

Chen W, Calco PA, Malide D, et al. (2001) A novel influenza A virus mitochondrial protein that induces cell death. Nat Med 7:1306–12

Clements ML, Murphy BR (1986) Development and persistence of local and systemic antibody responses in adults given live attenuated or inactivated influenza A virus vaccine. J Clin Micro 23:66–72

Clements ML, Subbarao EK, Fries LF, Karron RA, London WT, Murphy BR (1992) Use of single-gene reassortant viruses to study the role of avian influenza A virus genes in attenuation of wild-type human influenza A virus for squirrel monkeys and adult human volunteers. J Clin Micro 30:655–662

Cox NJ, Bender CA (1995) The molecular epidemiology of influenza virus. Sem Virol 6:359–370

Cox NJ, Brammer TL, Regnery HL (1994) Influenza: Global surveillance for epidemic and pandemic variants. Eur J Epidemiol 10:467–470

Cox NJ, Kitame F, Kendal AP, Maassab HF, Naeve C (1988) Identification of sequence changes in the cold-adapted, live attenuated influenza vaccine strain, A/Ann/Arbor/6/60 (H2N2). Virology 167:553–567

Edwards KM, Dupont WD, Westrich MK, Plummer Jr., WD, Palmer PS, Wright PF (1994) A randomized controlled trial of cold-adapted and inactivated vaccines for the prevention of influenza A disease. J Infect Dis 169:68–76

Enami M, Palese P (1991) High-efficiency formation of influenza virus transfectants. Journal of Virology. 65:2711–3

Fodor E, Devenish L, Engelhardt OG, Palese P, Brownlee GG and Garcia-Sastre A (1999) Rescue of influenza A virus from recombinant DNA. Journal of Virology 73:9679–82

Garcia-Sastre A, Egorov A, Matassov D, et al. (1998) Influenza A virus lacking the NS1 gene replicates in interferon-deficient systems. Virology 252:324–330

Govaert TME, Thifs CTMCN Masurel N, Sprenger MJW Dinant GJ, Knottnerus JA (1994). Efficacy of influenza vaccination in elderly individuals: a randomized double-blind placebo-controlled trial. JAMA 272:1661–1665

Gregory V, Bennett M, Orkhan MH, Al Hajjar S, Varsano N, Mendelson E, Zambon M, Ellis J, Hay A, Lin YP (2002) Emergence of influenza A H1N2 reassortant viruses in the human population during 2001. Virology. 300:1–7

Gruber WC, Kirschner K, Tollefson S, et al. (1993) Comparison of monovalent and trivalent live attenuated influenza vaccines in young children. J Infect Dis 168:53–60

Guo YJ, Li JW, Cheng I, et al. (1999) Discovery of humans infected by avian influenza A (H9N2) virus. Chinese J Exp Clin Virol 15:105–108

Herlocher ML, Clavo A, Maassab HF (1996) Sequence comparison of A/AA/6/60 influenza viruses: mutations which may contribute to attenuation. Virus Res 42:11–25

Hoffmann E, Krauss S, Perez D, Webby R and Webster RG (2002a) Eight-plasmid system of rapid generation of influenza virus vaccines. Vaccine 20:3165–3170

Hoffmann E, Mahmood K, Yang C-F, Webster RG, Greenberg HB and Kemble G (2002b) Rescue of influenza B virus from eight plasmids. Proc Natl Acad Sci USA 99:11411–11416

Hoffmann E, Neumann G, Hobom G, Webster RG and Kawaoka Y (2000a) "Ambisense" approach for the generation of influenza A virus: vRNA and mRNA synthesis from one template. Virology 267:310–317

Hoffmann E, Neumann G, Kawaoka Y, Hobom G and Webster RG (2000b) A DNA transfection system for generation of influenza A virus from eight plasmids. Proceedings of the National Academy of Sciences of the United States of America 97:6108–6113

Hoffmann E, Stech J, Guan Y, Webster RG and Perez DR (2001) Universal primer set for the full-length amplification of all influenza A viruses. Archives of Virology 146:2275–2289

Izurieta HS, Thompson WW, Kramarz P, et al. (2000) Influenza and the rates of hospitalization for respiratory disease among infants and young children. New Eng J Med 342:232–9

Jin H, Lu B, Zhou H et al. (2003) Multiple amino acid residues confer temperature-sensitivity to human influenza virus vaccine strain (FluMist™) derived from *ca* A/Ann Arbor/6/60. Virology 306:18–24

Johnson PR, Feldman S, Thompson JM, Mahoney JD, Wright PF (1986) Immunity to influenza A virus infection in young children: a comparison of natural infection, live cold-adapted vaccine, and inactivated vaccine. J Infect Dis 154:121–127

Karron RA, Steinhoff MC, Subbarao EK, et al. (1995) safety and immunogenicity of a cold-adapted influenza A (H1N1) reassortant virus vaccine administered to infants less than six months of age. Pediatr Infect Dis J 14:10–16

Kendal AP (1997) Cold-adapted live attenuated influenza vaccines developed in Russia: Can they contribute to meeting the need for influenza control in other countries? Eur J Epidemiol 13:591–609

Kendal AP, Maassab HF, Alexandrova GI, Ghendon YZ (1981) Development of cold-adapted recombinant live, attenuated influenza a vaccines in the USA and USSR. Antiviral Research 1:339–365

Khan AS, Polezhaev F, Vasiljeva R et al. (1996) Comparison of US inactivated split-virus and Russian live attenuated, cold-adapted trivalent influenza vaccines in Russian schoolchildren. J Infect Dis 173:453–456

Kilbourne ED (1969) Future influenza vaccines and the use of genetic recombinants. Bulletin of the World Health Organization 41:643–5

Klimov AI, Cox NJ, Yotov WV, et al. (1992) Sequence changes in the live attenuated, cold-adapted variants of influenza A/Leningrad/134/57 (H2N2) virus. Virology 186:795–797

Klimov AI, Egorov AY, Gushchina MI, et al. (1995) Genetic stability of cold-adapted reassortant vaccine strains before and after replication in children. J Gen Virol 76:1521–1525

Klimov AI, Kiseleva IV, GI Alexandrova, Cox NJ (2001) International Congress Series: Options for the Control of Influenza IV, Osterhaus A, Cox N, Hampson A, eds Excerpta Medica, Amsterdam, p955–959

Klimov AI, Rudenko LG, Egorov AY, Romanova JR, Polezhaev FI, Alexandrova GI, Cox NJ (1996) Genetic stability of Russian cold-adapted live attenuated reassortant influenza vaccines. In: Brown LE, Hampson AW, Webster RG (eds) Options for the Control of Influenza III, Elsevier Sciences B.V. pp129–136

Li S, Liu C, Klimov A, et al. (1999) Recombinant influenza A virus vaccines for the pathogenic human A/Hong Kong/97 (H5N1) viruses. Journal of Infectious Diseases 179:1132–8

Luytjes W, Krystal M, Enami M, Pavin JD and Palese P (1989) Amplification, expression, and packaging of foreign gene by influenza virus. Cell 59:1107–13

Maassab HF (1967) Adaptation and growth characteristics of influenza virus at 25°C. Nature 213:612–614

Maassab HF. 1968. Biologic and immunologic characteristics of cold-adapted influenza virus. J Immunol 102:728–732

Maassab HF, Kendal AP, Abrams GD, Monto AS (1982) Evaluation of a cold-recombinant influenza virus vaccine in ferrets. J Infect Dis 146:780–790

Meltzer MI, Cox NJ, Fukuda K (1999) The economic impact of pandemic influenza in the United States: priorities for intervention. Emerg Infect Dis 5:659–671

Mendelman PM, Cordova J, Cho I (2001) safety, efficacy and effectiveness of the influenza virus vaccine, trivalent, types A and B, live, cold-adapted (CAIV-T) in healthy children and healthy adults. Vaccine 19:2221–2226

Mills J, Chanock RM (1971). Temperature-sensitive mutants of influenza virus. I. Behaviour in tissue culture and in experimental animals. J Infect Dis 123:145–157

Murphy BR, Coelingh K (2002) Principles underlying the development and use of live attenuated cold-adapted influenza A and B virus vaccines. Viral Immunol 15:295–323

Murphy BR, Hinshaw VS, Sly DL, et al (1982a) Virulence of avian influenza A viruses for squirrel monkeys Infect immune 37:1119–1126

Murphy BR, Nelson DL, Wright PF, Tierney EL, Phelan MA, Chanock RM (1982b) Secretory and systemic immunological response in children infected with live attenuated influenza A virus vaccines. Infect Immun 36:1102–1108

Murphy BR, Park EJ, Gottlieb P and Subbarao K (1997) An influenza A live attenuated reassortant virus possessing three temperature-sensitive mutations in the PB2 polymerase gene rapidly loses temperature sensitivity following replication in hamsters. Vaccine 15:1372–8

Muster T, Subbarao EK, Enami M, Murphy BR and Palese P (1991) An influenza A virus containing influenza B virus 5' and 3' noncoding regions on the neuraminidase gene is attenuated in mice. Proc Natl Acad Sci U S A 88:5177–81

Neumann G, Kawaoka Y (2002) Synthesis of influenza virus: new impetus from an old enzyme, RNA polymerase I. Virus Research 82:153–158

Neumann G, Watanabe T, Ito H, et al. (1999) Generation of influenza A viruses entirely from cloned cDNAs. Proceedings of the National Academy of Sciences USA 96:9345–50

Neumann G, Watanabe T and Kawaoka Y (2000) Plasmid-driven formation of influenza virus-like particles. Journal of Virology 74:547–51

Neuzil KM, Mellen BG, Wright PF, Mitchel RF, Griffen MR (2000) Effect of influenza on hospitalization, outpatient visits, and courses of antibiotics in children. New Eng J Med 342:225–31

Nichol KL, Wuorenma J, Von Sternberg T (1998) Benefits of influenza vaccination for low- intermediate- and high-risk senior citizens. Arch Intern Med 158:1769–76

Nicholson KG (1998) Human influenza In: Nicholson KG, Webster RG, Hay AJ (eds), Textbook of Influenza Blackwell Science Ltd, Oxford, pp219–264

Nicholson KG, Tyrrell DAJ, Oxford JS, et al. (1987) Infectivity and reactogenicity of reassortant cold-adapted influenza A/Korea/1/82 vaccines obtained from USA and USSR. Bull WHO 65:285–301

Palache AM (1997) Influenza vaccines: a reappraisal of their use. Drugs 54:841–856

Palese P, Garcia-Sastre A (2002) Influenza vaccines: present and future. Journal of Clinical Investigation 110:9–13

Parkin NT, Chiu P and Coelingh KL (1996) Temperature sensitive mutants of influenza A virus generated by reverse genetics and clustered charged to alanine mutagenesis. Virus Res 46:31–44

Parkin NT, Chiu P and Coelingh K (1997) Genetically engineered live attenuated influenza A virus vaccine candidates. Journal of Virology 71:2772–8

Patriarca PA, Weber JA, Parker RA et al (1985) Efficacy of influenza vaccine in nursing homes: reduction in illness and complications during influenza A (H3N2) epidemic. JAMA 253:1136–1139

Peiris M, Yuen KY, Leung CW, et al. (1999) Human infection with influenza H9N2. Lancet 354:916–7

Powers DC, Murphy BR, Fries LF, Adler WH, Clements ML (1992) Reduced infectivity of cold-adapted influenza A H1N1 viruses in the elderly: correlation with serum and local antibodies. J Am Geriatr Soc 40:163–167

Rudenko LG, Alexandrova GI (2001) Current strategies for the prevention of influenza by the Russian cold-adapted live influenza vaccine among different populations. In: international Congress Series 1219: Options for the Control of Influenza IV, Osterhaus A, Cox N, Hampson A, eds Excerpta Medica, Amsterdam, p945–950

Rudenko LG, Arden NH, Grigorieva E, et al. (2001) Immunogenicity and efficacy of Russian live attenuated and US inactivated vaccines used alone and in combination in nursing home residents. Vaccine 19:308–318

Rudenko LG, Slepushkin AN, Monto AS et al. (1993) Efficacy of live attenuated and inactivated influenza vaccines in schoolchildren and their unvaccinated contacts in Novgorod, Russia. J Infect Dis 168:881–887

Salvatore M, Basler CF, Parisien JP, et al. (2002) Effects of influenza A virus NS1 protein on protein expression: the NS1 protein enhances translation and is not required for shutoff of host protein synthesis. Journal of Virology 76:1206–1212

Schickli JH, Flandorfer A, Nakaya T, Martinez-Sobrido L, Garcia-Sastre A and Palese P (2001) Plasmid-only rescue of influenza A virus vaccine candidates. Philos Trans R Soc Lond B Biol Sci 356:1965–1973

Simonsen L, Clarke MJ, Schonberger LB, Arden NA, Cox NJ, Fukuda K (1998) Pandemic versus epidemic influenza mortality: a pattern of changing age distribution. J Infect Dis178:53–60

Snyder MH, Betts RF, De Borde D, et al. (1988) Four viral genes independently contribute to attenuation of live influenza A/Ann Arbor/6/60 (H2N2) cold-adapted reassortant virus vaccines. J Virol 62:488–495

Steinhoff MC, Halsey NA, Wilson MH et al. (1991) The A/Mallard/6750/78 avian-human but not the A/Ann Arbor/6/60 cold-adapted, influenza A/Kawasaki/86 (H1N1) reassortant virus vaccine retains partial virulence for infants and children J Infect Dis 163:1023–1028

Subbarao K, Katz J (2000) Avian influenza viruses infecting humans. Cellular and Molecular Life Sciences 57:1770–1784

Subbarao EK, Kawaoka Y and Murphy BR (1993a) Rescue of an influenza A virus wild-type PB2 gene and a mutant derivative bearing a site-specific temperature-sensitive and attenuating mutation. J Virol 67:7223–8

Subbarao K, Klimov A, Katz J, et al. (1998) Characterization of an avian influenza A (H5N1) virus isolated from a child with a fatal respiratory illness. Science 279:393–6

Subbarao EK, London W, Murphy BR (1993b) A single amino acid in the PB2 gene of influenza A viruses is a determinant of host range. J Virol 67:1761–1764

Subbarao K, Chen H, Swayne D, et al. (2003) Evaluation of a genetically modified reassortant H5N1 influenza A virus vaccine. Virology in press

Subbarao EK, Park EJ, Lawson CM, Chen AY and Murphy BR (1995) Sequential addition of temperature-sensitive missense mutations into the PB2 gene of influenza A transfectant viruses can effect an increase in temperature sensitivity and attenuation and permits the rational design of a genetically engineered live influenza A virus vaccine. J Virol 69:5969–77

Subbarao EK, Perkins M, Treanor JJ, Murphy BR (1992) The attenuation phenotype conferred by the M gene of the influenza A/Ann Arbor/6/60 cold-adapted virus (H2N2) on the A/Korea/82 (H3N2) reassortant virus results from a gene constellation effect. Virus Res 25:37–50

Takeda M, Pekosz A, Shuck K, Pinto LH and Lamb RA (2002) Influenza A virus M2 ion channel activity is essential for efficient replication in tissue culture. Journal of Virology 76:1391–1399

Talon J, Salvatore M, O'Neill RE, et al. (2000) Influenza A and B viruses expressing altered NS1 proteins: A vaccine approach. Proceedings of the National Academy of Sciences of the United States of America 97:4309–4314

Tanaka T, Urabe M, Goto H and Tobita K (1984) Isolation and preliminary characterization of a highly cytolytic influenza B virus variant with an aberrant NS gene. Virology 135:515–523

Tobita K, Tanaka T, Odagiri T, Tashiro M and Feng SY (1990) Nucleotide sequence and some biological properties of the NS gene of a newly isolated influenza B virus mutant which has a long carboxyl terminal deletion in the NS1 protein. Virology 174:314–9

Tolpin MD, Massicot JG, Mullinex MG, et al. (1981). Genetic factors associated with loss of the temperature-sensitive phenotype of the influenza A/Alaska/77-ts-1A2 recombinant during growth in vivo. 112:505–517

Treanor JJ, Kotloff K, Betts RF, et al. (2000) Evaluation of trivalent, live, cold-adapted (CAIV-T) and inactivated influenza vaccine in prevention of virus infection and illness following challenge of adults with wild-type influenza A (H1N1), A (H3N2), and B viruses. Vaccine 18:899–906

Watanabe T, Watanabe S, Ito H, Kida H and Kawaoka Y (2001) Influenza A virus can undergo multiple cycles of replication without M2 ion channel activity. Journal of Virology 75:5656–62

Watanabe T, Watanabe S, Neumann G, Kida H and Kawaoka Y (2002) Immunogenicity and protective efficacy of replication-incompetent influenza virus-like particles. Journal of Virology 76:767–73

Webster RG, Bean WJ, Gorman OT, Chambers TM, Kawaoka Y (1992) Evolution and ecology of influenza A viruses Microbiol Rev 56:152–179

Wright PF, Webster RG (2001) Orthomyxoviruses. In: Knipe DM, Howley PM (eds) Fields' Virology 4th ed Lippincott-Raven, Philadelphia, pp 1533–1579

Subject Index

compiled by G. Neumann

Current Topics in Microbiology and Immunology

Volumes published since 1989 (and still available)

Vol. 240: **Hammond, John; McGarvey, Peter; Yusibov, Vidadi (Eds.):** Plant Biotechnology. 1999. 12 figs. XII, 196 pp. ISBN 3-540-65104-7

Vol. 241: **Westblom, Tore U.; Czinn, Steven J.; Nedrud, John G. (Eds.):** Gastroduodenal Disease and Helicobacter pylori. 1999. 35 figs. XI, 313 pp. ISBN 3-540-65084-9

Vol. 242: **Hagedorn, Curt H.; Rice, Charles M. (Eds.):** The Hepatitis C Viruses. 2000. 47 figs. IX, 379 pp. ISBN 3-540-65358-9

Vol. 243: **Famulok, Michael; Winnacker, Ernst-L.; Wong, Chi-Huey (Eds.):** Combinatorial Chemistry in Biology. 1999. 48 figs. IX, 189 pp. ISBN 3-540-65704-5

Vol. 244: **Daëron, Marc; Vivier, Eric (Eds.):** Immunoreceptor Tyrosine-Based Inhibition Motifs. 1999. 20 figs. VIII, 179 pp. ISBN 3-540-65789-4

Vol. 245/I: **Justement, Louis B.; Siminovitch, Katherine A. (Eds.):** Signal Transduction and the Coordination of B Lymphocyte Development and Function I. 2000. 22 figs. XVI, 274 pp. ISBN 3-540-66002-X

Vol. 245/II: **Justement, Louis B.; Siminovitch, Katherine A. (Eds.):** Signal Transduction on the Coordination of B Lymphocyte Development and Function II. 2000. 13 figs. XV, 172 pp. ISBN 3-540-66003-8

Vol. 246: **Melchers, Fritz; Potter, Michael (Eds.):** Mechanisms of B Cell Neoplasia 1998. 1999. 111 figs. XXIX, 415 pp. ISBN 3-540-65759-2

Vol. 247: **Wagner, Hermann (Ed.):** Immunobiology of Bacterial CpG-DNA. 2000. 34 figs. IX, 246 pp. ISBN 3-540-66400-9

Vol. 248: **du Pasquier, Louis; Litman, Gary W. (Eds.):** Origin and Evolution of the Vertebrate Immune System. 2000. 81 figs. IX, 324 pp. ISBN 3-540-66414-9

Vol. 249: **Jones, Peter A.; Vogt, Peter K. (Eds.):** DNA Methylation and Cancer. 2000. 16 figs. IX, 169 pp. ISBN 3-540-66608-7

Vol. 250: **Aktories, Klaus; Wilkins, Tracy, D. (Eds.):** Clostridium difficile. 2000. 20 figs. IX, 143 pp. ISBN 3-540-67291-5

Vol. 251: **Melchers, Fritz (Ed.):** Lymphoid Organogenesis. 2000. 62 figs. XII, 215 pp. ISBN 3-540-67569-8

Vol. 252: **Potter, Michael; Melchers, Fritz (Eds.):** B1 Lymphocytes in B Cell Neoplasia. 2000. XIII, 326 pp. ISBN 3-540-67567-1

Vol. 253: **Gosztonyi, Georg (Ed.):** The Mechanisms of Neuronal Damage in Virus Infections of the Nervous System. 2001. approx. XVI, 270 pp. ISBN 3-540-67617-1

Vol. 254: **Privalsky, Martin L. (Ed.):** Transcriptional Corepressors. 2001. 25 figs. XIV, 190 pp. ISBN 3-540-67569-8

Vol. 255: **Hirai, Kanji (Ed.):** Marek's Disease. 2001. 22 figs. XII, 294 pp. ISBN 3-540-67798-4

Vol. 256: **Schmaljohn, Connie S.; Nichol, Stuart T. (Eds.):** Hantaviruses. 2001, 24 figs. XI, 196 pp. ISBN 3-540-41045-7

Vol. 257: **van der Goot, Gisou (Ed.):** Pore-Forming Toxins, 2001. 19 figs. IX, 166 pp. ISBN 3-540-41386-3

Vol. 258: **Takada, Kenzo (Ed.):** Epstein-Barr Virus and Human Cancer. 2001. 38 figs. IX, 233 pp. ISBN 3-540-41506-8

Vol. 259: **Hauber, Joachim, Vogt, Peter K. (Eds.):** Nuclear Export of Viral RNAs. 2001. 19 figs. IX, 131 pp. ISBN 3-540-41278-6

Vol. 260: **Burton, Didier R. (Ed.):** Antibodies in Viral Infection. 2001. 51 figs. IX, 309 pp. ISBN 3-540-41611-0

Vol. 261: **Trono, Didier (Ed.):** Lentiviral Vectors. 2002. 32 figs. X, 258 pp. ISBN 3-540-42190-4

Vol. 262: **Oldstone, Michael B.A. (Ed.):** Arenaviruses I. 2002, 30 figs. XVIII, 197 pp. ISBN 3-540-42244-7

Vol. 263: **Oldstone, Michael B. A. (Ed.):** Arenaviruses II. 2002, 49 figs. XVIII, 268 pp. ISBN 3-540-42705-8

Vol. 264/I: **Hacker, Jörg; Kaper, James B. (Eds.):** Pathogenicity Islands and the Evolution of Microbes. 2002. 34 figs. XVIII, 232 pp. ISBN 3-540-42681-7

Vol. 264/II: **Hacker, Jörg; Kaper, James B. (Eds.):** Pathogenicity Islands and the Evolution of Microbes. 2002. 24 figs. XVIII, 228 pp. ISBN 3-540-42682-5

Vol. 265: **Dietzschold, Bernhard; Richt, Jürgen A. (Eds.):** Protective and Pathological Immune Responses in the CNS. 2002. 21 figs. X, 278 pp. ISBN 3-540-42668-X

Vol. 266: **Cooper, Koproski (Eds.):** The Interface Between Innate and Acquired Immunity, 2002, 15 figs. XIV, 116 pp. ISBN 3-540-42894-1

Vol. 267: **Mackenzie, John S.; Barrett, Alan D. T.; Deubel, Vincent (Eds.):** Japanese Encephalitis and West Nile Viruses. 2002. 66 figs. X, 418 pp. ISBN 3-540-42783-X

Vol. 268: **Zwickl, Peter; Baumeister, Wolfgang (Eds.):** The Proteasome-Ubiquitin Protein Degradation Pathway. 2002, 17 figs. X, 213 pp. ISBN 3-540-43096-2

Vol. 269: **Koszinowski, Ulrich H.; Hengel, Hartmut (Eds.):** Viral Proteins Counteracting Host Defenses. 2002, 47 figs. XII, 325 pp. ISBN 3-540-43261-2

Vol. 270: **Beutler, Bruce; Wagner, Hermann (Eds.):** Toll-Like Receptor Family Members and Their Ligands. 2002, 31 figs. X, 192 pp. ISBN 3-540-43560-3

Vol. 271: **Koehler, Theresa M. (Ed.):** Anthrax. 2002, 14 figs. X, 169 pp. ISBN 3-540-43497-6

Vol. 272: **Doerfler, Walter; Böhm, Petra (Eds.):** Adenoviruses: Model and Vectors in Virus-Host Interactions. Virion and Structure, Viral Replication, Host Cell Interactions. 2003, 63 figs., approx. 280 pp. ISBN 3-540-00154-9

Vol. 273: **Doerfler, Walter; Böhm, Petra (Eds.):** Adenoviruses: Model and Vectors in Virus-Host Interactions. Immune System, Oncogenesis, Gene Therapy. 2004, 35 figs., approx. 280 pp. ISBN 3-540-06851-1

Vol. 274: **Workman, Jerry L. (Ed.):** Protein Complexes that Modify Chromatin. 2003, 38 figs., XII, 296 pp. ISBN 3-540-44208-1

Vol. 275: **Fan, Hung (Ed.):** Jaagsiekte Sheep Retrovirus and Lung Cancer. 2003, 63 figs., XII, 252 pp. ISBN 3-540-44096-3

Vol. 276: **Steinkasserer, Alexander (Ed.):** Dendritic Cells and Virus Infection. 2003, 24 figs., X, 296 pp. ISBN 3-540-44290-1

Vol. 277: **Rethwilm, Axel (Ed.):** Foamy Viruses. 2003, 40 figs., X, 214 pp. ISBN 3-540-44388-6

Vol. 278: **Salomon, Daniel R.; Wilson, Carolyn (Eds.):** Xenotransplantation. 2003, 22 figs., IX, 254 pp.ISBN 3-540-00210-3

Vol. 279: **Thomas, George; Sabatini, David; Hall, Michael N. (Eds.):** TOR. 2004, 49 figs., X, 364 pp.ISBN 3-540-00534-X

Vol. 280: **Heber-Katz, Ellen (Ed.):** Regeneration: Stem Cells and Beyond. 2004, 42 figs., XII, 194 pp.ISBN 3-540-02238-4

Vol. 281: **Young, John A. T. (Ed.):** Cellular Factors Involved in Early Steps of Retroviral Replication. 2003, 21 figs., IX, 240 pp. ISBN 3-540-00844-6

Vol. 282: **Stenmark, Harald (Ed.):** Phosphoinositides in Subcellular Targeting and Enzyme Activation. 2003, 20 figs., X, 210 pp. ISBN 3-540-00950-7

Vol. 283: **Kawaoka, Yoshihiro (Ed.):** Biology of Negative Strand RNA Viruses: The Power of Reverse Genetics. 2004, 24 figs., approx. 350 pp. ISBN 3-540-40661-1

MIX
Papier aus verantwortungsvollen Quellen
Paper from responsible sources
FSC® C105338

If you have any concerns about our products,
you can contact us on
ProductSafety@springernature.com

In case Publisher is established outside the EU,
the EU authorized representative is:
Springer Nature Customer Service Center GmbH
Europaplatz 3, 69115 Heidelberg, Germany

Printed by Libri Plureos GmbH
in Hamburg, Germany